AF449158

D. Dietrich
P. Neumann
H. Schweinzer (eds.)

Fieldbus Technology

Systems Integration, Networking,
and Engineering

Proceedings of the Fieldbus Conference FeT '99
in Magdeburg, Federal Republic of Germany,
September 23–24, 1999

SpringerWienNewYork

Univ.-Prof. Dipl.-Ing. Dr. Dietmar Dietrich
Institut für Computertechnik

Ass.-Prof. Dipl.-Ing. Dr. Herbert Schweinzer
Institut für Elektrische Meßtechnik und Schaltungstechnik

Technische Universität Wien, Vienna, Austria

Prof. Dr.-Ing. habil. Peter Neumann
Institut für Automation und Kommunikation e.V.
Otto-von-Guericke-Universität Magdeburg
Magdeburg, Federal Republic of Germany

Printed by Novographic, Ing. Wolfgang Schmid, A-1230 Wien
Graphic design: Ecke Bonk
Printed on acid-free and chlorine-free bleached paper
SPIN 10744232

With 235 Figures

ISBN 3-211-83394-3 Springer-Verlag Wien New York

Preface

Although being a matter of course in science and industry, in the public opinion fieldbus technology is not always seen in its real importance. Terms like "home automation", "industry automation", etc. are not always perceived in a positive way: they sound ultramodern, mysterious, unreliable. Often conventional but well-known solutions instead of fieldbus technology are used. On the other hand, fieldbus technology saves money, systems are offering more services and test integration increases reliability.

Fieldbus technology is not an argument for marketing. However, its potential is enormous and unlimited although some believe realization will be a matter of the far future.

For several years, fieldbus systems have been seen as local installations of small interest in comparison with LANs, Internet, or mobile phones. This situation has changed: in automobiles, a fieldbus has to manage the communication of about one hundred nodes. Fieldbus systems are monitored via classical LANs, fieldbus systems in airplanes are used not only for control, but also for online testing, and in some building automation systems more than 10.000 nodes are connected by fieldbus systems. Great demands are arising with consequences in system complexity, reliability, availability and security, all being aspects relevant for scientific work. Activities in this field are numerous and mostly combined with other areas of scientific or engineering work. This is challenging and important because of the connection to different engineering disciplines.

There are slogans as "IT revolution in industry"[1], "the transparent factory"[2], or "concert of intelligent chips"[3]. In any way, our age is stamped by information technology. All spheres of our living are being influenced and changed by innovations of information technology. Mobile phones, teleworking, electronic commerce are revolutions in daily life. Similar to the industrial revolution which brought dramatic modifications, collecting, storing, transporting and computing of information leads to radical changes in human society.

In nature which is an important model for technical systems, primitive micro-organisms bear the information in their system structure. Animals being "more intelligent" are based on internal information systems, neuronal networks in the most complex form. Reception and communication with the external world is performed by use of highly sophisticated and adaptable sensor systems, but also internally a great number of sensors is used for process control of life functions.

---

[1] Die IT-Revolution in der Industrie, MegaLink 11/99

[2] Die transparente Fabrik; Messe Wien viet; Presse-Info, 18.6.1999

[3] Drössler, Ch.: Konzert der schlauen Chips, DIE ZEIT Nr. 24; 10.6.1999

Also fieldbus technology allows more intelligent systems. Smart sensors and complex sensor systems increase the functionality and the effectiveness of the system. In a weekly newspaper[4] a prognosis of the household of tomorrow is given where cooking, controlling of goods in stock and so on are performed by integrated transponders and neuronal algorithms. This leads to the following scenario:

- more and more smart sensors will be integrated in things of every day life and will be part of the fieldbus system,

- demand of controllers and software will increase dramatically to compute, filter and apply all of these data,

- various aspects of this new technology demand extensive analysis because numerous arising questions have to be answered by scientists,

- average life will change completely, a process which cannot be stopped.

Fieldbus technology carries an enormous part of this development without appearing in public like the Internet or the use of mobile phones. Moreover, this part will increase in the years to come. A clear example for this is the continuous change of mechanical systems by the use of fieldbus technology: when for the first time in Airbus A320, mechanical control of an airplane was replaced by "fly by wire", the public did not take notice of this, moreover passengers should not be informed about this fact. Nowadays, automobiles are in a similar situation. "Steer by wire" replaces the mechanical steering-gear by electronic control of every wheel, a technique more efficient and flexible, but also cheaper.

In industry and in the private sphere a similar development can be assumed. More and more, decisions for product development are influenced by the total costs which includes service and maintenance. Moreover, the competition of producers leads to increased features of the products. These are all arguments for an integration of fieldbus technology.

Large organizations like factories, hospitals, laboratories need transparency gained by automatic data acquisition. These data are not only directly related to the costs, but also to the personnel and to logistic parameters such as location, functionality, and status of maintenance of the equipment. Dynamic management of the facilities combined with data from processes and the personnel enables a more efficient and reliable operation of this complex system.

In the technical realization of such systems, problems are arising from the interconnection of different types of networks, especially fieldbus systems with LANs and the Internet. Furthermore, security is an important aspect. These all are subjects reflected in this book.

---

[4] Drössler, Ch.: Konzert der schlauen Chips, DIE ZEIT Nr. 24; 10.6.1999

Network interconnection is a technical and financial problem. Solutions are normally based on compromises and can be optimized by technical efforts. Security is a complex problem because of various aspects, especially the anxiety of the users. Intranet used in companies and banks is an expensive technique which shields against the Internet. On the other hand, an increasing number of systems can be read out and manipulated by remote control and remote service over networks which allow system access needing strict security provision methods. Several principles are known which have to prove their values.

Communication networks connect different spheres and in so far they help to overcome barriers[5]. They create new products, new markets and also new professions. Fieldbus technology is not at the front end of this trend because it acts in the background and in critical domains. Equipped with sensitive sensors, fieldbus technology becomes the backbone of many processes of our life.

This book is a collection of articles dealing with the matter of fieldbus technology. The articles were submitted to the fieldbus conference FeT'99 taking place in Magdeburg, Germany. Two very successful conferences before, the FeT'95 and the FeT'97, were held in Vienna, Austria. The articles were reviewed by the international program committee which decided to also include some high quality articles not presented at the conference.

The book comprises eleven chapters dealing with important aspects of fieldbus technology and reflecting areas of main activity in science and industry. These chapters also coincide with the sessions at the conference.

A number of people were strongly involved in the preparation and completion of this book: M. Wollschlaeger, Magdeburg, managed a main part of the communication with the authors to ensure the preparation of the articles in time. W. Kandler, Vienna, made the final check of the layout of the book which entailed a lot of corrections. We like to thank them very much. We also thank Springer-Verlag Wien for printing and binding the book in its well known high quality.

We would like to thank all those who helped in the organization of the conference: H. Mueller and H. Rosenzweig, VDI/VDE-GMA, who were responsible for the organization, Th. Sauter, W. Kandler and M. Wollschlaeger, who made preparations for the program committee meetings, and the program committee itself. We thank our authors and speakers who provided the valuable contributions for our stimulating discussions. We also thank our four invited speakers who followed our invitation.

---

[5] Dietrich, D.; Posta, R.; Schmalek, R.: Feldbussysteme - Synthese statt Konflikt; etz 1-2- 96, Jahrbuch elektrotechnik; VDE

VIII

Our aim is to reach a broad audience. Therefore the conference fees were kept moderate. This became possible only with the support of sponsoring organizations whose financial support we appreciate very much.

Magdeburg, September 1999

Dietmar Dietrich   Herbert Schweinzer   Peter Neumann

Supported by: ABB Automation Products, D-30179 Hannover

Endress + Hauser GmbH + Co, D-79689 Maulburg

Kist Europe, D-66123 Saarbruecken

Phoenix Contract GmbH & Co, D-32825 Blomberg

Siemens AG, D-90327 Nürnberg

Weidmueller Connext GmbH & Co, D-32760 Detmold

Contents

6 OLE for Process Control OPC 239

7 System Aspects 267

11 NOAH Esprit 26951 Project   413

Conference Chairman

Prof. Dr. P. Neumann, ifak, D

Chairman of the Program Committee

Prof. Dr. D. Dietrich, TU Vienna / ICT, A

Program Committee

Dr. M. Adams, Dr.Seufert GmbH, D
Prof. Dr. K. Bender, TU Munich / ITM, D
Dipl.-Ing. R. Bent, Phoenix Contact GmbH & Co., D
Prof. Dr. J. Böttcher, Univ. der Bundeswehr Munich, D
Prof. Dr. J.-D. Decotignie, CSEM Neuchatel, CH
Dipl.-Ing. A. De Decker, Siemens NV, B
Dr. Keith Dimond, University of Kent, GB
Prof. Dr. D. Dietrich, TU Vienna / ICT, A
Prof. Dr. K. Etschberger, stzp, IXXAT Automation GmbH, D
Josef F. Faller, Carinthian Tech Research, A
Prof. P. Fischer, FH Dortmund, D
Prof. Dr. W.A. Halang, Fernuniversität Hagen, D
G. Hodgkinson, PROFIBUS International, GB
Dipl.-Ing. J. Johansen, PROCES-DATA A/S, DK
Prof. Dr. K. Kabitzsch, TU Dresden, D
Prof. Dr. W. Kriesel, GMA, D
L. Liljegren, ABB, S
Dr. M. Merx, Weidmüller ConneXt GmbH&Co., D
Ing. W. Morrenth, Siemens AG, A
Dipl.-Ing. H. Müller, VDI/VDE-GMA, D
K. A. Myrvang, AD Elektronikk AS, N
Prof. Dr. P. Neumann, ifak Magdeburg, D
Prof. Dr. O. Nisamutdinov, TU Perm, RUS
Dipl.-Ing. P. Noury, CEGELEC, F
Dr. T. Sauter, TU Vienna / ICT, A
Ass.Prof. Dr. H. Schweinzer, TU Vienna / EMST, A
Dipl.-Ing. H.-J. Schweinzer, TU Vienna / ICT, A
Dipl.-Ing. H. K. Tronnier, EIBA, B
Dr. K. Watson, Fraunhofer-IITB, D
Dr.-Ing. M. Wollschlaeger, IPE, University Magdeburg, D
Dipl.-Ing. H. Zeltwanger, CAN-in-Automation e. V., D

Steering Committee

Prof. Dr. D. Dietrich, TU Vienna / ICT, A
Prof. Dr. P. Neumann, ifak, D
Ass. Prof. Dr. H. Schweinzer, TU Vienna / EMST, A

Dipl.-Ing. W. Kandler, TU Vienna / EMST, A
Dr. Th. Sauter, TU Vienna / ICT, A
Dr.-Ing. M. Wollschlaeger, IPE, University Magdeburg

Organisation

Dipl.-Ing. H. Mueller, VDI/VDE-GMA
Mrs. H. Rosenzweig, VDI/VDE-GMA

Index of Authors

Chapter 1: Protocol Specifications

Invited Paper presented in Plenary Session 1

Papers presented in Session 1.2

Further Paper

Main Paradigms as a Basis for Current Fieldbus Concepts

Jean-Pierre THOMESSE, Miguel LEON CHAVEZ

LORIA - INPL - ENSEM, 2, av. de la Forêt de Haye
F-54516 Vandoeuvre-lès-Nancy, France
Tel. (33) 3 83 59 57 46   Fax (33) 3 83 44 07 63
email: {thomesse,mleon}@loria.fr

Abstract. This paper discusses the main paradigms on fieldbuses from the applications requirements to the most common solutions derived to fulfill them. Temporal properties of the data to be exchanged are analyzed. Data are classified according to their criticality and their transmission frequency, for each of these types, the communication requirements are identified with emphasis on the real-time aspects. Two architectures are identified in order to design the applications, and an overview of their characteristics is presented. Main paradigms on fieldbuses are analysed at different levels of the architecture and in different contexts: synchronous vs asynchronous, periodic vs aperiodic, centralized vs distributed, static vs dynamic, deterministic vs nondeterministic. Several examples of the fieldbuses and the MAC protocols are analysed for each paradigm with regard to possible future fieldbuses.

1 Introduction

There exist several different fieldbuses from some years ago and they seem to be quite different from one another if their protocols are considered. But are they really different? Are the differences related to the services, to the quality of service, or only to some choices of protocol mechanisms? Are the differences result from the physical or from the MAC layer? From the application layer? From the definition of objects?

This paper tries to analyse some fundamental points to understand the differences of the fieldbus origins and to look at the future. Therefore the paper is entitled "Main Paradigms as a Basis for Current Fieldbus Concepts".

We consider the following paradigm pairs: synchronous vs asynchronous, periodic vs aperiodic, static vs dynamic, centralized vs distributed, deterministic vs nondeterministic. They are applied at different levels and in different contexts, and their meanings are very different from one scientific community to another.

To analyse these aspects, the paper will be organized as follows. In the first part, the problems of application architectures and application design will be analysed. This analysis will take into account, on one hand, the end-users requirements in order to identify the different traffics in relation with the architectures defined for such applications and, on the other hand, the main solutions proposed in order to design such architectures.

In the second part, this paper will recall the vocabulary and the meaning of the above paradigms and give examples of ambiguities and misconceptions associated to misunderstandings in order to look at the future fieldbuses. The problem of scheduling will be introduced as the solution to meet the timing constraints, and the open problem of joint scheduling (coordinated scheduling of messages and of tasks) will be presented.

2 Application requirements

Typically, fieldbuses are used in distributed applications to monitor and control a manufacturing process such as automotive industry, textile machinery, factory automation, semiconductor fabrication (chip-manufacturing), electronics manufacturing, foot and beverage, chemical processing, and so on.

The requirements of the manufacturing applications are directly deduced from the end-users ones, such as the field devices, the production machines controllers, the transport systems, the people who are in charge of the control, of the maintenance, of the production scheduling, of the technical management, etc.

The communication requirements of these applications can be characterized by both the criticality and the transmission frequency of the data.

In these applications there are data that have tight timing constraints, i.e., data that have a limited temporal validity. Hence, the response times of all the involved activities (production, transmission, consumption, computing, ...) in their processing must be small and bounded in order to efficiently monitor and control the proper operation of the manufacturing process and the quality of the products.

These data are called *critical data*, they can be either *soft* or *hard*, the latter being those in which the applications cannot tolerate any delay or loss in their processing.

In these applications there are also data without timing constraints. In this case, a delay of the respective processing has no effect on the proper operation of the manufacturing process itself nor on the quality of the products. These data are called *non-critical data* and are normally handled by background application tasks.

Another time-related aspect that concerns the data is the regularity and frequency at which they are produced and processed. Typically, the manufacturing applications know the current status of the manufacturing process by means of the data produced and transmitted both periodically and randomly by the sensors or higher level devices. In the same way, the applications control the manufacturing process by means of the control data transmitted to the actuators or control devices. In any case, when the transmission of such data is carried out with a constant frequency, they are called *periodic*, and they can be critical or not.

Aperiodic data are used to disseminate critical data in random emergency situations. Notice, however, that there might be non-critical data requested aperiodically, too, by a high-level supervision task. In some case it is possible to bound the minimum separation time between two consecutive aperiodic transmissions of the same source. This particular sort of aperiodic data is known as *sporadic*.

Figure 1 shows the temporal properties of data in manufacturing applications according to, on one hand, their criticality, and on the other hand, their transmission frequency. It can be noted that the data in the shaded area have two types of timing constrains, one due to the response times of their processing, which must be small and bounded, and the other due to their transmission frequency, which can be constant, random or with a known minimum separation time. Usually, these data are referred as *real-time data* and the data in the column of non-critical data as *non-real-time data*.

DATA	Critical		Non-critical
	Hard	Soft	
Periodic			
Aperiodic	**Real-time**		
Sporadic			

Fig. 1. Temporal properties of data in manufacturing applications.

In brief, with regard to the types of data, the communication requirements of the manufacturing and control applications can be summarized as follows:

- Differentiation between real-time and non-real-time data is essential [16]. The communication system should carry the data from source to destination without altering their designation.

- It should be possible to distinguish the relative urgency of data, basically between periodic and aperiodic hard-critical data.

- The communication system should guarantee that all hard-critical data transfers will be completed within a given time interval.

- The communication system should try its best effort to transfer the soft-real-time data, since minimizing the number of delayed or lost transfers will result in a proper operation of the manufacturing process in the long term, for example.

- Non-real-time data shall not jeopardize the delivery of real-time data.

3 Application architectures

Two approaches have been proposed in order to design the manufacturing applications: Event-triggered and time-triggered architectures [18]. In *event-triggered* architecture, the occurrences of events (significant state changes) in the manufacturing process initiate the appropriate application activities. A *time-triggered* architecture is driven by periodic observations of the state of the manufacturing process and starts actions as a consequence of the analysis of these states.

It is interesting to note that these approaches are well adapted for designing the control and manufacturing applications based on the transmission frequency of the data, i.e., a lot of fieldbuses have been designed to transmit firstly the periodic traffic. Regulators or controllers and PLC softwares are often based on the sampling theory, the time-triggered approach is then well adapted for designing these applications, in which the aperiodic traffic is handled by one or more periodic servers. But considering all the information as periodic leads to overload the communication channel. So, other mechanisms, corresponding to the event-triggered approach, have been proposed to handle the aperiodic traffic. Nevertheless, a single combined architecture is needed.

With regard to the criticality of the data, the time-triggered approach requires a detailed design phase, during which the maximum response times of all activities must be established and calculated, and appropriate execution schedules are constructed. In this way, the temporal behavior of the application can be predicted in order to guarantee that the timing constraints induced by the manufacturing process be met.

On the other hand, in the event-triggered approach, some events can be predictable, applying known physical laws, and others are random. In the first case, it is possible to anticipate such events and reserve resources to process and transmit them. In the second case, the random events can only be predicted probabilistically. Typically, the signaling of the events is realized by an interrupt mechanism, the architecture then dynamically determines the execution order of the application tasks and the transmission order of the messages. In this way, depending on the event sequences and the duration of the activities to process them, several schedules are possible, which may or may not meet the timing constraints induced by the manufacturing process.

These architectures can be modeled either as synchronous or asynchronous systems. Ideally, the time-triggered architecture recognizes a single interrupt signal [20]: the ticks generated by the local clock at predetermined points in time. The input data are then polled, stored in memory giving an image of the manufacturing process. An application task may detect the occurrences of events by the change of values between consecutive samplings, performs its algorithm and computes its outputs. A distributed application must then synchronize all the local clocks to represent a globally synchronized time.

The event-triggered architecture is designed to react upon external events, these are mapped on a set of interruptions and then the architecture dynamically determines the task to be run and the message to be sent or received.

Theoretically, in a *synchronous message-passing system* [4], the computation is performed in rounds. At the beginning of a round, each processor sends messages, and waits to receive messages that were sent by its neighbors in this round. Upon receiving these messages, the processor performs some internal operations and then decides what messages to send in the next round. In an *asynchronous message-passing system*, messages incur an unbounded and unpredictable (but finite) delay, and processors take steps at arbitrary rates. There are also intermediate models of partially synchronous systems, where there are various limitations on message delivery time and processors' step time.

But synchronous and asynchronous are two qualifiers used to characterize systems, message-passing models, programming languages, operating systems, and also network communications, the next section will analyse.

4 Synchronous vs asynchronous

4.1 Network communications

In network communications, synchronous and asynchronous have different meaning according to the layers of the OSI Reference Model.

At the physical layer, in the *asynchronous communication*, the receiver node synchronizes its receiving logic with that of the sender node only at the beginning of a new message. In *synchronous communication*, the receiver node resynchronizes its receiving logic during the reception of a message to the ticks of the sender's clock [19].

At the MAC layer, synchronous and asynchronous are usually synonyms of periodic and aperiodic messages [23]. A synchronous message is characterized by its constant inter-arrival times, which is the time at which the message becomes available for transmission. An asynchronous message arrives at stochastic inter-arrival times.

4.2 System models

In a *synchronous system* [13], there is a known fixed upper bound Δ on the time required for a message to be sent from one processor to another and a known fixed upper bound Φ on the relative speeds of different processors. In *an asynchronous system* no fixed upper bound Δ and Φ exist. If there is a fixed upper bound Δ on the time for messages to be delivered, then the communication is said synchronous. And if there is a fixed upper bound Φ on the rate at which one processor's clock can run faster than another's, then the processors are said synchronous. Moreover, it is easy to extend these models to allow processors (as well as communication) to be partially synchronous. That is, Φ (Δ) can exist but be unknown, or Φ (Δ) can be known but actually hold only from some time onward.

It can be noted that this definition takes into account, on one hand, a temporal requirement of the manufacturing applications, the bounded transmission times, and on the other hand, the clock synchronization requirement of the time-triggered architecture.

4.3 Programming languages and operating systems

From the programming languages and operating systems point of view, message-passing statements are synchronous and asynchronous depending on whether their execution could cause a delay [3]. Synchronous and asynchronous are so sometimes synonyms of blocking and nonblocking, respectively. A statement is nonblocking if its execution never delays its invoker; otherwise the statement is said to be blocking.

With synchronous message-passing [6], the sender is blocked until the receiver has accepted the message (explicitly or implicitly). Thus the sender and receiver not only exchange data, but they also synchronize. With asynchronous message-passing, the sender does not wait for the receiver to be ready to accept its message. Conceptually, the sender continues immediately after sending the message. The implementation of the language may suspend the sender until the message has at least been copied for transmission, but this delay is not reflected in the semantics. Between these two extremes there is buffered message-passing [2], in which the buffer has finite bounds and hence sender delays only when the buffer is full.

4.4 Communication models

The communication models for distributed systems may be characterized as synchronous or asynchronous (blocking or nonblocking). For example, the point-to-point model can be both synchronous and asynchronous; the rendezvous model is synchronous; the remote procedure call model is synchronous; the one-to-many messages model (multicast and broadcast) is typically asynchronous; the generative communication model, in which the processes share a single memory space called tuple space, is similar to asynchronous message-passing model.

There is another communication model, the client-server model that promotes modular, flexible and extensible system design [1]. In this model, the server renders a service to a client. The client requests a service by sending a message to the server. The server repeatedly receives a request for service from a client, performs that service, and returns a completion message to the client [3]. A client interface specifies the individual services or operations supported by the server. Client can only request services that conform to the client interface for the given server.

The client-server communication model can be implemented with two point-to-point messages: In one way, from the client to the server, as either synchronous or asynchronous message-passing; In the other way, from the server to the client, as asynchronous message-passing. But, at the programming language level, a single higher level construct is easier to use and more efficient to implement. Two such constructs are rendezvous and remote procedure call [6].

However, the client-server model is a good one to establish the relationship between pairs of entities and their exchanges. Thus for example, the notions of client and server apply to a number of different pairs of entities: users, process, layers, and so on.

This model is used in modern operating systems [25], where a usual approach is to implement most of the operating systems functions in user process. To request a service, a user process (now called client process) sends the request to a server process, which then does the work and sends back the answer. In this model, all the kernel does is handle the communication between clients and servers. By splitting the operating system up into parts, each of which only handles one facet of the system, such as file service, process service, terminal service, or memory service, each part becomes small and manageable. Furthermore, because all the servers run as user-mode processes, and not in kernel mode, they do not have direct access to the hardware

The client-server model can also be applied to the layers of the OSI Reference Model, where every layer is a client of its underlying layer, and each one is a server of the layer immediately above it. Furthermore, the relationship layers at the same level can be modeled by this model.

5 Periodic vs aperiodic

Usually, control and manufacturing applications are composed of a a set of tasks processing the data from and to the manufacturing process. As section 2 has analysed, the data can be characterised by their criticality and their transmission frequency.

Although many of the options taken in the protocols of each layer may influence the ability of the communication system to meet the requirements of these applications, the Medium Access Control (MAC) protocol has the utmost importance in this matter because it is responsible for the scheduling of the network traffic [21, 23]. This scheduling establishes the transmission order of the messages and, therefore, it directly influences the response time of the communication system.

Several MAC protocols have been proposed in order to meet the communication requirements of these applications, and some solutions have been implemented in existing fieldbuses to handle both periodic and aperiodic traffic [26].

For example, *WorldFIP* [29] uses the producer-distributor-consumer model and defines one macro-cycle that is composed of micro-cycles, themselves composed of a periodic part and an aperiodic part, and possibly a special part to adjust the constant duration of the cycle. Each variable is identified by a unique label and each node can be a producer and/or consumer of one or more variables.

During the periodic cycle, the bus arbiter reads a list of periodic variables and injects each variable label onto the network, each variable has only one producer. Consumers needing to utilize the variable, alerted by the label, store and use the value broadcasted by the producer.

When an aperiodic variable or message are produced, the producer utilizes the response, of a received periodic variable label, to request its transmission. The bus arbiter stores this request and after completing the periodic cycle, adds all aperiodic variables that have been requested and they are then handled in the same way as periodic variables, where the relevant producers reply current values.

In *Profibus* [12, 11] the nodes are organized as masters and as slaves (I/O devices). The master nodes form a logical token ring. When a master receives the token, it can communicate with all slave nodes in a master-slave relationship and with all master nodes in a master-master relationship. The token rotation time depends on the configured token holding time of each master.

Slave nodes do not have bus access rights and can only acknowledge received messages or send messages to the master upon request. However, slave nodes can spontaneously transfer events to the master with a diagnostic message, then the master explicitly acknowledges alarms received from the slave.

The *Controller Area Network (CAN)* protocol [7, 15] does not make any differentiation between periodic and aperiodic traffic, each message in the network has an identifier field that is used as priority. The length of the identifier in the standard format is 11 bits and in the extended format, 29 bits.

At network operation, when the bus is detected as idle each node begins to transmit its highest priority message while monitoring the channel. The most significant bit of the identifier field is transmitted first. The message is coded so that if a node transmits a recessive bit (logical 1) but monitors a dominant bit (logical 0) then a collision is detected. The node knows that its message is not the highest priority message, stops transmitting and waits for the bus to become idle. Meanwhile, the node becomes a receiver of the message. Each receiver performs an acceptance test to determine whether the data are relevant. If so, the data are accepted; otherwise they are ignored.

Tindell et al. [28, 27] present an analysis that bounds the response time of all CAN messages, including the lowest priority message. Navet et al. [24] propose an analysis by providing an analytical method for computing the worst case deadline failure probability. These analyses make CAN suitable as a bus for hard real-time data.

DeviceNet and *Smart Distributed System* (SDS) are two CAN-based control networks.

In the *Foundation Fieldbus* there is a bus scheduler, called the Link Active Scheduler (LAS), similar to the bus arbiter of the WorldFIP, that has a list of transmit times for all periodic messages in all the nodes. When it is time for a node to send a periodic message, the LAS transmits a Compel Data message to the node. Upon reception, the node broadcasts or "publishes" the data in its buffer to all the nodes. Any node configured to receive the data is called a "subscriber".

If there is no time to do anything before next Compel Data message, the LAS transmits idle messages while waiting, otherwise it passes a token to each node. When a node receives it, the node is allowed to send its aperiodic messages until it has finished or until the maximum token holding time has expired, whichever is the shorter time. If the node has nothing to send, it immediately returns the token to the LAS.

6 Centralized vs distributed

Centralized and distributed are two qualifiers often used to characterize user applications or programs, operating systems, and so on. But they are also used to qualify MAC protocols.

For example, a *distributed program* [2] is a concurrent (or parallel) program in which tasks communicate by message passing. The name results from the fact that such programs are typically executed on distributed architectures - such as multicomputers or network computers - in which processors do not share memory. A distributed program can, however, also be executed on a shared memory multiprocessor or even on a single processor.

A *distributed operating system* [25] is one that appears to its users as a traditional uniprocessor operating system, even though it is actually composed of multiple processors. In a true distributed operating system, users should not be aware of where their programs are being run or where their files are located; that should all be handled automatically and efficiently by the operating system. In contrast, a network operating system, the users are aware of the existence of multiple computers, and can log in to remote machines and copy files from one machine to another. Each machine runs its own local operating system and has its own user or users.

The dichotomy of centralized or distributed qualifies also the way to take a decision, to access a resource, an objet, at any level. The question here is whether the decision making process should physically reside in a single processor (*centralized*) or whether the work involved in making decisions should be physically distributed among the processors (*distributed*).

The decision is taken from a knowledge of the state which is locally available to the deciding process. The state is composed of local elements and, possibly, images of remote elements.

Casavant and Kuhl [9] take into account two fundamental components when considering the decision making policy of a system: responsibility and authority. If *responsibility* for making and carrying out policy decisions is shared among the entities in a distributed system, the system is *distributed*. If *authority* is distributed to the entities of a resource management system, the system is *decentralized*. This differentiation persists in many other organizational structures. Any system which possesses decentralized authority must have distributed responsibility, but it is possible to allocate responsibility for gathering information and carrying out policy decisions, without giving the authority to change past or make future decisions.

In the OSI Reference Model, the MAC protocol is an algorithm used by the nodes to manage the access to and use of a communication channel. Since the nodes are distributed, if each node runs the same algorithm to access the channel, the MAC protocol is then distributed as well. Note that all the nodes run the same algorithm, but no one has the authority to change it.

A example of centralized protocol is *roll-call polling*, where one node sends a message to each node in turn, inquiring whether or not the node has anything to send. If not, the node sends back a special poll reject message. In the *distributed polling* protocol, the node which has just transmitted polls the other nodes to pass on the access rights.

Some examples of centralized and distributed fieldbuses are the followings: The *CAN* protocol is distributed. Each node knows its own state, from the PDU to be sent, and the channel state, from the signals on the cable. If the channel is sensed idle, the node transmits its package.

The *WorldFIP* protocol is centralized. There is one node, the distributor or bus arbiter, having a polling table which defines, at any moment, the communication requirements of all nodes. The table is defined at network configuration time, and several different tables can exist, but one being active at any given time.

The *Foundation Fieldbus* is centralized. Two types of nodes are defined: Link Masters and Basic Devices. The former are capable of becoming the Link Active Scheduler (LAS), the latter do not have this capability.

Other possibilities exist, as in *Profibus*, where the decision making process is distributed amongst the master nodes and it is centralized on each master node with respect to the slave nodes.

7 Static vs dynamic

Static and dynamic are two qualifiers used to characterize scheduling algorithms. But they are also used to qualify MAC protocols.

Task scheduling algorithms, themselves, can be static or dynamic [10]. A static approach calculates schedules for tasks at pre-run-time and requires the complete prior knowledge of the tasks' characteristics. A dynamic approach determines schedules for tasks at run-time and allows tasks to be dynamically invoked.

Although static approaches or cyclic executives have low run-time cost, they are inflexible and cannot adapt to a changing environment. Any change in the set of tasks can require drawing and fully testing a new time-line. They are not well suited to handling mixtures of periodic and aperiodic tasks [17]. In short, they provide highly deterministic yet inflexible behavior [5].

In contrast, dynamic approaches or best-effort scheduling involve higher run-time costs, but they are flexible and can easily adapt to changes in the environment, and, at best, allow only probabilistic predictions of run-time performance.

Static and dynamic qualify also priority assignment in task scheduling. In static-priority scheduling there is a fixed base priority associated with each task, although the task may temporally acquire a higher active priority at run time. Static-priority policies include rate monotonic and deadline monotonic scheduling.

Dynamic-priority scheduling policies assign variable priorities at run time, these policies include earliest deadline first and least laxity scheduling, and can lead to better processor utilization than static-priority ones, but are harder to implement [14].

On the other hand, a MAC protocol can be considered as two processes [23]: an access arbitration process and a transmission control process. The former determines when a node can send a message over the channel. The latter determines how long a node can continue to send messages over the channel.

As above, these two processes can be static or dynamic. The static process determines time or access duration at network design or network configuration time. The dynamic process determines time or access duration at network operation time.

For example, in the *Time Division Multiple Access* (TDMA) protocol the access arbitration and the transmission control processes are static. In this protocol each node periodically uses the channel for some fixed amount of time.

In the *Token Bus* protocol (IEEE 802.4) these processes are static, there is a special frame, called token, that circulates around the ring. A node cannot transmit until it first receives the token from its predecessor, then the order of transmission is fixed according to the node position in the ring and each node possesses the token for some fixed amount of time.

In the *Token Ring* protocol (IEEE 802.5) the processes are dynamic, there is a priority field into the token that controls the transmission rights and is examined by each node whenever it arrives. The node inserts the higher priority amongst their waiting messages, if and only if this priority is higher than the inserted priority in the token. A node can only seize the token and send its message when the token comes with the requested priority and after that it has made a round through all nodes. In this way, the transmission order is defined by the priorities of the messages and not by the position of the nodes in the ring. A node can send at least one message and can continue if there are no other higher priority messages in the network or until a fixed token holding time.

In the *CSMA/DCR* (Carrier Sense Multiple Access with Deterministic Collision Resolution) protocol [22, 8], the access arbitration and the transmission control processes are dynamic. At network configuration time, each node has been assigned an index or address which is used by the protocol to implement the binary tree search algorithm.

As long as no collision occurs, CSMA/DCR behaves exactly like CSMA/CD. Upon collision occurrence, CSMA/DCR iteratively dichotomizes the set of message sources, i.e., the set of the nodes is divided into winners and losers. The former consists of all the nodes having addresses higher than or equal to the average of the addresses. The loser nodes cease emitting until all frames of the winner nodes are transmitted.

The process of division continues until all branches of the tree have been analysed: each branch of the tree of addresses is analysed until a node successfully answers (without collision) or until none node answers (a branch is abandoned as soon as none node answers at the end of a given time). Having examined a given branch, the algorithm DCR passes to the branch immediately to the left.

8 Deterministic vs nondeterministic

The determinism is based on the capability to predict something in the future from the knowledge of the past and of the present. Even if this dilemma has been studied by a lot by philosophers and scientists for several centuries, it is a subject of discussion in computing systems, in networks and especially in fieldbuses.

The determinism doesn't exist in general unless some hypotheses are explicitly stated, particularly in terms of reliability or dependability, of bounded delays, of absence of errors... However one can read sentences such as "this protocol is a deterministic one". Let us recall the discussions on Ethernet and the token techniques at the beginning of the eighties. many times we have seen the confusion between the "determinism" of a MAC protocol and the "determinism" of a received frame?

This paradigm must in its essence be applied to the application behavior whatever the environment may be. It is always required that the whole application behavior be deterministic as much as possible. But what is a deterministic behavior? Whatever the environment may be? In all the possible event sequences? That's the problem of reliable or of fault tolerant systems design. A misconception is to say that static processes or synchronous systems automatically lead to deterministic systems.

9 Conclusion

This contribution has studied the main paradigms of the communication systems designed for fieldbuses with examples taken from the standards. Fieldbuses are all synchronous regarding the existence of delays. Except for CSMA/DCR, they are essentially static with a very low dynamic part. They are deterministic if several assumptions are made. The problem is that they are made only implicitly. Most of the fieldbuses combine in a different manner the periodic and aperiodic traffic management. In all cases, time-triggered systems are privileged to the event-triggered ones.

Is this solution definitively the right one? If some smart sensors or actuators are considered, we may imagine that the periodic traffic will decrease with the integration of periodic algorithms into the devices and that the aperiodic traffic increase related to abnormal situations and to the event-triggered approach.

In the context of robustness or fault-tolerance issues, the time-triggered approach has been chosen essentially because:

- The systems are relatively static. This means that the devices or the nodes are known at the configuration stages and it has been possible to define static mechanisms.

- This approach simplifies the protocols in the sense that the receivers know when a message must be received. In the event-triggered approach, the transmission control is made by the receiver rather than by the sender.

- This approach is used even in so called distributed MAC protocols, as in *CAN*, but it is at the application process level in order to be able to prove some properties and the proofs are possible regarding some limitations.

We can consider the evolution of the protocols in other applications, for example, the multimedia applications. Where the transmission control is one of the WANs, these new applications introduce timing constraints on the traffics, which become similar to the fieldbuses ones, and the solutions are based on more dynamic mechanisms, but always by techniques of resource reservation.

An important goal for the future is a convergence between these two types of solutions, leading to the definition of a simple communication stack with:

- A common kernel managing dynamically time-critical and not-time-critical traffics.

- An adaptation of the lowest level to the environmental conditions (IS, data rate, length,...).

- An adaptation of the application level to the application objects, statically or dynamically managed, created, destroyed, etc.

References

1. R.M. Adler. Distributed coordination models for client/server computing. *IEEE Computer*, 28(4):14-22, April 1995.

2. G.R. Andrews. Paradigms for Process Interaction in Distributed Programs. *ACM Computing Surveys*, 23(1):49-90, 1991.

3. G.R. Andrews and F.B. Schneider. Concepts and Notations for Concurrent Programming. *ACM Computing Surveys*, 15(1):3-43, 1983.

4. H. Attiya. *Parallel and Distributed Computing Handbook*, chapter Distributed Computing Theory, pages 127-159. McGraw-Hill, 1996.

5. N.C. Audsley, A. Burns, R.I. Davis, K.W. Tindell, and A.J. Wellings. Fixed priority pre-emptive scheduling: An historical perspective. *Real-Time Systems*, 8:173-198, 1995.

6. H.E. Bal, J.G. Steiner, and A.S. Tanenbaum. Programming Languages for Distributed Computing Systems. *ACM Computing Surveys*, 21(3):261-322, 1989.

7. Bosch,R. GmbH. *CAN Protocol Specification V2.0 (A,B)*, 1992.

8. J. Boudenant, B. Feydel, and P. Rolin. An IEEE 802.3 compatible deterministic protocol. In *IEEE INFOCOM'87*, pages 573-579, Sn. Francisco, USA., April 1987.

9. T.L. Casavant and J.G. Kuhl. A Taxonomy of Scheduling in general-purpose Distributed Systems. *IEEE Transactions on Software Engineering*, 14(2):141-154, Feb. 1988.

10. S.C. Cheng, J. Stankovic, and K. Ramamritham.. *Hard Real-Time Systems: Tutorial*, chapter Scheduling Algorithms for Hard Real-Time Systems: A Brief Survey, pages 150-173. IEEE, 1988.

11. DIN, German Draft Standard 19245-4. *PROFIBUS-PA, Profibus for Process Automation*, 1996.

12. DIN, German Standards 19245-1 to 19245-3. *PROFIBUS, Process fieldbus*, 1990.

13. C. Dwork, N. Lynch, and L. Stockmeyer. Consensus in the presence of partial synchrony. *Journal of the ACM*, 35(2):288-323, April 1988.

14. C.J. Fidge. Real-Time Schedulability Tests for Preemptive Multitasking. *Real-Time Systems*, (14):61-93, 1998.

15. ISO DIS 11898. *Road Vehicules-Interchange of Digital Information-Controller Area Network (CAN) for High Speed Communication*, 1992.

16. ISO TC 184/SC5/WG2 Architecture and Communications. *User requirements for system supporting time critical communications*, 1992.

17. M.H. Klein, J.P. Lehoczky, and R. Rajkumar. Rate-monotonic analysis for real-time industrial computing. *IEEE Computer*, pages 24-33, january 1994.

18. H. Kopetz. Should Responsive Systems Be Event-Triggered or Time-Triggered? In *IEICE Trans. on Information and Systems*, volume {E76-D}, Inst. of Electronics, Information and Comm. Engineers, Tokio, Japan, Nov. 1993.

19. H. Kopetz. *Real-Time Systems, Design Principles for Distributed Embedded Applications*. Kluwer Academic Publishers, 1997.

20. H. Kopetz and G. Grünsteidl. TTP - A Protocol for Fault-Tolerant Real-Time Systems. *IEEE Computer*, 24(1):14-23, 1994.

21. J.F. Kurose, M. Schwartz, and Y. Yemini. Multiple-Access Protocols and Time Constrained Communication. *Computing Surveys*, 16(1):43-70, 1984.

22. G. Le Lann and P. Rolin. *Procédé et dispositif pour la transmission de messages entre différentes stations à travers un réseau local à diffusion*, 1984. French patent no. 8416957.

23. N. Malcolm and W. Zhao. Hard Real-Time Communication in Multiple-Access Networks. *Real-Time Systems*, 8:35-77, 1995.

24. N. Navet, Y.-Q. Song, and F. Simonot. Worst-case deadline failure probability in real-time applications distributed over CAN (Controller Area Network. *Journal of Systems Architecture*, to appear in 1999.

25. A. Tanenbaum. *Modern Operating Systems*. Prentice Hall, 1992.

26. J.P. Thomesse. A Review of the FieldBuses. *Annual Reviews in Control*, (22):35-45, 1998.

27. K. Tindell, A. Burns, and A. Wellings. Calculating Controller Area Network (CAN) Message Response Times. *Control Engineering Practice*, 3(8):1163-1169, 1995.

28. K. Tindell, A. Burns, and A.J. Wellings. Analysis of Hard Real-Time Communications. *Real-Time Systems*, 9:147-171, 1995.

29. prEN 50170, WorldFIP. General Purpose Field Communication System. 1995.

Communication Profile for Embedded Networks

Holger Zeltwanger

CAN in Automation (CiA), Am Weichselgarten 26, D-91058 Erlangen, Tel. +49-9131-69086-0 (Fax –79), Email: headquarters@can-cia.de, URL: www.can-cia.de

Abstract. Embedded networks are used for machine or system internal communication between module and sub-systems. One of the most important embedded networks is Controller Area Network (CAN). In the early days of CAN, the embedded system designers have developed their own higher-layer protocols. Today, most of the system integrators like to use standardized application layers and communication profiles to reduce design efforts. CANopen is a family of specifications including communication profiles, frameworks, and device profiles for embedded networks based on CAN.

1 CANopen Specifications

CANopen is a family of specifications covering a communication profile as well as several device, interface and application profiles. The communication profile is based on the CAN data link layer and the high-speed physical layer as specified in ISO 11898. CANopen was pre-developed in the ESPRIT project called ASPIC under the chair of Bosch. In the year of 1994, CANopen was handed over to the CAN in Automation (CiA) international users and manufacturers. CiA Members had reviewed the CANopen documents and published them in 1995. The current specification (version 4.0) is available by CiA (DS-301). At first CANopen was used by European machine manufacturers as a standardized embedded network. Typical examples are bending machines, textile machines, packaging machines, and injection molding machines. In the meantime, CANopen has also been used as an „open" network in applications such as production lines, inventory systems, and power plants. However, CANopen is not only used in industrial automation, there are also applications in medical equipment, maritime electronics, off-road vehicles, railways, and military systems. The CANopen communication profile was submitted to CENELEC for European standardization. More than 10 companies offer CANopen source code libraries, and several companies provide different configuration, development, and analyzing tools.

2 Standardized Embedded Networking

CANopen was originally developed as a standardized embedded network with high flexible configuration capabilities. Most of the CAN applications are embedded networks. In the early days of CAN, each designer of a CAN-based embedded network has developed his own protocol or has implemented additional functionality in the application. CANopen unburdens the developer from dealing with CAN-specific details such as bit-timing and implementation-specific functions. CANopen provides standardized communication objects for real-time data (Process Data Objects), configuration data (Service Data Objects), and special functions (Time Stamp, Sync Message, and Emergency Message) as well as network management data (Boot-Up Message, NMT Message, and Error Control Message).

Besides the standardization of communication objects, CANopen specifies the description of application objects. In the CANopen device profiles, there are defined application objects to achieve a partly interchangeability of CANopen devices. CiA has published device profiles for generic I/O modules, drive and motion controllers, human machine interfaces, and encoders. Under development are device profiles for measuring devices and closed-loop controllers, proportional hydraulic valves, door control units, and railway brakes. Other device profiles in preparation include specifications for diesel engines, joy sticks, medical-specific devices, etc. There is also an application profile for passenger information systems for public transportation available.

For special application fields CiA is developing frameworks. The CANopen framework for programmable devices as well as the interface profile for IEC 1131-compliant devices are already published. The framework for safety-relevant communication based on CANopen will be available by end of this year. Other frameworks under development are the framework for maritime electronics and the framework for medical systems.

To guarantee interoperability, there is a CANopen conformance test tool available. CiA uses the very same tool for certifying CANopen devices and implementations. The first modules have passed the conformance and are already certified.

3 Object Dictionary and Communication Objects

In each decentralized control application there are different communication objects required. In CANopen all these communication objects are standardized and well described in the Object Dictionary. The CANopen Object Dictionary is accessible by a 16-bit index and in the case of Arrays and Structures additionally by an 8-bit sub-index. The use of an object dictionary allows specifying objects independent of the CAN identifiers. That is, why CANopen can distinguish more than 2048 objects even if CANopen is based on standard CAN frames with 11-bit identifiers.

The CANopen Object Dictionary also describes all application objects of the device. Application objects may be specified by a CANopen Device Profile or by an Application Profile. In addition, a device manufacturer may define non-standardized application objects, but then this device will not be interchangeable with one of the same class from another company. For standard devices only the range 6000h to 67FFh is mandatory in order to guarantee that they fit into a Multiple Device Module. The entry 67FFh is reserved for the device type entry of a certain device. The object at index 67FFh and multiples with an offset of 800h describe the type of each virtual within one physical device and its functionality. This technique allows implementing 8 virtual devices in one physical CANopen module.

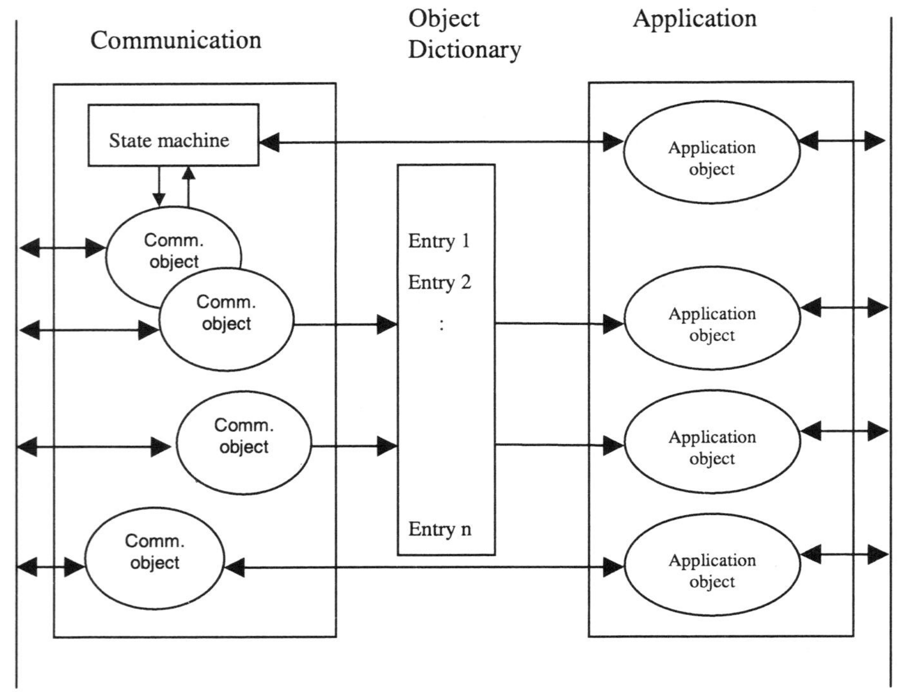

Fig. 1. CANopen Device Model

There are four classes of communication objects standardized in CANopen. Process CANopen Data Objects (PDO) are mapped to a single CAN frame using all 8 bytes of the data field to transmit application objects. Each PDO has a unique identifier and may be transmitted by only one node, but it can be received by more than one (producer/consumer communication). PDOs may be transmitted in different modes that is driven by an internal event, by an internal timer, by remote requests and by a sync message received from a specific node. The default mapping of application objects as well as the supported transmission mode is also described for each PDO in the Object Dictionary. PDO identifiers have a high priority to guarantee good real-time performance. If real-time performance is required the system designer can configure an inhibit-time for each PDO. The inhibit-time forbids this object to be transmitted within a specific time. So he can design a deterministic PDO behavior on more than one object. PDOs are transmitted with no confirmation.

Which application objects are transmitted within a PDO is defined in the PDO Mapping Object. It describes the sequence and length of the mapped application objects. The CANopen device profiles specify the default PDO Mapping. A device that supports dynamic mapping of PDOs must support this during the Pre-Operational

State. If variable mapping during Operational State is supported, the SDO Client is responsible for data consistency.

The second class of communication objects are Service Data Objects (SDO) transmitting configuration data, which is sometimes longer than 8 bytes. The SDO transport protocol allows one to transmit objects of any size. The first byte of the first segment contains the necessary flow control information including a toggle bit to overcome the well-known problem of double received CAN frames. The next three bytes of the first segment contain index and sub-index of the Object Dictionary entry to read or write. The last four bytes of the first segment are available for configuration data. The second and the following segments using the same CAN identifier contain the control byte and up to seven bytes of configuration data. The receiver confirms each segment, so that there is a peer-to-peer communication (client/server). In the future CANopen will allow also a fast SDO transfer confirming not only each segment, but also the complete object.

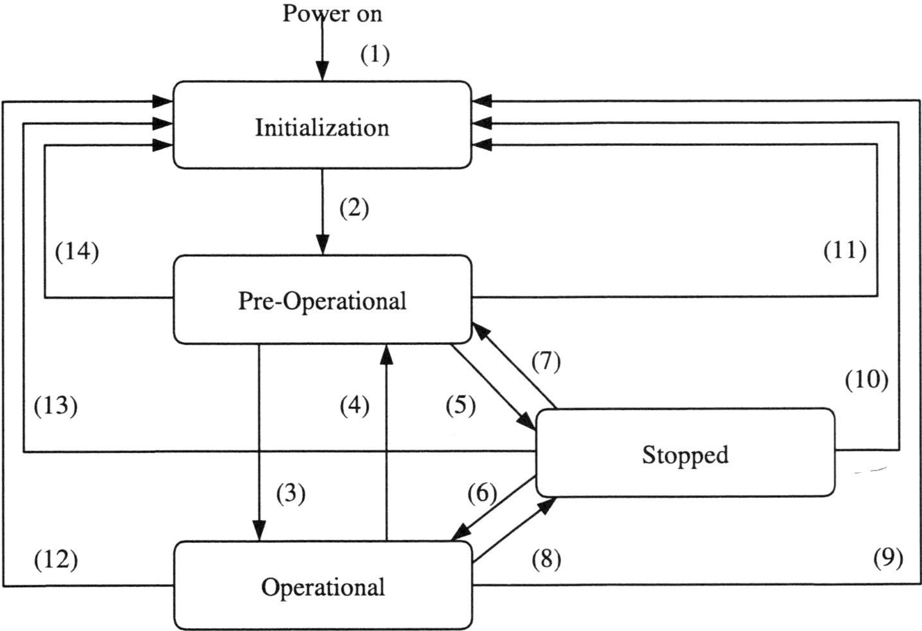

Fig. 2. NMT Slave State Machine

The third class of communication objects is the network management objects: Boot-up object, Network Management object, and Error Control object. The Boot-up object has the same identifier as the Error Control object and is transmitted after Initialization before the node is set into Pre-operational state. The Network Management object is the highest prior object in a CANopen network. The network management master node transmits it. With this object the NMT slave state machine is controlled. There are defined different states: Initialization, Pre-Operational, Operational, and Stopped. After power on each CANopen node is in the Initialization State and transits automatically to the Pre-operational state. In this state Sync objects and Error Control is provided, also the transmission of SDOs is allowed. If the NMT master has set one or more nodes to the operational state they are allowed to transmit and receive PDOs. In the Stopped state no communication is allowed except NMT objects.

The Initialization state is divided in three sub-states in order to enable a complete or partial reset of a node. In the Reset_Application sub-state the parameters of the manufacturer-specific profile area and the standardized device profile area are set to their default values. In the Reset_Communication sub-state the parameters of the communication profile area are set to their power-on values. The third sub-state is the Initialization State, which a node enters automatically after power on or after reset communication or reset application. Power-on values are the last stored parameters.

The Error Control object indicates, which state a node has. This object can be requested remotely by the NMT master module (Nodeguarding) or can be transmitted by the NMT slave modules periodically. Transmission cycle times as well as expected reception cycle times can be configured.

CANopen defines also three specific objects for synchronization, emergency indication, and time stamp transmission. The Sync Object is broadcast periodically by the Sync Producer. This object provides the basic network clock. The time period between Sync messages is defined by the Communication Cycle Period Object, which may be written by a configuration tool to the application devices during the boot-up process. There can be a time jitter in transmission by the Sync Producer due to some other objects with higher priority identifiers or by one frame being transmitted just before the Sync Object. The Sync Object is mapped to a single CAN frame with the identifier 128. By default, the Sync Object does not carry any data, but it can have up to 8 bytes of user-specific data.

Emergency Objects are triggered by the occurrence of a device internal fatal error situation and are transmitted from an emergency client on the concerned application device. This makes them suitable for interrupt type error alerts. An Emergency Object may be transmitted only once per 'error event'. As long as no new errors occur on a device no further Emergency Object must be transmitted. Zero or more Emergency Object consumers may receive these objects. The reaction on the emergency consumers is not specified. CANopen defines several emergency error codes to be transmitted in the Emergency Object, which is a single CAN frame with 8 data bytes.

By means of the Time Stamp Object a common time frame reference is provided to application devices. It contains a value of the type Time-of-Day. This object transmission follows the producer/consumer push model. The associated CAN frame has the identifier 256 and a data field of 6 bytes length.

4 Pre-Defined Identifier Allocation

As CAN is a communication object (COB) oriented network, where each COB has one or more associated identifiers which specifies its priority implicitly, the allocation of identifiers to the COBs is an essential issue in the system design. In order to reduce configuration effort for simple CANopen networks a mandatory default identifier allocation scheme is defined. These identifiers are available in the Pre-Operational State and may be modified by means of dynamic distribution. A CANopen device has to provide the corresponding identifiers only for the supported communication objects. The default profile ID-allocation scheme consists of a functional part, which determines the object priority and a module-ID-part, which distinguishes between devices of the same functionality. The ID-allocation scheme corresponds to a pre-defined master/slave connection set and allows peer-to-peer communication between a single-master device and up to 127 slave devices. It also supports the broadcasting of non-confirmed NMT, Sync and Time-Stamp Objects. The pre-defined master/slave connection set supports one Emergency Object, one SDO as well as up to 4 Receive-PDOs and up to 4 Transmit-PDOs, and Error Control (Boot-up) object.

To optimize the identifier allocation the system designer may change the allocation of identifiers. This can be done statically, which means that the identifiers are fixed during the run-time of the system. Dynamic distribution is that the identifiers are distributed via the CAN network using the SDO services. The dynamic distribution is the preferred method in CANopen-based systems.

In order to transfer data items to the correct location in the network it is sometimes necessary to link together Transmit-PDOs and Receive-PDOs, which are transmitted from one device and received from other ones. This is the case in systems, where the default identifier distribution is not sufficient. In the process of PDO Linking, the system designer must take care that the Transmit-PDOs have unique identifiers and that the data length of linked PDOs are identical.

References

1. ISO 11898 (1993): Road vehicles – Interchange of digital information – Controller area network (CAN) for high-speed communication

2. CiA DS-301 (1999): CANopen communication profile

Shared Data on InterBus

H. Schweinzer, W. Kandler

Institute for Electrical Measurement and Circuit Technology (E354)
University of Technology, Vienna
Gusshausstrasse 25, A-1040 Vienna, Austria

Abstract. InterBus is a fieldbus system with cyclic operation making use of a total frame protocol. It is very well suited for real time applications. The additionally defined PCP (peripherals communication protocol) is designed for the exchange of numerous data without hard real time demands, eg. parameters and program downloads. „Shared Data" is a new approach presented in this paper which enhances the features of PCP. It offers a data channel on InterBus generally useable both for tasks similar to PCP and for temporary real time connections among a small number of partners. The shared data channel allows data transfers marked with specific transfer codes directing data to a predefined number of modules. An immediate read out of data of modules is possible as well as a transfer of multiple data within the same execution cycle.

1 Introduction

InterBus is a fieldbus system standardized in the European standard EN 50254-2 [3]. Basically InterBus is a master slave system which means that the central module, the master, makes use of several individual connections to slave modules in parallel which are cyclically updated in real time. Physically InterBus is realized as a ring where the whole data frame is shifted through all slaves. Nevertheless data exchange is only performed between the master and each slave module.

The purpose of this paper is to describe an expansion of InterBus which allows to define a data channel which can be used commonly by several modules making communication more flexible. In normal configurations of InterBus, an increase of the number of communication partners and their associated ports of communication also increases the time of the operation cycle proportionally. This results from the total frame protocol of InterBus which separately reserves an adequate amount of I/O-data according to the transfer need of each module.

It is a totally new perspective at InterBus to allow several modules to access to the same I/O-space. The allocation of this I/O-space at InterBus called „shared data channel" (SD channel) has to be done in the usual way by a special InterBus slave module. All the InterBus slaves together define the cycle time which is not increased by additional modules that only operate on the SD channel.

Requesting access to the SD channel, modules have to reserve this data frame for themselves for a certain amount of time or operation cycles respectively. During the reserved cycles, these modules place write data on the SD channel. Other modules are allowed to listen to these data in parallel. Because of the changing assignment of the SD channel to different modules and the generally changing content of the data, it is clear that it makes no sense to repeat any I/O-operation on the SD channel periodically for every cycle as it is done on the normal InterBus. On the other hand, a fast communication can be realized on base of the InterBus cycle which is flexible and expansible nearly without limits.

The method offers the opportunity to transmit process data commonly available for all modules taking part in the SD communication. Applications for this form of communication are characterized by demands in flexibility of installation and usage of the communication which are especially testing, delivery of continuous test data, visualization and voice communication. Also the communication with PCP (Peripherals Communication Protocol) as defined within the InterBus definition can be realized using the SD channel.

The Peripherals Communication Protocol (PCP) is designed for the exchange of numerous data without hard real time demands. It enables the transfer of parameters or programs to InterBus slaves without reserving more than a few bytes of data exchange area in each slave taking part in this service. However, this reserved bytes in the slave allows a permanent data exchange with the InterBus master which probably is not necessary for this mode of communication.

The SD channel offers an alternative for the realization of the PCP service. The channel has to be reserved by the master which can perform transfers individually coded to affect the correspondent slave. Choosing this method, internal bytes for PCP communication of InterBus slaves can be saved and, on the other hand, be used to realize the more flexible SD channel. The size of the total frame of InterBus data may remain about the same as before.

2 Principle

InterBus is a single master fieldbus system using a total frame protocol. Similar to a shift register, within a bus cycle every piece of data is transferred through all modules. Output data are shifted from the master to the slaves while input data are shifted from the slaves to the master within the same bus cycle. The last module in the chain of modules recognizes the miss of an outgoing cable which would connect to a following module. It connects the end of the cable to a backgoing cable which is included in the bus cable and forms a ring leading back to the master [1].

The SD channel is realized by an InterBus module attached to the end of the bus cable reserving the SD frame slot. Being the last module on the bus, its input data are the first ones running at the backgoing path and can be detected simultaneously everywhere on the return cable (Fig.1).

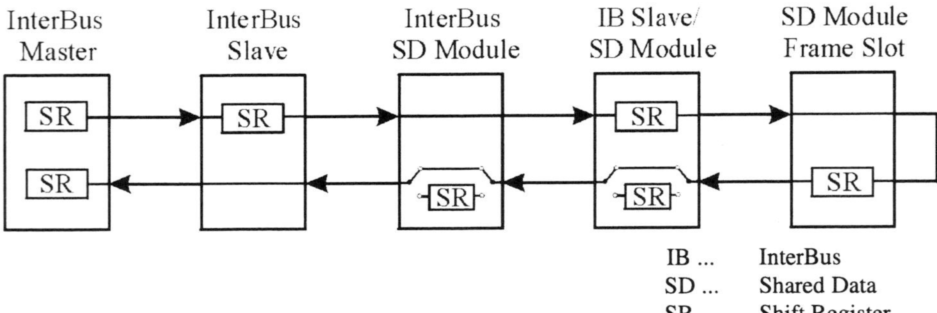

Fig. 1. Principle

Primarily, this last module reserving the SD frame slot has the task to copy its output data as input data for the next bus cycle into the frame. This action is necessary to close the ring and to continue the round trip of the data back to the transmitter of shared data. Nevertheless this last module can also be included in the SD communication as a fully operating module .

Modules accessing the SD channel have to be inserted in the backgoing cable of InterBus. Expense and functionality of the module interface are similar to a standard InterBus slave. However, operating with an internal InterBus shift register with zero length, all SD modules are logically at the same position at the ring while physically one after the other. They all can manipulate the shared data easily because the SD frame slot is located at the beginning of every data sequence. SD modules have to listen to the stream of InterBus data where the end of each cycle is clearly marked by a check sequence. Shared data are the first ones at the beginning of a new data sequence. At any physical location in the return cable of InterBus it is possible to read the SD channel and also to modify its data.

Writing data by a SD module at any physical location is an online modification of the frame data. Although having a shift register length of zero, the realization of the interface is only slightly different from a normal InterBus slave interface. The incoming data stream will be recorded in an internal shift register of the module while the outgoing data stream will be switched from incoming data to internal data of the module which will be sent instead. Of course, this switching must also be done during the check sequence at the end of the bus cycle. Any modification of the shared data implies a checksum correspondingly calculated in the SD module which has to be transmitted to the InterBus master.

The main assumption for performing a write operation is to ensure that the SD module is allowed to do so. Only one or two writing modules within one cycle enable a proper data transmission on the SD ring. A protocol of using the SD channel guarantees this condition. Moreover, two priority levels are realized to meet real time requirements.

3 Protocol

The protocol for accessing the SD channel aims at a flexible use of this common data area. It includes information about the size of the SD channel and number and types of the modules of the given configuration. Modules or groups of modules are addressed by transfer codes. A token is used to avoid collisions in write access.

3.1 Arrangement of the shared data channel

An information set on the SD channel consists of a two byte header and n byte transparent data (Fig. 2). The receipt of a set token bit by a module requesting channel access passes control of the current information transfer to this module. If this module for instance plans to place its write data on the SD channel it replaces any received token byte with zeros. This behavior ensures that the module is the only one selected by receiving the token and allows an immediate placement of the write data following the token byte.

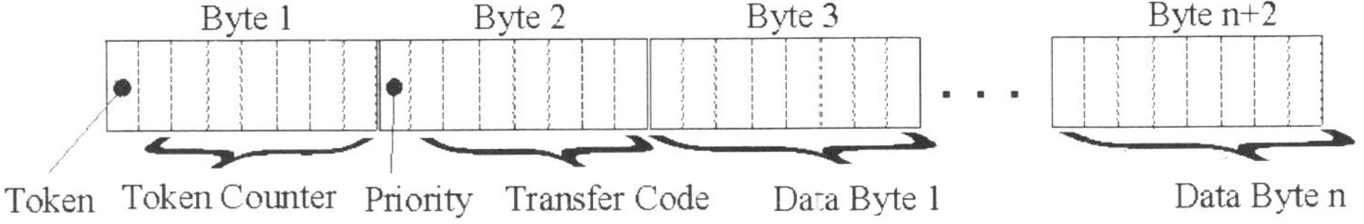

Fig. 2. Shared data frame

The token counter is a number already passed with the last data transfer provided by the actual sending module. In the beginning it is set to an initial value indicating the number of cycles after which modules with higher priority have to be activated. This activation is signaled by setting the priority bit after the token counter has reached zero again. A set priority bit disables every module with lower priority to take the token in the next cycle. A module with higher priority is allowed to take the token and to use the bus as long as it wants. Together with the last data transmission, this module defines by means of the value of the token counter the number of bus cycles after which the modules with higher priority have to be activated again. In the next cycle following the last data transfer the token (and the value of the token counter again) is passed. A receiving module can take the token also in a situation where write data were prepared just immediately before passing the token. The token count has to be decremented every bus cycle which is performed under control of this module.

The transfer code characterizes the meaning of the packet of transparent data which follows. Certain codes are fixed and further codes can be defined at run time as necessary. Fixed transfer codes are used for initialization and configuration of the SD channel where the message length of the SD frame is defined. At this time, a certain minimum length is necessary for the communication. Transfer codes can also be predefined to realize connections to certain modules similar to normal addresses.

3.2 Initialization

The number of InterBus Slaves and the number of their I/O-bytes result in total in a specific bus cycle time of the InterBus varying with each application. The SD channel needs an additional frame slot increasing the cycle time of InterBus too. This byte number has to be defined according to the transfer need of the SD modules and is primarily controlled by modules with real time demand using the higher priority bus access level. On the other hand, the total number of SD modules does not influence the cycle time of InterBus. Because of different application demands the frame slot of the SD channel should be configurable. The total byte number has to be set at the last InterBus module which is the organizer of the SD channel. This byte number must be written to all SD modules within the initialization phase.

It is necessary to know which modules can take part in SD communication. In a configuration phase each SD module has to fetch the bus token at least once to transfer its individual parameters (name, type, names of variables, subaddresses, and so on) in a broadcast transfer to all modules in the ring. This allows each listening module to initiate a connection and further on to perform data transfers with identified modules if necessary.

3.3 Normal Operation

Some applications, for instance the recording of received test variables during a given time, have a well defined demand for continuous transfers on the SD channel. Moreover, real time requirements can make it necessary to perform these transfers at higher priority level. A module with such a communication demand first has to fetch the token to establish the connection to the partner module by an initial dialog. Local buffering in the transmitting module enables it to collect an adequate number of data bytes which afterwards are sent in several succeeding bus cycles. Within the last data cycle the token count is sent as a preview to the number of cycles which can be used by lower priority modules until at least the control of the SD channel has to be returned to the higher priority modules. With the begin of the next cycle the token will be sent. Getting the token a module is informed about the maximum number of cycles the bus may be held. If the token is passed after a smaller number of cycles to another module the remaining number of cycles is sent again as a preview token count with the last data cycle. If a module is performing the allowed maximum number of data cycles the priority bit in the second byte of the header has to be set again combined with the last data transfer. This effects in activating a waiting module with higher priority. Especially the module having defined the initial value of the token count may regain control of the SD channel. This strategy guarantees the real time access for one module at the time. Several modules working in parallel with higher priority have to coordinate their activities and to manipulate the token count in a cooperative way.

3.4 Reduction of transfer delays by optional use of a "hardware bridge"

Concerning data transfers at the SD channel, InterBus looks like a closed ring. The token will be passed in the direction of the data shift process and rotates in the ring. In every case, a full rotation in the ring causes a minimum delay of one InterBus cycle.

This results from the need of buffering the shared data in the last InterBus slave, the SD organizer.

Standard implementations of InterBus masters take one bus cycle to process the new input data. During this processing time the next InterBus cycle is already running transferring data previously processed. This results in an additional delay of two bus cycles for a data packet or the token to be passed on to the SD organizer. Being sent by the SD organizer as its new input data in the next cycle, this data packet or token returns to the sending modules after a delay of three bus cycles.

This great delay time can be reduced to its minimum value of one cycle utilizing a "hardware bridge" bypassing the master in an external or internal realization (Fig.3). The hardware bridge has to copy the SD frame slot directly in the output data of the master. This effects in the direct transfer of these data to the SD organizer. It is possible to do so because the shared data are the first data incoming at the InterBus master. On the other hand, the first two output bytes of the master are forming the loop back word which is well defined. While sending of the loop back word takes place, the hardware bridge is able to receive and buffer the first two bytes of shared data. Following the loop back word, the output data directed to the SD organizer are sent by the master which will easily be replaced by the hardware bridge with the already received SD input data. Therefore the hardware bridge has to buffer two bytes only. Of course the bridge has to be configured according the size of the SD channel. Realizing the bridge by external hardware, the SD output data of the InterBus master are overwritten in any case. Thus the standard master can not be included in the SD communication. If the master should be able to participate with the SD communication the functionality of a SD module has to be added independently to the master. This can be realized more effectively by an internal realization of the hardware bridge.

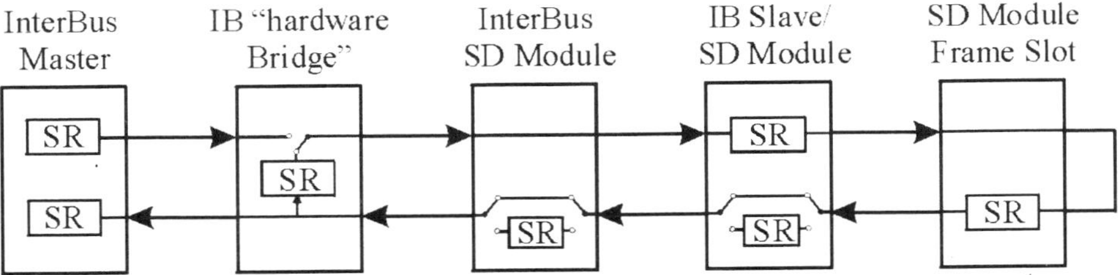

Fig. 3. Option: hardware bridge

4 Requirements for shared data modules

If a module has the task to access the SD frame slot of the total frame of the InterBus it has to fulfill the following requirements at least partially (Fig.4):

- recognize the corresponding position of shared data within the frame (leading data of the cycle)
- listen to and shift in the flow of data (read)

28

- modify the data flow by shifting out internal data (write, write and read)

- calculate a CRC (cyclic redundancy check) information

- interpret address and control information according to the protocol

In contrast to a standard InterBus slave module the data shift register of a SD module is not included in the shift path at any time. The output shift register will only be activated by a multiplexer if the output data have to be placed on the SD channel (Fig.4). Normally the incoming input data will pass the module directly. The protocol execution which is controlled by an interpretation of the header bytes of shared data has to guarantee the placement of every byte at the right time.

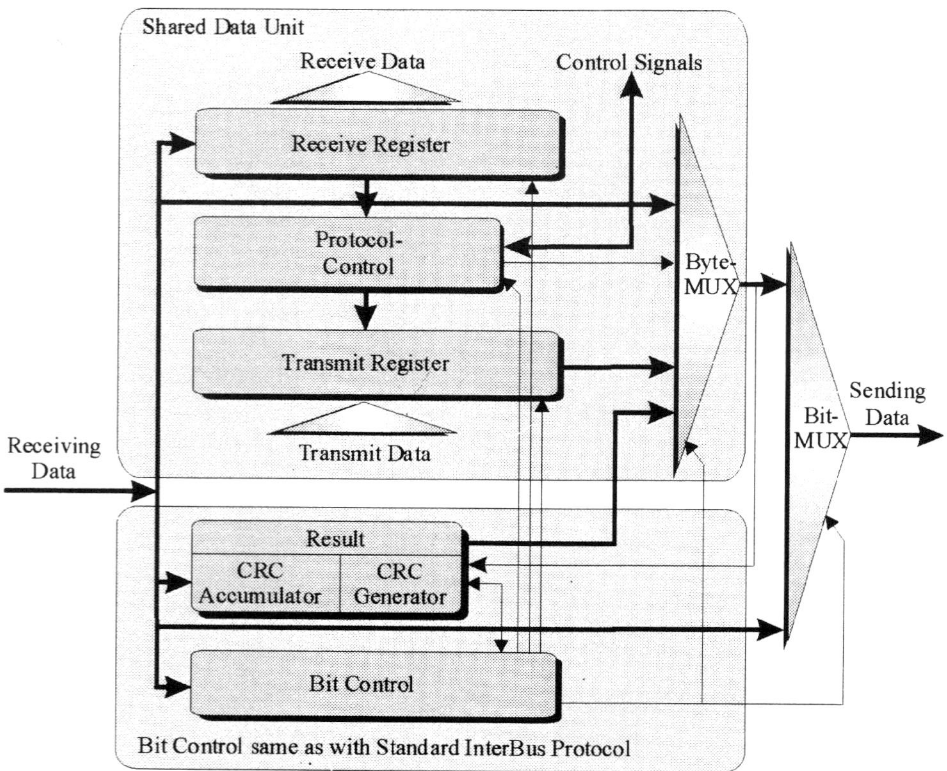

Fig. 4. Block diagram of a Shared Data module

A basic assumption for inserting data in the SD channel is that receiving and interpreting a incoming byte may lead to an immediate placement of internal data on the SD channel beginning with the following byte. This allows the following applications:

- If an SD module waiting for a token receives the token bit set in the first incoming byte, the module has to transmit its own output data immediately beginning with the second byte of the SD channel. The first byte of the own output data (the second byte of shared data) includes the transfer code (and the priority bit) which characterizes the meaning of the transfer for all other modules. In any case, waiting for a token the module has to replace the first byte by a zero byte.

- If an SD module, which does not have the token, receives a transfer code in the second byte of shared data which has to be replied to with internally prepared data, the module has to transmit its own output data immediately beginning with the third byte of the SD channel. This transfer type allows both receiving and transmitting data with the same operation (write and read) which can be used for bidirectional data transfer between two modules within one rotation at the ring. In the lack of internally prepared data, the module has to respond with data bytes with zeros to the transfer code.

A principal consequence for the construction of a SD module is that the analysis of the token bit (set / reset) and of the receipt of some specific transfer codes has to be made by hardware to yield the necessary reaction speed. However, the implementation cost of the hardware solution remains small. The main part of more complicated analysis and preparations of the protocol can be performed by software actions because the necessary information is previewed one bus cycle ahead. This software analysis can be executed within the exactly defined bus cycle time and is especially concerned with

- interpreting and calculating the token count which has to be transmitted with own data transfers,

- deciding to call the modules with higher priority by setting the priority bit within the next data transfer.

5 Summary

The presented method of transmitting commonly accessible data in a SD channel of the total frame protocol expands the universality of InterBus. Contrary to the standard behavior of InterBus, possible connections between modules are not totally available in parallel. On the other hand, a great number of participants can be included in the system without increase of cycle time of the bus and without reconfiguration of the InterBus installation. This allows to enlarge the number of services with improve test, diagnoses and further utilities at the bus.

References

1. Alfredo Baginski, Martin Müller: InterBus-S: Grundlagen und Praxis, Hüthig Verlag Heidelberg, 1994.

2. Phoenix Contact (Hrsg.): Grundkurs Sensor/Aktor- Feldbustechnik, Vogel Buchverlag, 1997.

3. High Efficiency Communication Subsystem for Small Data Packages, volume 2: InterBus-S, prEN 50254-2, 1997.

4. L.M.Casey, R.C.Dittburner, N.D.Gamage: FXNET – a Backbone Ring for Voice and Data, IEEE Communication Magazine, vol. 24, no. 12, dec. 1986.

Adding Multi-Master Capabilities to Interbus-S

Salvatore Cavalieri, Antonio Consoli, Orazio Mirabella

University of Catania, Faculty of Engineering
Istituto di Informatica e Telecomunicazioni
Viale A.Doria, 6 95125 - Catania, Italy

Abstract. Fieldbus technology is achieving more and more attentions throughout all the different sectors of industrial automation. They were born as local area networks dedicated to applications in the area of data acquisition and control of sensors and actuators inside the factory. Nowadays there are many fieldbus systems commercially available, and each of them is peculiar for particular kinds of applications. In this scenario the Interbus-S system is playing the role of a main character. Born under the name of the "sensor-actuator" bus, it has been appreciated by control systems designers and technicians essentially for two important qualities. Firstly, the reliability of its open-architecture, being supported by a variety of products coming from different manufacturers. Secondly, for the efficiency of its own protocol, that makes it a very interesting suite for real time applications. Notwithstanding, the architecture suffers the centralization of the mono-master arbitration scheme, which make it unsuited for a wide range of applications. Maintaining full compatibility with the existing standard, we have studied, defined and implemented a protocol extension which overcomes these limitations and makes it ready for new scenarios.

1 Introduction

The transmission performance of a fieldbus network in process control applications is measured in relation of two different classes of data. The first class is concerned with process data: they are pieces of information not longer than a few bits, have a direct effect on the process to be controlled, they are generated periodically and they have to be delivered on a real-time basis. The output of a temperature sensor, the set point for a controller or the status of a switch are examples of process data. The second class is composed by parameters (or messages) and they have characteristics totally different from process data. They are usually long bit strings, even several Mbits long, they do not have strict deadlines and they are not periodic. Parameters are used to monitor and program intelligent devices. While process data do not require special error control or security mechanisms, parameters need to be transmitted with the implementation of more complex protocols for flow control, error control and error recovery. It is desirable that a network for process control guarantees deterministic delivery of process data even when large block of parameters are transmitted over the same cable. This is a main problem that all Fieldbus protocol must take into account. Another problem is the protocol efficiency, that is, the ability to exploit the bandwidth available.

Most Fieldbusses use a Data Link protocol that makes use of a frame for each information to be transferred. The need of transferring a lot of redundant information (i.e. headers, trailers, addresses, control fields, etc.) wastes a consistent amount of bandwidth. The Interbus-S Fieldbus has the ability to solve the two problems outlined above [1].

Interbus-S is based on the summation frame protocol, which is a variation of the TDMA (Time Division Multiple Access) transmission procedures. According to TDMA, every device is allocated a time slot with duration proportional to the word length it has to transmit, according to its function. This makes it simple to calculate the transmission time as the sum of all the time slots. In this way, the protocol can guarantee the delivery time of each periodic information, thus satisfying the main requirement of a process control system. Moreover, Interbus-S is often referred to as one-total-frame protocol. In fact all the data to be transmitted from all the devices are inserted into a one frame. This frame is prepared in the outgoing interface of the master and then it is shifted along all the peripherals connected to the network (slaves) until it comes back to the master, through its ingoing interface.

The use of a single frame as a carrier for all information, save a significant portion of bandwidth, this way providing Interbus-S a much higher efficiency if compared with other existing Fieldbusses.

Even if these two features make Interbus-S very attractive for many applications, a significant limit exists. The adoption of a centralized management, based on a Master-Slave communication concept, limits the kind of information exchange the system can perform, by preventing two devices to directly exchange information.

In this paper, we propose an extension of Interbus-S protocol that can overcome these limitations still maintaining full compatibility with the existing protocol.

2 Some Remarks on Interbus-S Protocol.

In Interbus-S, a network is configured as a ring, with the two extremities connected to the ingoing and outgoing interfaces of the master [1]. Every slave is coupled to the bus by means of shift registers. Shifting on these registers the data of the frame makes transmission possible.

There are two types of frames in the Interbus-S protocol: the identification frame and the data frame. The identification frame is used by the master to acquire knowledge of network configuration, i.e. number, type and loop position of the slaves, data width, duration of cycle scan time, and so on. Data frames interest process and parameter data transmission during normal scan cycles.

Every data frame prepared by the master is composed by a loopback word, marking the start of the frame and user data for input devices (actuators). When the frame arrives to the master after a scan cycle, it is composed by data from output devices (sensors) or process image, and the loopback word. This is possible because in every slave as soon as data is received from the ingoing interface, output data replace input data on the frame while this is shifted along the ring. The scan cycle terminates with the transmission of a CRC field and a control word, which represent, together with the loopback word, the only overhead added by the DLL. This makes Interbus-S one of the most efficient networks in terms of information bit per transmitted bit. By incorporating all the data from peripherals into one total frame, overhead is simply reduced to the minimum. And since frames take always the same time to be scanned, Interbus-S can guarantee deterministic delivery times. Provided that the last cycle is completed without errors, process data is delivered as soon as the frame finishes its loop, and the user can calculate cycle scan time basing on the characteristics of the network. Interbus-S can also transmit large blocks of parameter data without affecting the times of process data delivery. In fact cycle times do not change even when large blocks of parameters are transmitted together with process information. This is possible because parameter data follow a dedicated path during information transfer. In every scan cycle there is a certain amount of time slots reserved for message transmission.

3 Why a Multi-Master Scheme.

One of the present limits of Interbus-S architecture is its strong centralization, due to the master/slave arbitration scheme. Apart from being a central fault point, the master is the only station where the control algorithm can be situated. This scheme is well suited for simple control applications, where a single PLC interacts with I/O devices. For more complex applications, this scheme may not be efficient. Fig.1, for example, shows an application with two PLCs performing two separate control applications over the same network. PLCs communicate between each other and with a SCADA device (Supervisory Control and Data Acquisition). This control scenario can not be implemented using an Interbus - based solution. The only solution allowed by the Interbus-S architecture, shown by Fig.2, consists of connecting the two PLCs as single slaves. Unfortunately this scheme does not allow direct communication between the PLCs, but only between a PLC (slave) and the SCADA (master), and that is why this does not meet our needs.

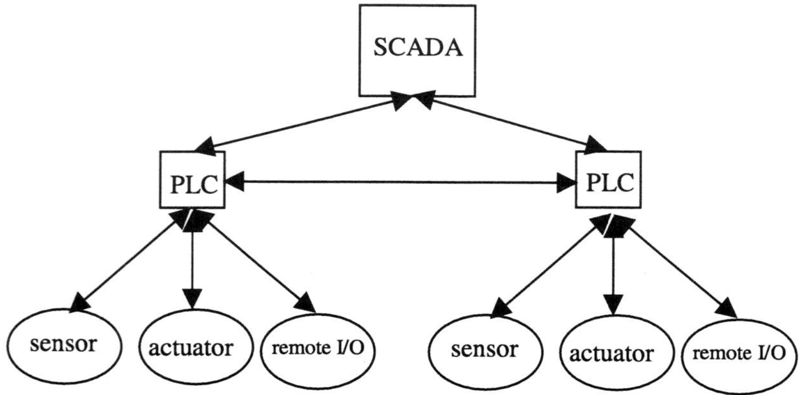

Fig. 1. Control Scenario.

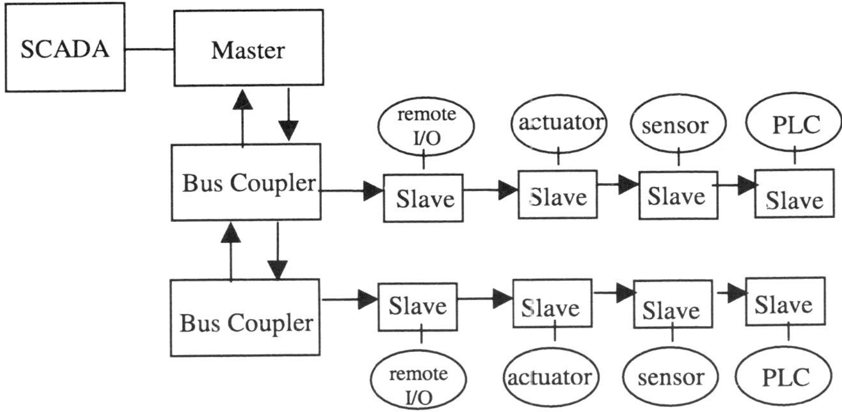

Fig. 2. Interbus-S Solution.

When there is the need to share devices on the regulation loops, the only way to do that is to duplicate devices. It would be desirable for certain kinds of applications to have more than one master over the same network, and a different arbitration scheme, in order to make it possible for different masters to concurrently establish a communication with a set of shared devices. Fig.3 shows how the system architecture can be modified if a Multi-Master approach is used. Each Master manages its own devices maintaining, at the same time, the possibility of exchange data with the other Masters.

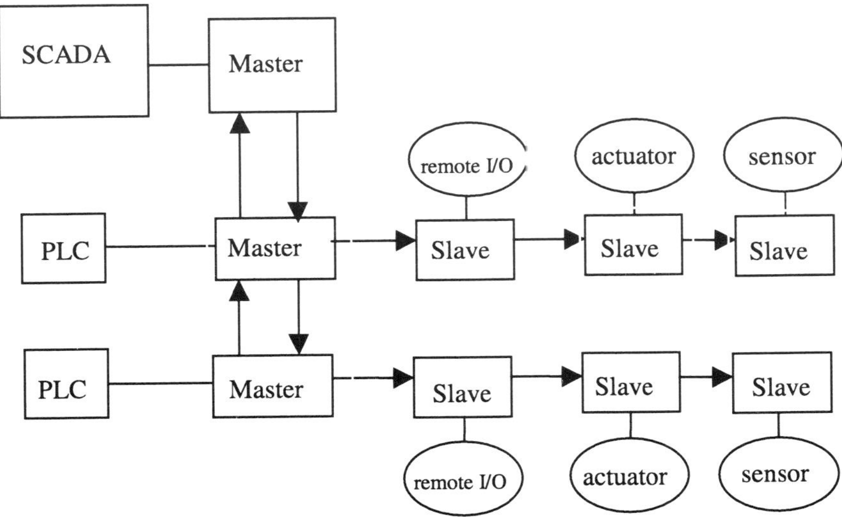

Fig. 3. Proposed Multi-Master Solution

4 The Multi-Master Approach.

The idea of our work is to add multi-master capabilities to the Application Layer protocol defined in the stack of Interbus-S [1]. The Application entity should only be introduced in those stations able to behave as a master, and in the actual DLL master, which from now on will be called scheduler, acting as a real scheduler of the bus. Master functions will only be achieved at the application interface of those devices endowed with the new layer, while the DLL of the network will continue to consider those stations as normal slaves. The main advantage of this scheme is that we still have full compatibility with all the existing slaves, which do not need to be changed.

The way in which the new functions are introduced is very simple. Master stations requiring to manage a loop with a subset of existing devices will have to establish a connection with the scheduler of the bus, which still acts as the only master at the DLL layer. The implementation provides a service by which a master can require the active configuration of the loop to the scheduler, which has achieved it by means of the traditional identification procedure. Master stations are still considered normal slaves at the DLL layer. Process data for input devices are sent from the master to the scheduler through the established connection during a normal scan cycle. The scheduler, in the subsequent scan cycle, will insert them in the time slot of the destination devices. At the same time, the scheduler will collect data from output devices using a first scan cycle. This information will be delivered to the involved masters in the next data cycle. Communication among masters and slaves is possible as long as the scheduler keeps memory of the connections by means of internal tables.

The central point of this new architecture is the scheduler. Being the protocol entity residing on the DLL master, it is encharged of routing the messages belonging to active connections through the DLL interface of the master. Since there might be several active connections at one time, each belonging to a different master, the scheduler makes it possible that data belonging to each connection actually are received by all input stations involved in a connection. If one piece of data has to be received by several input stations, the scheduler has to replicate it into the next transmitted frame as many time as the number of stations having to receive it.

The main functions performed by the scheduler are: managing of asynchronous service requests, collecting process data from output devices and transmitting process data to input devices. These actions correspond to three different protocol entities inside the scheduler, acting separately and simultaneously. They are the Application Base Protocol Machine (ABPM), the Application Receive Protocol Machine (ARPM) and the Application Transmit Protocol Machine (ATPM). They interact with the application entities of the master stations, called Application Master Protocol Machines (AMPMs).

The ABPM responds to the requests of asynchronous services coming from external application entities, that is, those of the master stations (AMPMs). Those services include the establishment of a new connection between a master and the scheduler, the registration of a master in the scheduler tables, and other configuration and management services. The ABPM can negotiate connections with masters wishing to communicate with slave stations. When a master wants to require access to a number of network devices (input, output or input/output stations), it formulates a CR_APDU (Connection Request Application PDU), which contains a connection identifier and a list of all the devices required for the connection. The devices required for a connection must be slave stations. They do not have an application entity and do not take part to the negotiation phase. Therefore the scheduler grants access to slave devices without asking their permission. For every device it requires from the scheduler, the master specifies the loop position inside the network ring, the device code and the type of access it requires. The device code is a 16-bit field which describes the type of the slave, data width and direction, plus other relevant information concerning the device itself. Every physical device, except the DLL master, has its own device code. The type of access (input, output or input/output) must be compatible with the class of the device the master is requiring. For instance a master can require input access for input or input/output stations, but not for output only devices. For every connection established with a master, the scheduler adds a record inside the connection table, which stores information concerning every active connection in the network.

The connection table has a row for every active master inside the network. In the columns there are specified the loop position and the device code of the master station which manages the connection, a priority for the connection, an identifier assigned to the connection by the scheduler and a pointer to the list of all the slave devices which perform communication inside this connection. Some of the fields might be left bank until the master has required a connection. Rows are added to the table whenever a new master becomes active. Masters become active when they invoke the registration service.

Connection management is only one of the tasks performed by the ABPM. Another task is related to the registration of the masters on the network, through which a station informs the scheduler that it is a master and that it can manage connections with a certain priority. A higher priority has to be assigned to those masters which perform more critical regulating loops. A master performs registration on the server transmitting a MR_APDU (Master Register Application PDU).

The last service the ABPM implements is that by which every master can ask the scheduler the active configuration of the network. This information is transmitted by the scheduler using an appropriate APDU called LC_APDU (Loop Configuration Application PDU).

The Application Receive Protocol Machine (ARPM) is an entity which listens on the network and collects all the piece of process data belonging to the connections that are active at a certain time. It is passed the data received by the DLL in the last scan cycle and formulates appropriate transmission requests according to the information stored in the connection table. Transmission requests are inserted into a transmission queue (TQ) with their priority, which is that of the master leading the connection. In particular, the ARPM formulates a transmission request (TR) for every input or input/output device belonging to one connection.

Transmission Requests are processed by the ATPM, which extracts them from the TQ according to their priority, and formulates the DLL service primitives requests for their transmission. The ATPM also optimizes transmission combining together all the transmission requests in order to reduce the number of bus cycles and minimize the number of DLL frames to be transmitted. In this way process data transmission is delayed by one scan cycle time. In fact transmission of a piece of information coming from a slave devices need to be collected by the scheduler in a first scan cycle, and then delivered to a master in the subsequent scan cycle. At the same time, data transmitted from a master to a slave need to be transmitted to the scheduler first, and then they will be routed to the receiving slave stations. This process includes the possibility to share devices among several masters, through data replication. Output data collected from a sensor might be transmitted from the scheduler to all the stations that require them.

The last entity defined in the application layer is the one performing the master protocol functions, that is the AMPM. The first action performed by a master is to registrate to the scheduler. In this way, the scheduler knows that the master exists and may require access to slave devices and process data. A master registrates itself transmitting a MR_APDU (Master Register Application PDU). In order to require access to a subset of devices, a master has to know actual network configuration. For this purpose, it produces and transmits a LR_APDU and is responded with a LC_APDU by the scheduler, supplying all the information required. Once it knows the kind and number of the devices in the network it can formulate connection requests, using CR_APDUs. Connection is granted when a CC_APDU is received. Masters can require, handle and cancel connections as long as they are in the registered status. They exit from this status only if they explicitly ask for de-registration or if network configuration changes. In this case the scheduler cancel their entry in the connection table and they have to register and ask for loop configuration again.

Fig.4 shows how the control scenario depicted in Fig.1 can be realized with our solution. As can be seen, now the master devices can communicate with their slaves, through the scheduler.

5 Conclusions

In this paper we have proposed a particular strategy able to transform the Interbus-S protocol (which is intrinsically Master-Slave) into a Multi-Master protocol, this way making it ready for new applications and improving its potential market. The protocol extension has been obtained by adding suitable application processes in those nodes which must perform as Master. Doing this we have obtained several useful results. First, all application protocol functions have been implemented using parameter messages supported by the DLL protocol. In this way zero overhead has been added to process data exchanged by masters and slave devices. In other words, all the protocol PDUs travel through the parameter channel and do not affect process data transfer. As a result, implementation of the regulating loop is completely transparent. It does not have to worry about how many masters will share a device. Information transfer has been delayed by one scan cycle time, which is very little, considering band efficiency provided by the protocol

Second, we have full hardware and software compatibility with existing slave stations, which can be used in the new framework without any change. Master boards can be added a piece of software in order to produce scheduler stations. Application masters can be implemented using existing stations coupled with a microprocessor system performing the action of the master entity. The only requirement is that master stations are devices provided with parameter channel.

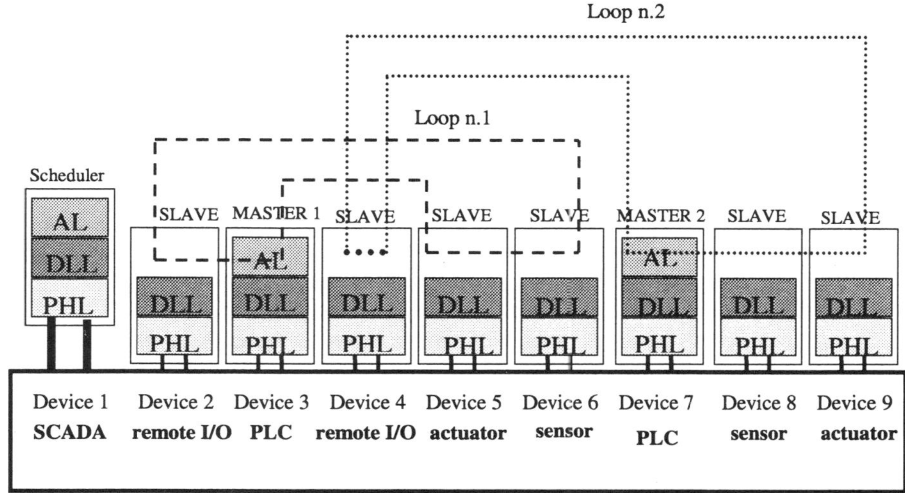

Fig.4. An Example of the Proposed Solution

References

1. CENELEC, EN 50254 Volume 2 "High Efficiency Communication Subsystem for Small Data Packages".

Telecontrol Standard IEC 60870-6 TASE.2 Globally Adopted

Karlheinz Schwarz

Schwarz Consulting Company (SCC)
Im Eichbaeumle 108, D-76139 Karlsruhe, Germany
schwarz@scc-online.de, www.scc-online.de

Abstract. The Information exchange has taken on increasing importance for the restructured energy industry. A standard communications protocol that powers this information exchange, even between incompatible communication systems, enables improved system performance, often accompanied by dramatic cost savings. Developed collaboratively by a team including energy companies, vendors, and EPRI, the Telecontrol Application Service Element or TASE.2 (IEC standard 60870-6 TASE.2 – also known as the Inter-Control Center Communications Protocol or ICCP in the United States) facilitates seamless exchange of time-critical data over local and wide area networks. TASE.2 is the most capable, widely adopted open communications protocol available to the electric power industry today. TASE.2 defines open functions and an object model for application data. TASE.2 supports the integration of instrumentation and control into the corporate wide information and data processing in any application domains, e.g., utilities, manufacturing, or process control.

1 TASE.2 is based on proven technology

The energy markets are characterized by the deregulation worldwide. Production, transport, distribution and sale of electric energy were carried out by a single enterprise till now. In future, these units will form independent companies and compete with each other. Thus, it is becoming possible that a company sells energy of power stations belonging to different companies.

Worldwide, the electric power industry uses more operating data than perhaps any other industry. Energy companies, transmission companies, grid operators and other industry parties obtain a constant flow of information - monitoring and control system data, scheduling data, energy accounting data, and system operator messages. Seamless communication of this information to appropriate locations smoothes the process of generating, transmitting, and distributing the world's electricity. To achieve this objective, in the past energy companies developed multiple communications protocols to facilitate exchange of these data from one location to another. Unfortunately, these protocols were often developed on an as-needed basis, leading to a proliferation of proprietary, incompatible protocols.

At the same time, the power industries in many countries began undergoing fundamental changes. For instance, the United States has witnessed an increase in cooperative enterprises such as power pools and regional centers. More than ever, U.S. industry participants needed the seamless exchange of data but found themselves hampered by the technical and economic limitations of incompatible communications protocols.

In Europe, the European Union Directive 96/92 EU on the Internal Market for Electricity calls for the opening of transmission systems throughout much of the subcontinent for marketplace use by February 1999. The Union for the Coordination of Production and Transmission of Electricity (UCPTE) is responsible for implementing this requirement. Clearly, the need for a standard communications protocol transcends national boundaries. With most energy management system (EMS) vendors offering their products internationally, a globally-recognized standard protocol became necessary. TASE.2 was developed as an open interface by energy utility providers in cooperation with manufacturers of control systems.

The design goals of TASE.2 were:

- higher safety of the plant,
- lower costs of components,
- reduced costs for installation and operating,
- shorter times for planning, design and installation,
- simplified selection of the devices and systems,
- increasing interoperability,
- lower training costs,
- higher usage of the operating resources,
- vendor independency,
- more support by the supplier, and
- use of generally available industrial solutions.

TASE.2 is part of the Utility Communications Architecture (UCA version 2). The IEEE SCC 36 (Utility Communications Architecture, UCA) has unanimously decided in April 1999 to publish the UCA Version 2 specification as IEEE Technical Report (IEEE TR 1550) in July 1999. The Utility Communications Architecture (UCA) is a standards-based approach to utility communications which provides for wide scale integration at reduced costs, and which solves many of the most pressing communications problems for today's utilities. The UCA is designed to apply across all of the functional areas within the electric, gas, and water utilities.

2 Overview about the TASE.2

TASE.2 comprises three documents:

- IEC 60870-6-503: TASE.2 Services and Protocol
- IEC 60870-6-802: TASE.2 Object Models
- IEC 60870-6-702: TASE.2 Application Profile

The standard has a three-step number. The proceeding number 60870 stands for "Telecontrol Equipment and Systems". The second number is the 6, it stands for the project "Telecontrol Protocols compatible with ISO and ITU Standards". At third place are the individual documents, both of the series TASE.1 (ELCOM based) and of the series for

TASE.2 (ICCP based). The documents IEC 60870-6-1, IEC 60870-6-2, IEC 60870-6-3, and IEC 60870-6-4 define common parts of the two series. The part IEC 60870-6-503 contains the core functionality of TASE.2. The service and protocol definitions (IEC 60870-6-503) are based on MMS (ISO/IEC 9506). MMS and the underlying protocols are be mapped to ISO transport protocol or to TCP/IP (RFC 1006 or RFC 1070).

The TASE.2 model defines the communication between two control centers. For the communication, TASE.2 models a control center as a TASE.2 server in which the objects (application data) and the defined functions are located. Another control center (the TASE.2 client) accesses that server. The TASE.2 client invokes a function which is then executed in the TASE.2 server. The TASE.2 server answers with the corresponding actions and return codes.

TASE.2 client and server describe the locality of the data (TASE.2 objects) and of the functions (TASE.2 services) and the direction in which a function operates. The function Read Variable, for example, operates from the client to the server, i.e. the client reads the values from the server. Real implementations (computers) generally support both tasks (mostly even on a single connection). Thus, one control center e.g. can read data from another one and vice versa.

TASE.2 offers a series of functions which are generally demanded in the process data communication. The following list shows the functions. The corresponding TASE.2 objects are listed in parenthesis:

- **setting up of connections** between partners with authentication and password protection (Association),
- **access only to data which are unlocked** (Bilateral Table),
- **definition of names and structures** of application data, for example single point information, double-point information with/without quality information/time stamp (Data Value, Data Set),
- **one-time inquiry** of messages, status information (Data Value, Data Set),
- **spontaneous transmissions** of simple or complex information (Transfer Set),
- **notifications** of a change and event indications of arbitrarily simple or complex information with or without buffering (Transfer Set),
- **general interrogation** of arbitrarily simple or complex information (Transfer Set),
- **transmission of equidistant values** in real time or from archives, with/without buffering in the data source (Transfer Set),
- **exchange of long data areas** (Transfer Set),
- **remote definition of groups** with assigning of application-oriented names (Data Set),
- **remote configuration** of the reporting behavior of the data source, for example: log-on and log-off, cyclic transmissions, transmission at change of value, acknowledgment of critical information (Transfer Set),
- **control of devices** with exclusive access right (Device),
- **program control** (Program),

- **exchange of information for deregulated energy markets** between control centers and between control centers and power stations or substations.

3 TASE.2 products

A high quality of the standard was obtained by standardizing and implementing at the same time. IEC 60870-6 TASE.2 is implemented in different forms such as:

- Integrated into a control system.
- As gateway to the port of a proprietary control system (front-end processor to existing control centers).
- As software for example under Unix or Windows 3.1, 95 or NT (OEM-Products).

Among other, the following manufacturers offer one or several of these solutions:
- ABB Systems Control,
- Bailey Network Management Systems,
- CAE,
- Cegelec ESCA,
- Cycle Software, Inc.,
- GE-Harris Energy Control Systems,
- Landis & Gyr,
- QEI,
- Repas AEG,
- Siemens Power Systems Control,
- SISCO,
- Tamarack,
- Valmet.

At first these implementations were intended for the communication between control centers. In the meantime, implementations for smaller devices (MMS for embedded Systems) have been developed, for the most important MMS functions:

- Implemented base models of MMS: Association control, VMD Support, Read, Write, Information Report, Domain Management, Event Management (basic).
- Conformant to the requirements of the EPRI MMS forum (for ICCP, ...).
- Standard ANSI C code.
- Code Space: 9 kByte to 25 kByte.

This MMS solution is already integrated in several devices.

Shortly after having published the TASE.2 standard as IS (International Standard, IEC 60870-6 TASE.2), far more than 100 projects had already been carried out. Also, TASE.2 is more and more being used for the coupling of stations and power substations as well as substations with power control systems.

4 UCPTE Real Time Information Communication Architecture

Electricity market deregulation as imposed by the European Commission has come in force in April 1999. A result of this will be the growth of energy exchanges over the international high-voltage network (380 kV). At present this network is already intensely operated and the members of UCPTE (Union pour la Coordination de la Production et le Transport de l'Energie Electrique - Union for Coordination of Generation and Transmission of Electrical Energy in Western Europe) are fully responsible for energy exchange and power system security. In this changing electricity market, measures must be taken to avoid propagation of incidents throughout the European network. This requires also the putting into place of a network to automatically exchange the relevant data.

At present the international high-voltage grid of UCPTE is interconnected by AC-links with CENTREL and also with NORDEL and Great-Britain by some DC-links. More than 10% of the total generated energy of UCPTE is already exchanged and the members of UCPTE are fully responsible for the management of energy exchange and power system security. Today plenty of information is already exchanged more or less on a bilateral basis in order to make possible the international energy exchange - but there is until now no common UCPTE-information exchange system. At present, practically all information exchanged between UCPTE partners is by facsimile or by telephone.

The member countries of UCPTE are: Austria, Belgium, Croatia, France, Germany, Greece, Italy, The Netherlands, Portugal, Spain, Switzerland, and The Federal Republic of Yugoslavia. At present, few UCPTE partners exchange real-time telcontrol information using this recommendation, other partners use different communication technologies in complex and inadequate ways.

By the liberalization of the electricity market in Europe an important increase of the energy exchanges and energy exchange players is expected. Following the extension of UCPTE's synchronism area, combined with the significant evolution experienced in new powerful protocols since 1982, it has appeared necessary to review the 1982 recommendation. The objective is to define an information exchange system between the European Transmission System Operators (TSO) which enables an automatic exchange of relevant net-work security data in real time or close to real time.

Three protocols were considered particularly suitable for this type of data exchange: ELCOM-90, TASE.1 and TASE.2. For reasons of openness and interoperability the TASE.2 protocol was finally recommended for data exchanges within the UCPTE (this is also the protocol adopted by IEC as the International Standard IEC 60870-6 TASE.2). The low-level protocol will be TCP/IP, installed in a meshed network dedicated to this project. The UCPTE partners wishing to automatically exchange data with other partners have been recommended to use this standard.

The aim of an UCPTE working group is to draw up a comprehensive technical specification describing :

- the implementation of a TASE.2 interface

- a list of available object-models
- the nomenclature rules for these objects.

The specification has been accepted by all the UCPTE partners at the beginning of 1999 so as to ensure full international compatibility.

The UCPTE ad-hoc group was created in September 1997 to define the infrastructure of such a real time information exchange system. This new process data network is named URTICA: UCPTE Real-Time Information Communication Architecture. This "ad-hoc working group" was assisted by an "operating subgroup" for the functional analysis and an "experts subgroup" for technical aspects.

The URTICA-System is not a public information system for all energy exchange players on the market because the technical requirements to exchange financial and economic data are completely different. Nevertheless a special link from URTICA to such a commercial system is planned because a small part of the information can also be important in the commercial system. In order to manage the exchange of purely economic financial and commercial data (prices, ex-change contracts, orders etc.) a separate, more open information exchange network system will be created.

The objective of URTICA is the definition of the specifications for a common architecture for inter control center real-time communication between UCPTE partners based on open standards. It bas been strongly recommended that CENTREL follows the proposed URTICA specification to allow a seamless communication throughout the western Europe synchronized transmission network. This common architecture enables each of the partners to achieve the necessary data exchanges with any other partner without having to invest in costly individual hardware and software solutions. These specifications cover all current needs to exchange power system security related data:

- real-time network security-related data:
- voltage, active and reactive powers;
- breakers and disconnectors status.

non real-time data (optional):

- schedules of energy exchanges;
- energy accounts: metering data, exchanges accounts, imbalances, ...
- messages between operators (i.e. outage information) , electronic mail;
- network models.

This specification does not impose which data must be exchanged between UCPTE partners to ensure network security. The basic principle of data exchange between two partners is based on a bilateral agreement: only the two partners are concerned by the data they want to exchange.

The recommendation does not prevent using any particular solutions already in place as a result of existing bilateral agreements. However, each partner is expected to implement the recommendation without unnecessary delay as soon as one of his partners notifies him his wish to exchange with him data via the recommended technology. Any other new individual approach would not be welcomed.

5 Early TASE.2 adoption in the USA

While numerous TASE.2 implementation projects are underway in the United States and abroad, a few serve to illustrate the benefits TASE.2 users can realize. An early TASE.2 adopter in the United States, the New York Power Pool (NYPP), completed implementation of a TASE.2-compliant communications system. A consortium of the seven investor-owned utilities of New York state and the New York Power Authority, the NYPP was operating a proprietary communications protocol that had limited capabilities. NYPP recognized that a standardized communications protocol that expanded the pool's capabilities and enabled real-time exchange of data would best serve its members in the changing business environment.

Because of the protocol's standardized nature, the NYPP can now utilize the most advanced telecommunications technologies, such as frame relays and ISDN lines, to expedite data transmission. The lower initial cost of the TASE.2-compliant system, com-pared to a proprietary system, provided immediate saving - estimated at $300,000. In addition, the pool's recurring communication costs, such as telephone charges, will be cut in half, saving NYPP an estimated additional $780,000 over five years. Moreover, the system will also provide a communications gateway into the United States for Hydro Quebec, one of the Northeast United States' major power providers.

These savings are typical of the early adopters of TASE.2 in the United States during 1995 and 1996. In 1997, competition and standardization reduced TASE.2 system costs even more – by as much as a factor of four! This price reduction occurred as vendors sold TASE.2-compliant systems as a fully developed standardized product. "Further reductions of more than an additional 40% are feasible in 1998 and beyond," says EPRI's David Becker, "as computer hardware costs decrease and communications system software is increasingly run on relatively low-cost operating systems such as Windows NT."

As a result of these advancements, the cost of transmitting information between power industry sites has decreased substantially. "TASE.2 has been the catalyst for this cost reduction, illustrating the value of standardization," says Becker. "The user, vendor, and technical community are all benefiting."

The collaborative efforts that have produced the TASE.2 standard are reaping rich rewards for energy companies worldwide. The protocol has gained widespread acceptance over the past year, with numerous vendors offering TASE.2 products. There are an estimated 150-200 completed or current implementations of TASE.2-compliant systems in the United States.

Without TASE.2 utilities would need to establish a variety of independent grow-as-you-go, point-to-point links. Since all major EMS vendors have adopted TASE.2, utilities can use the same protocol to communicate between all of their control areas and members, regardless of the EMS equipment they use.

6 What Is Next?

In a recent talk, EPRI's David Becker sees a continued rollout of more site implementations of TASE.2 from multiple vendors due to real-time data needs and the low costs of implementation in the USA. Europe is following close behind and Asia and Central America are beginning implementations.

In the United States, EPRI has been a catalyst for the development and adoption of TASE.2. A global organization, EPRI is available to work with adopters of TASE.2 compliant communications systems throughout the world to meet their unique needs. EPRI also maintains support of a USA TASE.2 technical working group to review any new issues, monitor interoperability, and review proposed enhancements.

References

1. Telecontrol equipment and systems – Part 6: Telecontrol protocols compatible with ISO standards and ITU-T recommendations – Section 503: Services and Protocol (ICCP Part 1) IEC 60870-6-503, 1998

2. Telecontrol equipment and systems – Part 6: Telecontrol protocols compatible with ISO standards and ITU-T recommendations – Section 702: Application Profile (ICCP Part 3) IEC 60870-6-702, 1998

3. Telecontrol equipment and systems – Part 6: Telecontrol protocols compatible with ISO standards and ITU-T recommendations – Section 802: Object Model (ICCP Part 4) IEC 60870-6-802, 1998

4. Becker, Gerhard; Gärtner, W.; Kimpel, T.; Link, V.; März, W.; Schmitz, W.; Schwarz, K.: Offene Kommunikationsplattformen für die Leittechnik nach IEC 870-6 am Beispiel der Netzleittechnik, etz-Report 28, VDE-Verlag Berlin, 1998

5. Becker, Gerhard; Gärtner, W.; Kimpel, T.; Link, V.; März, W.; Schmitz, W.; Schwarz, K.: Open Communication Plattforms for Telecontrol Applications – Benefits from the New Standard IEC 60870-6 TASE.2 (ICCP), etz-Report 32, VDE-Verlag Berlin, 1999

Chapter 2: Network Interconnections

Papers presented in Session 1.3

Opportunities and Problems of Wireless Fieldbus Extensions
Rauchhaupt, L.; Hähniche, J.

The Use of Radio Technologies in the Fieldbus Area – Using Interbus as an Example
Peter, M.

Internet Technologies and Fieldbuses
Laîné, T.

Further Papers

Security Architecture for Field Area Networks Connected to Internet
Gordeev, M.

P-NET-Management über das Internet
Ayal, M.; Knizak, M.; Sauter, Th.

Wireless Data Transfer System for Oil Drawing Plants
Perez, A.A.; Reguera, V.A.; Paliza, F.A.

Opportunities and Problems of Wireless Fieldbus Extensions[*]

Lutz Rauchhaupt, Jörg Hähniche

Institut für Automation und Kommunikation e.V. Magdeburg
Steinfeldstraße 3
D-39179 Barleben
Tel.:   +49 39 20 38 10-67
FAX: +49 39 20 38 11 00
E-mail: lra@ifak.fhg.de; haeh@ifak.fhg.de

Abstract. Many examples are known for the influence of technologies, originally made for consumer market, to industrial automation. The digital wireless communication technologies, mostly used for mobile telephony, are one of the most rapidly growing branches today. This paper shows that there is an actual need to introduce this technologies to the industrial automation. Many cases can be listed where today widely used fieldbus systems advantageously can be extended by wireless components. But the requirements of these different communication systems are very different as discussed in the paper. Radio technologies are introduced and a scenario is discussed which uses DECT to extend Controller Area Networks. Opportunities and problems of wireless fieldbus extensions are shown.

1 Introduction

An increasing influence of the commercial market can be monitored in the field of industrial automation. This fact is driven by the high number of customers together with resulting low prices and leads to an increasing usage of commercially available products in this kind of application field. Examples for this development are the following:

1. Control Area Network (CAN):
    CAN has been developed for cars. Meanwhile chips are available from many manufacturer and are integrated in an amount of microcontrollers. The application field has been extended from car to automation systems. For this purposes higher layer protocols have been specified.

2. Open interface technology:
    The open interface technology for office applications like text processing, table calculation as well as data base systems was driven by the requirement, to

---

[*] This project is supported by the German Federal Ministry of Education, Science, Research and Technology (BMBF).

exchange common data between this different applications. Examples for this dynamic data exchange are CORBA (Common Object Request Broker) and OLE (Object Linking and Embedding). The OPC (OLE for Process Control) technology forms the basis for future DCS systems in the industrial automation.

3. Ethernet:
   From the beginning Ethernet was developed for the connection of high end computer systems. Due to the high transmission rate, the simple installation as well as the low cost components of this technology is very successful on the market. This is one reason why more and more industrial applications use also this technology to exchange data between control systems.

4. Personal Computer (PC):
   The increasing performance and the availability make the PC technique interesting for industrial applications. Future developments of real-time operation system and the improving reliability of the PC will move the tasks of Hard-PLC to so called Soft-PLC based on the PC.

The above considered examples show the growing influence of commercial products and developments in the field of industrial automation. As we would like to show you later the adoption is not done without any risk.

The wireless communication has been improved enormous in the last few years. This technology will play an important role in the whole life. For this reason we would like to discuss the opportunities and problems of the extension of fieldbus systems by wireless communication.

2 Different Approaches for Communication Systems

Two kinds of Communication Systems are used in the "Public World": wired communication systems like Local Area Network, Wide Area Network or Metropolitan Network, and wireless communication systems like Mobil Phone. The requirements on the mentioned communication systems results mainly from the office applications. A high data throughput is required in this field of applications. The used communication systems are characterized by the following issues:

- File transfer with data packages of several kBytes;
- Environmental conditions: low influences of EMV but no explosive atmosphere;
- System reaction time in seconds;
- Errors during the data transmission are tolerated by repetitions without influencing the real time requirements;
- Using of the communication systems in three-dimensional limited area with extensions by repeater, bridge, router or gateways; delays are tolerable;
- High number of connected devices;
- only few types of different devices (PC or Workstations);
- Ease to install - plug and play;
- Transmission rate in office at least 10Mbit/s;

50

The mobile phone communication is characterized by the quality of a connection and the time to realize the connection. Errors during the mobile communication are tolerable and have effects on the quality of the connection only but will be accepted.

In the meanwhile more and more wireless communication will also be used in the office application because of more flexibility. In order to use the wireless technology instead of wired communication in office application more effort has been done to maximize the transmission safety.

In contradiction to the "Public World" the industrial applications have harsh requirements on the digital data communication. The digital communication systems has to realize a reliable and safety data transmission without any loss of data. The following requirements exist for a digital communication system in industrial applications in detail:

- High dynamic data exchange between decentralised devices;
- Usually short data packages;
- Deterministic bus access for each connected device;
- Remaining bit error probability $<10^{-6}$;
- Using in hazardous environmental conditions;
- Short system reaction time (<1ms in manufacturing automation)
- Limited number of connected devices;
- Many different types of devices (sensors, actuators, controller, hand-held);
- Using of the communication systems in three-dimensional limited area with extensions by repeater, bridge, router or gateways; delays are not tolerable;
- Ease to install - plug and play;
- Transmission rates up to Mbit/s;

The above mentioned requirements make it difficult to use the same technology of the "Public World" in the industrial applications. The available technology for data transmission of the "Public World" is used only in some application fields today. The most problems which have to be overcome are:

- the system reaction time,
- the reliability of the data transmission as well as
- the deterministic, predictable data transmission for each connected device.

However some efforts have been done and will be done in order to use the available chip technology of the "Public World" in field of industrial applications.

3 Motivation for Wireless Fieldbus Extensions

The extension of wired fieldbus systems by wireless components can be useful if one of the following conditions are given:

- a single device, far away
- difficult geographical conditions or difficult installation conditions

- mobile or portable devices
- temporary installed devices
- redundancy and safety

Because a line of side can not be guaranteed between the fieldbus devices in industrial applications infrared solutions are not possible in most cases. So we have to have a look at the technologies, regularities and solutions of radio transmission.

4 Approaches for Solution

4.1 Radio Technologies

The pre-condition for future-oriented solution concepts for wireless fieldbus data transmission is the use of digital radio technologies. So we have to taken into consideration following standards or technologies:

Digital European Cordless Telecommunications (DECT) [1]

DECT is an European standard for cordless phones. It is a cellular system with an indoor transmission range of 50m and an outdoor transmission range of 300m per cell. Nevertheless the equipment is relatively simple and not expensive.

Global System for Mobile Communications (GSM) [2]

The European GSM standard is the most important mobile communication technology all over the world. It is a cellular system with a transmission range of up to 100km per cell used for public phone. The services are put in disposal by providers. Further specifications are in progress.

Industrial Scientific Medical (ISM) Bands [3]

Different frequency bands are released under this label by the regulation bodies of many countries for free use. Some constraints are specified like transmission power and signal coding. You have many degrees of freedom but also many things to do to get an information infrastructure.

IEEE 802.11 [4]

This standard specifies a wireless kind of Local Area Networks (LAN) using the ISM bands. The transmission range is similar to that of DECT but the transmission rate is considerable higher. It is to be expected that devices and software concerning this standard will be available in a big assortment because of the constantly growing PC market and the growing degree of networking.

IEC 1158-2 [5]

Within this standard an air interface is being specified. Due to the consideration of national standards and the restrictions by the regulation bodies DECT and IEEE 802.11 standard are took into account. The standardization is not finished yet.

52

Trans European Trunked Radio Access (TETRA) [6]

A digital communication system especially for administration, police, fire department, logistic enterprises is being specified with that standard. Frequency channels are placed at disposal to providers. The transmission rate is very low. But if such systems already exist e.g. in harbors, airports or big companies it could be advantageous to use the given infrastructure for fieldbus data transmission, too.

HIPERLAN I [7]

HIPERLAN I is an existing standard similar to IEEE 802.11 using a higher frequency band. The implementation is regarded as very complicated. There are no devices available on the market yet.

Universal Mobile Telecommunication Systems (UMTS) [8]

A world-wide communication standard for mobile systems of the third generation is planed with UMTS. The target is to integrate the functionality of DECT, GSM and TETRA in one standard.

Wireless Asynchronous Transfer Mode (WATM) [7]

The wireless version of broad band ATM networks is being specified with Wireless ATM (HIPERLAN II). The aim is to get a fast wireless communication system for voice, picture and video transmission.

4.2 Scenario for Wireless Fieldbus Extension

Figure 1 shows a structure of a fieldbus system with wireless extensions. The numeration of the fieldbus devices means that all devices belong to one and the same fieldbus network. That means the discussion goes not about the wireless connection of two independent networks, but about one fieldbus network with wireless connections to devices or sub-networks. The boxes without numbers represent the connection between wired and wireless transmission. They have not to be fieldbus devices with a fieldbus address. This is what we call transparent transmission.

Regarding the Controller Area Networks (CAN) a transparent transmission via the physical medium is not possible, because the simultaneous transmission and reception of one and the same bit as defined in the CAN standard (ISO 11898) is not possible using radio transmission. Furthermore the intended usage of available low-priced radio devices demands another approach.

One of the most interesting radio technologies is DECT, because of the easy use and the low prices for equipment. DECT is a cellular system consisting of a fixed part (base station BS) and mobile parts (usually hand held phones). Time division multiple access and time division duplex is used to control the access to the air interface within a cell. Twelve time slots for down-link (from BS to Mobile) and twelve time slots for up-link (from Mobile to BS) are specified in one time frame of 10ms. So up to 12 mobiles theoretically can communicate with a base station. Frequency division multiple access and frequency hopping is used to get the best quality of service. Four protected frames

8 byte each can be transmitted in a time slot. This is suitable for wireless communication e.g. for CAN. An interworking unit connects the wired (CAN) and wireless (DECT) communication above the data link layer (see figure 2). Each CAN-Object has up to 8 bytes data and 2(8) bytes descriptors (11(29) bit identifier). So more than one CAN-Object can be transmitted in one time slot but they have to be buffered in the radio devices.

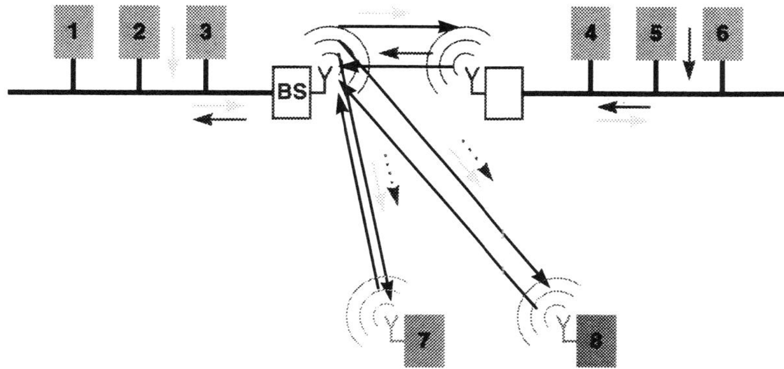

Fig. 1. fieldbus system using wireless extensions

In the down-link frame the CAN-Objects of the left sub-network are transmitted by the base station to all other radio devices. The base station does not filter any CAN-Objects. The transmission is transparent. In the up-link frame the base station gets the CAN-Objects of all sub-networks. The CAN-objects are transmitted in the left network and during the next down-link frame to all other radio devices.

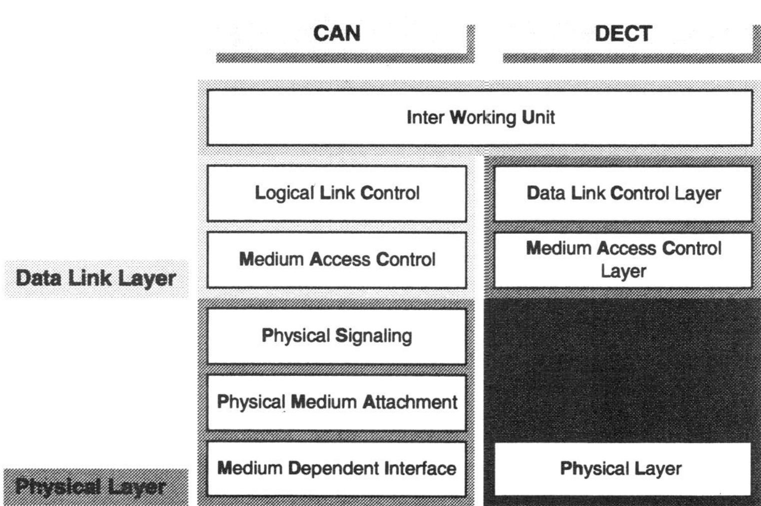

Fig. 2. communication layers used by DECT and CAN

In this scenario a determined transmission over the air interface is guaranteed but limited by the 10ms frame and the 32 bytes per connection. So the data throughput of the DECT system determines the possible baud rate in the CAN. The baud rate of such a system hardly exceeds 10kBit/s.

Because of the limited data length (32 bytes per connection) and the very low data throughput DECT is not suitable for fieldbus systems like PROFIBUS or INTERBUS. For this bus systems approaches with the wireless LAN standard IEEE 802.11 are advantageous. But the medium access control sublayer of IEEE 802.11 has to be changed for a better data throughput. The protocol data unit (PDU) and the access control has to be more adapted to the needs of the fieldbus system. Regarding CAN this is particularly important because its media access control method can't be used for radio transmission.

5 Summary

This paper has shown that available wireless technologies are suitable for industrial communication systems. DECT can be used e.g. for CAN. The frequency hopping method fits to the hazardous requirements in the field of industrial automation. Other technologies are possible (e.g IEEE802.11) but adoptions are necessary and this adoptions have to be done in different ways. The growing improvement of this new wireless technology like increasing of the data throughput or reliable data transmission will overcome the mentioned problems. Thus a wide employment of the wireless communication systems will be possible without limitation by industrial requirements.

References

1. DECT Forum: DECT - The standard explained. Solothurn, Switzerland, 1997

2. David, K.; Benkner, T.: Digitale Mobilfunksysteme. B. G. Teubner Stuttgart, 1996

3. ETS 300 328 Radio Equipment and Systems (RES); Wideband transmission systems; ISM Technical characteristics and test conditions for data transmission equipment operating in the 2,4 GHz ISM band and using spread spectrum modulation techniques. 1996

4. IEEE802.11: IEEE standard for Wireless LAN Medium Access Control (MAC) and Physical Layer (PHY) specification. New York, 1997

5. DIN IEC65C/155/CDV (Entwurf) Feldbus für industrielle Leitsysteme. Ergänzung 3 zu IEC 1158-2 : 1993; Spezifikation der Bitübertragungsschicht (Physical layer) und Definition deren Dienste (IEC65C/155/CDV : 1996)

6. Cox, S.: Land Mobile Radio. Eurotrade Communication, Vol. 2, No. 9, S. 22, 1997

7. Walke, B.: Mobilfunknetze und ihre Protokolle - Band2. B. G. Teubner Stuttgart, 1998

8. Walke, B.: Mobilfunknetze und ihre Protokolle - Band1. B. G. Teubner Stuttgart, 1998

The Use of Radio Technologies in the Fieldbus Area

-Using INTERBUS as an Example-

Michael Peter

Concept Development
Phoenix Contact GmbH & Co., Blomberg, Germany
Email: mpeter@phoenixcontact.com

Abstract. Until recently, data transmission by radio in the fieldbus area was inconceivable. The available radio technologies were too susceptible to interference, cumbersome, expensive and slow. However, new radio technologies such as DECT or the 2.4 GHz ISM band used in 802.11 offer promising new possibilities for fieldbus links which were only previously made possible using costly connection technology. This offers the user both straightforward installation and a higher degree of flexibility, without increased costs. Through the joint project supported by BMBF (German Federal Department of Education, Science, Research and Technology), "Wireless Fieldbuses in the Production Environment" (FUNBUS), and cooperation with the technical college FH-Lippe, initial experiences of using radio technologies on Interbus have been collated from various viewpoints and a product-related prototype has been developed for use on Interbus.

1 Introduction

Serial data exchange over fieldbus systems using traditional transmission media such as copper wires or fiber optics has proved safe and reliable in automation technology. However, for many applications, connecting two devices using cable-based media is not without its problems. A great deal of effort is often required when establishing a physical connection between two devices; there may be rough ground, rotating devices or devices moving at different levels or on free paths over large distances. In these cases trailing cables, slip rings or infrared transmission paths are used which are time-consuming, expensive, require a great deal of maintenance and are affected by environmental influences. In these cases, the use of radio technology offers a higher degree of freedom and is also maintenance-free and cost-effective. However, this technology is still traditionally considered susceptible to interference and unsafe. Industrial users are thus reluctant to use it. However, the radio technologies available today enable the creation of safe and reliable connections between fieldbus devices.

2 Requirements of Fieldbus Technology for Radio Transmission

A fundamental requirement of the use of radio technology in fieldbus systems is that the basic characteristics of the fieldbus are supported: safety of data transport, determinism, easy configuration and monitoring as well as clear and rapid error diagnostics are some of the points which are important here.

INTERBUS, standardized in EN 50254 Volume 2 [1], is a synonym in the fieldbus area for easy configuration, clear and rapid diagnostics, deterministic data transport and safe data transmission which must be reflected by radio technology.

Other important characteristics which a radio system designed for industrial use must display are low costs, subscriber freedom (especially in the use of radio components), flexibility of use and ease of installation. This also includes the use of radio transmission methods without special knowledge of radio technology. These requirements must also be met, so during development greater emphasis is being placed on user-friendliness at low cost.

3 Different Radio Technologies

In principle any radio technology is suitable for the transmission of data. Recent examples of these technologies are GSM, DECT and the ISM (Industrial, Scientific, Medical) bands, including the 2.4 GHz band which is also used in IEEE802.11 [2], to name but a few. However, the aforementioned requirements cannot be met by all radio technologies. The radio systems available on the market are not necessarily free of costs (e.g., GSM), can only transmit at low data rates (e.g., 433 MHz systems) or have long delays due to their principles of operation (e.g., DECT).

This means that evaluation criteria must be set for radio technology. These criteria comprise both the requirements of the fieldbus system and the various characteristics and marginal conditions for radio technologies. They include: ease of operation, cost, resistance to interference, license freedom, possible data rates, future availability (standards), ranges, compatibility, availability of the necessary frequency bands in the desired customer countries and existing distribution. To attain a clear capability profile, criteria evaluation is required. Important aspects can already be seen in the demands made on radio technologies by fieldbus systems. The use of radio technologies should be cost-effective and should not require further administrative effort or subsequent costs which are, for example, commonly charged by network providers (provider charges). They should also be able to transmit a high data rate effectively and to guarantee specified low transmission times to maintain determinism.

Closer examination of these characteristics reveals that only a few radio technologies meet the most important requirements, or have the opportunity to meet them through development or modification [3]. Those worth mentioning are DECT and the 2.4 GHz ISM band. While the transmission characteristics and possible options are specified in the DECT standard, in the 2.4 GHz ISM band only certain marginal conditions must be observed so that it can then be used without additional costs. In addition to determining the transmission technology (spread spectrum transmission) and the maximum transmitter power (100 mW EIRP), the radio system must be approved by the relevant regulating authority. In this frequency range, frequency spread technologies are used which have been developed to reduce susceptibility to interference. In both systems the user has no subsequent costs except for those incurred by operating the devices. The future looks good for both systems, because DECT and IEEE802.11 are either already established standards or will soon be established.

The best technological options are currently offered by the 2.4 GHz ISM band, which enables the development of individual radio transmission systems subject to certain conditions. It is this technology which forms the basis of the research into replacing the Interbus cable with a radio path.

4 Possible Applications

Interbus provides various starting points for the use of radio technologies. Depending on the marginal conditions required in the application, the most suitable selection may be decoupling data from an Interbus device or replacing the Interbus cable. For example, for small quantities of data where handover and roaming are desirable, decoupling data from one or more devices over radio is an option. Depending on temporal requirements, radio systems such as DECT can be used which already offer the aforementioned characteristics as standard. However, for the exchange of large quantities of data with no separate control on the receiver side, the replacement of the cable by a radio path is a better option. The requirements of the radio system vary widely between the two options. While for the decoupling of data from an Interbus device only this data - at the lowest required cycle time - is decoupled and transmitted, when the cable is replaced the entire data flow for a cycle is transmitted full duplex over the radio path. This results in higher demands on the radio technology.

The starting point for the work carried out was the "transparent replacement" of the cable, i.e., the cable between two devices was removed and replaced by a radio path as shown in Fig. 1. Ideally the user will not notice whether a radio path or a cable is being used. This means that the same demands are made of the radio path, and the operation of the radio path is set up in the same way as for a wired transmission.

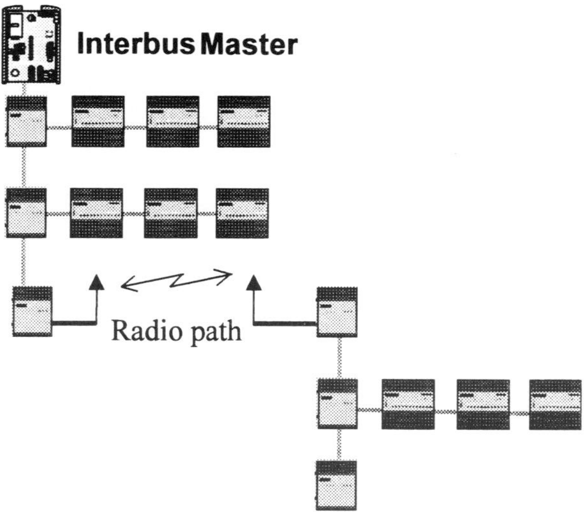

Fig. 1. The use of radio technology in the fieldbus area in using the example of Interbus

A commercial product which transmits data in the 2.4 GHz band using the Direct Sequence Spread Spectrum method was used for this implementation. In the 2.4 GHz band the modulation is specified by frequency spread technology. This spreading offers a higher level of redundancy compared with conventional modulation methods and thus ensures safer transmission. The module used (a commercial product) is able to transmit data semi-duplex at a maximum data rate of 2 Mbps over the radio interface. However, as Interbus transmission is full duplex, this full duplex transmission must be converted to a time duplex transmission with a correspondingly higher data rate on the radio interface. This creates additional bus delays, based on the fact that during transmission from one side to the other the system must wait for the radio interface to be enabled. During this delay, the data arriving over the connected interface is collected there. The delay comprises the duration of the preamble required for radio transmission (approximately 128 bit periods at a typical rate of 1 Mbps/s), the data to be transmitted by radio (which has been received during the delay) and a period for the switchover from receive to send mode - as shown in Fig. 2.

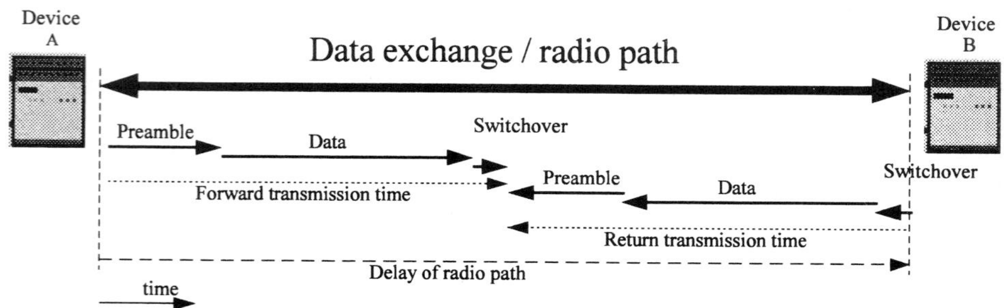

Fig. 2: Time behavior on the radio path when an Interbus cable is replaced

Fig. 3 shows this relationship in form of a message sequence chart.

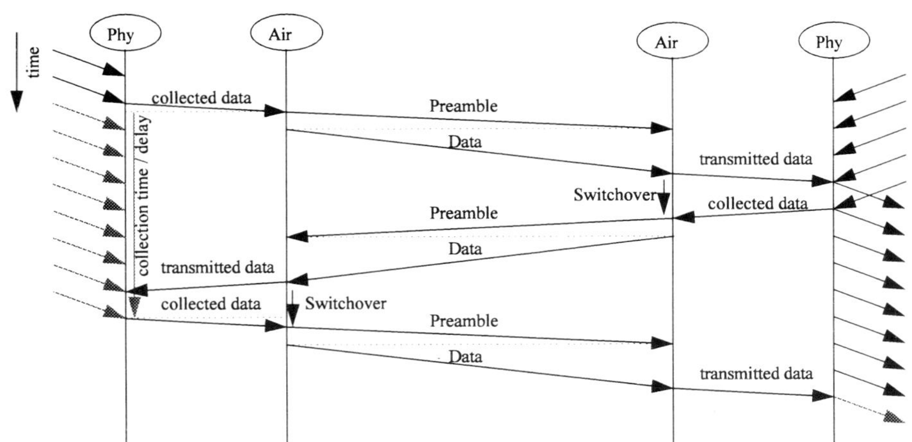

Fig. 3: Timing shown in a message sequence chart

5 Trials / Field Tests

If relatively noise-immune cable transmission is to be replaced by radio transmission, the question of reliability is important. In the past, radio technologies have not enjoyed the best of reputations in terms of immunity to interference [4]. In this area, radio technologies have recently been developed with a strong emphasis on the noise immunity aspect. To verify this, various radio systems were tested under industrial conditions and under different types of interference. The systems tested were the DSSS (Direct Sequence Spread Spectrum) and FHSS (Frequency Hopping Spread Spectrum) spread modulation methods used in the 2.4 GHz band, a DECT radio system and a radio system operating in the 433 MHz ISM band. To gain an impression of the susceptibility to interference of these systems in an industrial environment, they were tested under various sources of interference and obstacles which are typical in industry to determine their error behavior. Environmental conditions which were active sources of interference included galvanizing systems, welding systems, eroding systems and machine tools; passive sources of interference were racks in warehouses and large distances in assembly workshops. These tests showed that all except for the 433 MHz system offer a high level of resistance to interference. The possible ranges within industrial buildings are over 70 m, although the 433 MHz system and the 2.4 GHz DSSS system tested lag behind with lower ranges, which can be explained for the DSSS system by a high degree of sensitivity to multiway distribution. This result (Fig. 4)was however improved considerably by the use of antenna diversity, i.e., the use of two receiver antennae.

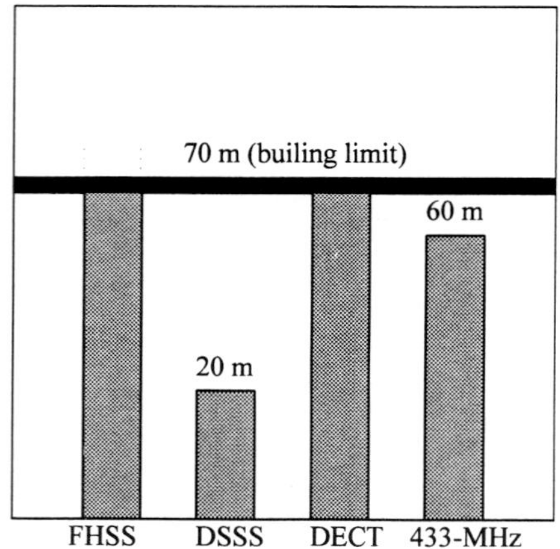

Fig. 4: Example of possible ranges within industrial buildings

6 Outlook

The use of radio technologies in the fieldbus area is currently still in its initial phase. However, great efforts are being made in this area to develop even more attractive and higher-performance transmission systems using new technologies. Higher data rates with improved noise immunity and appropriate transmission ranges, automatic functions such as passing on from stations and addressing several stations, alongside falling costs for the transmission system will lead to a strong increase in the use of radio technologies in the fieldbus area. However, to ensure the safe use of radio technologies, test profiles must be specified which can be used to measure the suitability of developed components under predetermined environmental conditions and thus make it possible to precisely determine their suitability for use.

References

1. EN 50254 „High Efficiency Communication Subsystem for Small Data Packages, Volume 2 (INTERBUS)", October 1998

2. IEEE 802.11 - Part 11: „Wireless LAN Medium Access Control (MAC) and Physical Layer (PHY) Specifications", Jun 1997

3. „Wireless Fieldbusses in the Production Environment (FUNBUS)", BMBF-Project 02PV4062, 1997 - 2000

4. D. Fischer, J. Konopka, „Development and Construction of a Wireless Subscriber Link for INTERBUS", Fachhochschule Lippe (Prof. Meier), Dissertation March 1999

Internet Technologies and Fieldbuses

Thierry LAÎNÉ

ALSTOM Technology
5 Av Newton, F-92140 CLAMART - France
Tel: +33 (0)1 46 29 12 59, Fax: +33 (0)1 46 31 75 45
Email: *thierry.laine@techn.alstom.com*

Abstract. Facing the rapid changes brought in the industry automation world by the wide spreading of Internet technologies, this paper describes what will be the impact of the use of the internet technologies as an improvement of the facilities that were initially brought by fieldbus at the lower levels of the industrial control systems.

1 Introduction

Computer technology, which is the basis of modern control systems, continues to develop at a rapid pace. Before the automation industry has time to draw the full benefits from one generation of technology, the next one has arrived, with its array of innovations that require the current offer to be reconsidered.

Faced with these extremely swift technological changes, users no longer always have the time to invest themselves and assess the pertinence of these new technologies. This causes them to simply follow the opinions of system manufacturers. However, for this reason, they expect effective solutions that are less expensive to buy and maintain.

Naturally, they want to reap the benefits of these new technologies without experiencing the inconveniences resulting from the introduction of such major innovations. Users also attempt to diversify their supplier base to take full advantage of these advances. Therefore, they expect flexible and open systems. What lessons can we learn from these current trends with regard to system improvements and their capacity to handle new functions, allowing more knowledge of, and control over, manufacturing processes.

2 Main directions

Not all technological breakthroughs have the same impact. They must be evaluated over time. Only those which will significantly influence the basic structure of control systems and their integration in the company must be retained.

2.1 Fieldbuses, local networks

The «fieldbus» concept took a long time to emerge. However, today it has become critical in all industries and service sectors. The offer has become more rational, focusing on a few major solutions that have become competitive. The prime qualities of fieldbuses are their speed in transmitting information, their reliability, and their immunity to industrial disturbance. These qualities, which have made fieldbuses successful, will probably help them endure as the natural link to processes. However, the demand for improved passband technology will continue to grow. With regard to local networks, which are older, Ethernet, with its TCP/IP stack, has established itself over solutions standardized by the ISO. However, it is a technological minimum which does not in itself allow open systems to be built.

2.2 Radio communication

Radio communication, long reserved for slow-speed auxiliary communication, is changing its face with the emergence of GSM-Data and high-speed communication. Today, the GSM-Data service allows speeds of 9.4 kbits/s, and will soon offer speeds of 32 kbits/s or 64 kbits/s. This will be suitable for slow supervision from very remote sites. With the freeing of the 2.5-gigaHz band and the adoption of the UMTS international standard, communicating at a speed of 2 megabit/s should be feasible over short distances, i.e. inside a workshop, or over long distances using the infrastructures which will accompany the development of the second-generation GSM.

2.3 Web technologies

This term covers at once the Internet concept itself, the technologies its comprises, and the navigator concept. These technologies have swept over the world of office systems and unified behavior with regard to information processing, in particular through the client/server concept. The mechanisms proposed offer a high level of flexibility for managing information, and can offset the lack of an application layer in Ethernet-based local networks.

These main directions will generate profound changes in control systems. At this stage, we can already examine the first major trends:

- Decentralization of processing and cooperation between equipment of different types.
- Integration of intelligent equipment to meet the special needs of the application and for setup purposes.
- Standardization and non-dedication of certain system components to increase openness.

3 Cooperation Decentralization

For more than twenty years, people have been talking about «distributed control systems» (DCS). In fact, the distribution aspect is a very minor one. In general, all the processors are the same, and they are often installed in the same protected geographic area. They only differ from multiprocessor systems through their serial bus link.

Current technological advances allow us to go much further by giving field equipment a certain degree of intelligence, i.e. by making them programmable for the function requested in the application in question.

This requires that fieldbuses be capable of sending - in addition to the program's critical time data - other information of various types (programs, configurations, the results of self-validation functions, etc.) in a reliable manner, and without disturbing the process.

This decentralization allows information to be processed at the proper level. For example, a motor can be controlled (torque, speed, position) entirely within the variable speed drive, despite its complexity. Likewise, statistical measurement calculations (average value, standard deviation and autocorrelation) will be drawn up in the sensor.

However, calculations involving several devices (control loops, measurement intercorrelations, etc.) will be located in the programmable controller. In this case, a more suitable name for the latter would be «automation controller.»

3.1 Communications

Integrating the functions implies that communication links be created according to need. However, shortages must not occur, even during a strong demand for communication. In some case, especially for communication with the process, for control functions, exchanges can be neither late nor early, since processing - and of course the process - depends on them. In such cases, we stress that communication sizing and the choice of protocols are critical at each level of the system architecture.

3.1.1 Fieldbuses

Despite efforts at standardization, it has become apparent that no universal network will emerge, but rather a small number of solutions. And even if a standardized network were defined, the right profile would still have to be chosen. Today, there is a wide selection, and each PLC or system manufacturer offers several solutions. It is therefore necessary to make the right choice to meet the needs of the application.

The initial setup difficulties have been gradually mastered through the use of tools offering users advanced means to make setup easy and quick - on condition that certain rules and limitations be observed.

On the other hand, strong competition often results in the use of streamlined transmission mechanisms designed for less demanding tasks in terms of the environment, integrity, reliability and longevity. This could damage user confidence, as they will encounter the same problems they experience today in their industrial use of «computer» networks - the very problems the fieldbuses were supposed to solve.

Much has been said, but there are a few key criteria to be taken into account:

1. Cost, which must include all related costs, including cabling and maintenance accessories.
2. Solutions that match the need of the application in question, avoiding excessive and untimely generalizations. Why, for example, choose a network with a theoretical throughput of 100 megabyte/s when an effective throughput of 1 megabyte/s would be largely sufficient?

3. Consistency between different levels. Complex bridges - which do not guarantee end-to-end transmission, and which require special message exchange configurations - should be avoided.
4. Effective performances in the system configuration chosen. For example, a 10 megabyte/s network may not be suitable because it cannot send the same data to several devices, whereas a network with a slower throughput can.
5. The capacity to separate real-time and service traffic, in order to ensure service quality in both cases.
6. Easy setup.

Too many different solutions still coexist, and this situation prevents progress on the issue of interoperability between devices which need to cooperate. However, an increase in the need for passbands and the transmission of increasingly complex information has been noted.

The standardization of IEC 61158 makes the issue somewhat clearer, but it comes too late, on a market that is already very rigid. Each major manufacturer has already installed its solution and does not want to migrate. Given this situation, the market could be tempted by another path originating in the computer world: Ethernet.

3.1.2 Ethernet

Clearly, Ethernet has established itself in many fields, beginning with office systems and all intra-company links. Some see it as the universal network that could replace the fieldbus, even if it would mean using the fast 100 megabyte/s version of Ethernet.

Let us examine its strengths and weaknesses:

1. The connection cost for terminal equipment is low for the 10 Base T, but this is not the case for routers, switches and other network accessories required to sort traffic and avoid unnecessary overloads.
2. Throughputs can be very high - up to 100 megabytes/s - but are such throughputs necessary, and can the field equipment support them. Moreover, when Ethernet is made more robust and therefore reasonably time-probable, its efficiency drops to 20% or even 10%, whereas for fieldbuses it remains higher than 50%.
3. Ethernet is very fast. However, «critical time» information transfers must not be deferred by others who have been initiated first and who are using the line at the time of the transmission. Ethernet handles neither priorities nor pre-empting of certain communications.
4. The cost of connecting and cabling an Ethernet network is low, but how will the cables behave in a difficult environment.
5. How robust is Ethernet in disturbed industrial environments? How will economic redundancy be feasible?
6. Radio communication will be the communication mode of the XXIst century. It develops according to special protocols, and allows links to be developed that compete with those made with Ethernet.

More fundamentally, in the minds of many, Ethernet represents a universal network, whereas in fact it is only one part of the communication protocol. What is actually being referred to are the protocols that use Ethernet, mostly TCP/IP. In fact, it is those layers and the higher layers which are crucial for the future of communication, since they structure exchanges in so-called «client/server» form. So it would be entirely possible to use different protocols beneath these layers to access the medium.

Today, it is already possible to use the following instead of Ethernet:

- ATM for long-distance links,
- GSM and, in three to four years, UTMS for radio links,
- satellite links for high throughputs,
- certain fieldbuses to manage and maintain field equipment.

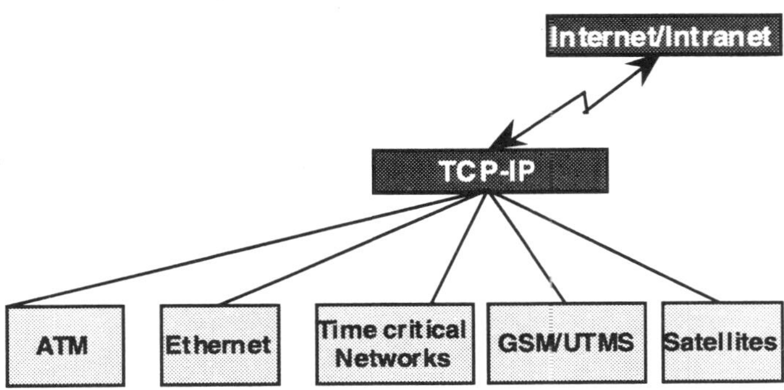

Fig. 1. The ascendancy of IP

The meeting point of all these protocols is IP (Internet Protocol), which is the unifying factor of low-level protocols. Given the use of IP, is would be logical to expect that it will remain a gateway to all communication protocols to be developed over the next ten years.

3.1.3 Coexistence of fieldbuses and Ethernet networks

Given the situation, it is legitimate to wonder about the future of fieldbuses. Ethernet is not a universal solution, but it could compete with certain fieldbuses. But TCP/IP does not have the features needed to carry critical time information. The conventional communication stack is not sufficient. There are two possible ways to get around the problem:

- make fieldbuses TCP/IP-compatible,
- built a special protocol above Ethernet in place of TCP/IP, and seek compatibility at another level.

The second approach, currently being investigated by the Fieldbus Foundation, is an attempt to install the upper layers of Fieldbus IEC 61158 on Ethernet. This would correct some of TCP's flaws, but it would not be sufficient to compete with fieldbuses. However, this second approach does not follow the current technological tide since it proposes a client-server message exchange service that is not recognized by Web technologies. This creates a discontinuity in supervisory-related transmissions, and makes it hard to ensure all the exchanges of a distributed multi-supplier application.

Inversely, by creating a TCP/IP gateway on the data link layer of a fieldbus, Web type technologies can be used up to the field equipment level, without having to cross special bridges. Obviously, in this case, critical time communications that only concern equipment connected to the same fieldbus would continue to use the fieldbus' dedicated communication stack.

The second approach, whose prime goal is to make Ethernet futureproof and isolate the production cell from Web technologies, could become an obstacle to system openness. Following the first approach, which is more rational, would require that only those fieldbuses that can support TCP/IP in addition to their own stack would adapt to a broad dissemination of Web technologies. In this case, only WorldFIP or IEC 61158 type networks would have an increased chance of fitting into this trend.

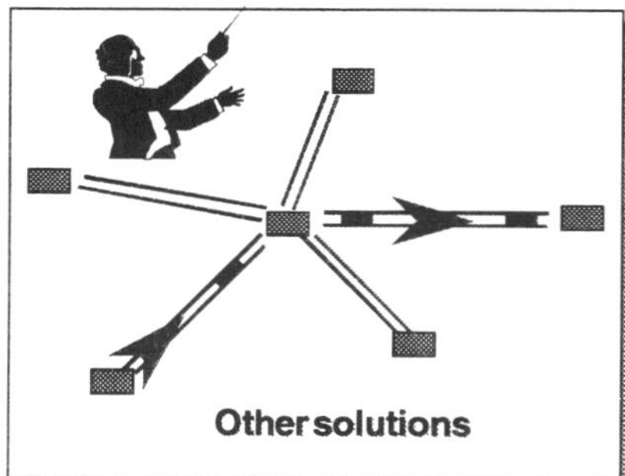

- Two service qualities
- Separate management of qualities

- One service quality
- Volume management of flows

Fig. 2. Cohabitation and service quality

These networks bring unique characteristics to the Web technologies that allow innovative solutions to be built for a wide range of fields: rail transportation, urban transportation, transportation infrastructures, telecommunications control, aerospace, road traffic management, aeronautics, tools for fundamental physics, etc.

4 Globalization

The Internet, and more specifically the Web technologies, are in the process of completely changing business computing. Their advantages open interesting doors for industrial computing, mostly by simplifying the integration of all the tasks that go into making a decision in all the departments of a company.

It is especially interesting to not have to worry about the future use of information when an application is being developed, and to format information simply according to its level of accessibility. The client navigator does the sorting based on the search criteria used at the time. Integrating information becomes easier, and information searches can be modified as desired without conducting prior studies.

With these unique factors, control systems can be considered from a new angle, and real-time control tasks and service tasks such as maintenance, logistics and technical management can be merged at all levels.

4.1 The positioning of the Internet

Could Web technologies replace current communication and automation control technologies? Before they could, Internet would have to overcome some of its limitations. Some limitations are temporary, such as the lack of security and limited performances. A great deal of work has been done in these areas, and solutions are beginning to be implemented. Other limitations are more inherent in the technology itself, such as lack of network management, lack of real-time and pre-emption capacities, lack of mechanisms for regularly updating shared information, and software that is often complex and voluminous.

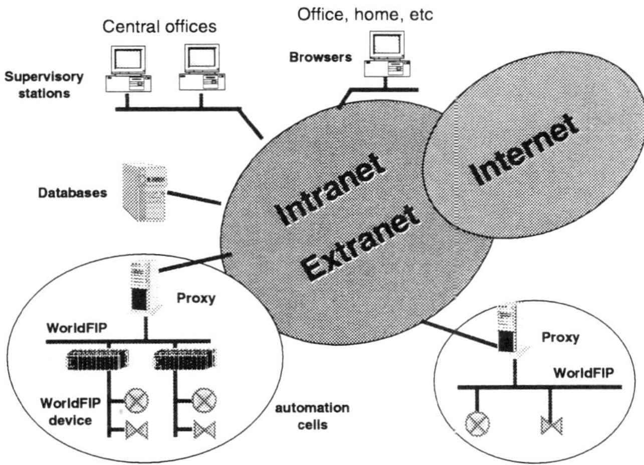

Fig. 3. Impact of Internet on Control Systems

Some fieldbuses, such as WorldFIP, offer qualities lacking in the Internet and provide the extensions needed to access terminal devices. Furthermore, higher speeds of around 25 Mbits/s would be quite feasible.

4.2 The FIPWEB server concept

The WorldFIP messaging service, which complies with standardization IEEE 801.2 for data link services, can provide TCP/IP with the same services as Ethernet while continuing to ensure the transportation of critical time information. End-to-end connections where the FIP/Ethernet coupler would act as the «bridge» (in the ISO sense of the term) would be feasible. In this manner, the service quality provided would be the same as for a homogeneous network.

Therefore, devices connected to WorldFIP are visible from the Internet network, and have a real IP address. This allows them to be interrogated using any navigator connected to the plant's intranet; they can be monitored, diagnosed and adjusted remotely, as long as the user has the required privileges. One can envisage the day - once security procedures allow - when you will be able to call on suppliers or experts to keep the equipment at its highest performance level. This microserver concept is the key to integration of the Web and control systems. It combines the best of both solutions to meet the needs of users.

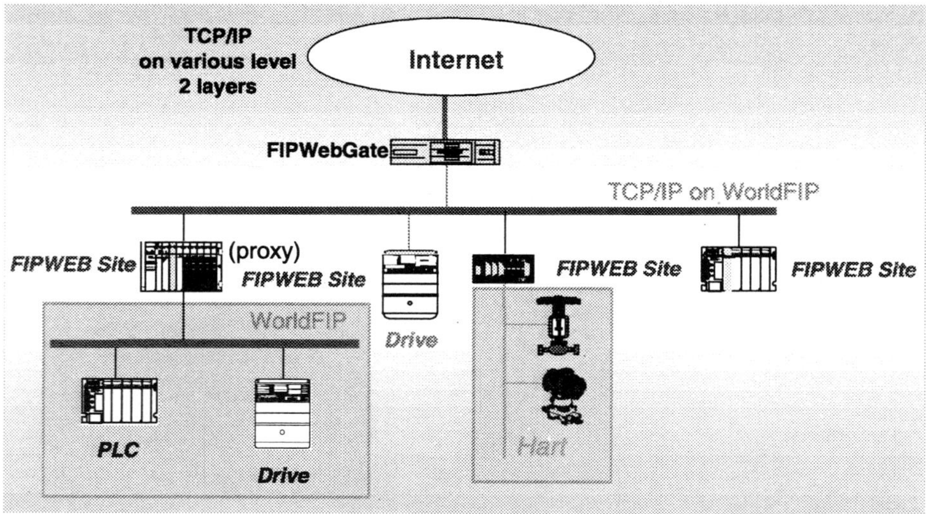

Fig. 4. Concept of Web server on WorldFIP

5 Conclusion

The tools and technologies are available in industry to meet the needs of users wishing to manage their productive assets in an optimum manner. This concept is called «asset management.» However, while such tools exist, they are not naturally integrated together. Global integration that would be valid for all applications is out of the question, and work on system design will always be necessary, and will in fact become more important if the spectrum of functions to be handled expands. Standardization will not solve all system problems, but it will provide a basis for systems engineers to design powerful architectures at a lower cost, and that are optimized for the customer's application.

The fact remains that such tools will call into question the traditional organization of companies and the idea that an item of information belongs to a specific framework. Future systems applying the Fieldbus/Web combination will be able to offer «the right information at the right place to the right person at the right time.» Only companies that head in this direction will take advantage of the richness of these systems. For example, it will be tempting to be able to have a process assessed by the right expert, without the latter having to leave his office, or for a supplier to monitor the evolution of the products he has delivered, and to advise users on the maintenance to be performed in order to keep the products in optimum operating condition. Users will be able to focus their efforts on production by optimizing the invested capital.

Security Architecture for Field Area Networks Connected to Internet

Mikhail Gordeev

Institute of Computer Technology
Vienna University of Technology
Gusshausstrasse 27-29
A-1040 Vienna, Austria
e-mail: mischa@ict.tuwien.ac.at

Abstract. It is already a reality that data security aspects start to play a dominant role almost in every fieldbus application. Most of all, it concerns fieldbus-Internet applications. The major goal of this paper is to propose and to describe a security architecture for such applications. The papers covers its main concepts and implementation basis. An undivided attention is paid at standardization and conceptual issues. The paper does not provide an universal solution, which can be directly applied to an application of any kind. It focuses at major concepts and guidelines, which should lead to secure fieldbus-Internet applications.

1 Risks in fieldbus – Internet applications

Since last decade, a role of the Internet is rapidly increasing in all areas of modern technology. World of fieldbus system is also not an exception. Realities of today dictate a strong need in open, flexible and safe fieldbus – Internet applications. As in any other type of Internet applications, data security aspects play in them a dominant role. You do not have to be an expert in Internet technology or data security to understand it. The world of Internet contains a big variety of risks and therefore it is really important to undertake certain measures in order to minimize or even to avoid them.

Almost in any fieldbus – Internet application we often deal with security sensitive information. On one hand, some data might have a confidential or private content. On another hand, some data might have a high relevance to the functionality of the overall application (e.g. control parameters or commands). Therefore there is a rather high risk that a certain party or a person will try to make attempts to gain an *unauthorized* access to it. It can be accomplished by a variety of different attacks or combination of them. Fig.1 demonstrates main categories of attacks, which can occur in any fieldbus – Internet application. Each of them can be based on a number of different actions [2].

Espionage is a passive action. Its goal is to intercept a data flow between communicating parties and just "listen" it. Manipulation and sabotage are active actions. The first of them can be targeted at data deleting, insertion, replaying and modification. The second one disables a communication process partly or completely.

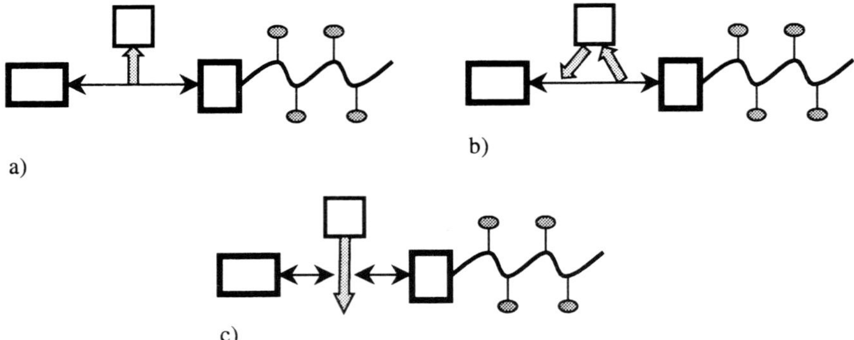

Fig. 1. Main categories of attacks: a) espionage, b) manipulation, c) sabotage

Today, there are a lot of solutions, which try to overcome security problems described above. They vary from simple password based access control mechanisms up to strong cryptographic protocols [3]. However each of these approaches has some disadvantages. Passwords are not reliable because they can be easily stolen, simple given to unauthorized persons, and finally cracked. Even usage of pure cryptographic protocols causes serious drawbacks like a need to implement some mechanisms for encryption keys distribution and management, to secure keys and cryptographic routines, etc.

2 Essential data security concepts

First of all, it very important to define a meaning of the word "security". In the area of information technologies, the term "security" is used in the sense of minimising the vulnerabilities of the information system assets and resources [1]. In other words, it means that all system weaknesses, which could be exploited to violate a system or the information it contains, must be minimised or even eliminated.

In general, three major requirements to data security can be formulated:

- *Availability*. It means that data must be always available when they are needed.

- *Integrity*. It means that data must not be altered or destroyed in an unauthorised manner.

- *Confidentiality*. It means that information is not made available or disclosed to unauthorised individuals, entities or processes.

The first one is one of the most difficult case in any fieldbus – Internet application. In most cases, it is almost impossible to figure out why certain data or services are not available. It could happen due to an ordinary hardware or network failure, or due to some indented actions (attacks). Therefore most of existing solutions does not provide any protection against it. This case is also out of the scope of this paper.

In order to provide a fieldbus – Internet application an intended level of security, it is necessary not only to take the requirements mentioned above into an account. The whole application and its environment must be carefully analysed in order to identify all sources of potential threats to its security. It is also necessary to define a significance of the application data and their protection levels. Usually it is done by risk analysis and management techniques [4].

All this information is used to define the application *security policy*. Security policy contains an informal description of the security principles and practise to be used. Essentially, a security policy states in general terms what is permitted and what is not permitted in the field of the application security during its life cycles. It must be defined for any application.

3 Major requirements to the security architecture

Modern fieldbus applications dealing with Internet are complex and heterogeneous. An adequately high level of their security can and must be reached *only* by combining different techniques, mechanisms and means. For fieldbus – Internet applications this combination is secure software, secure hardware[1], secure Internet protocols, cryptographic protocols. We shall pay an undivided attention at them in the next paragraphs of this paper.

It is mostly impossible to formulate "an universal solution" that can be directly applied to any fieldbus – Internet application and guarantee its security. These applications are different and they operate in different environments. Their security can be build up by following essential guidelines and principles. We can formulate the following major requirements to the security architecture:

-   solid and sufficient security policy

-   combination of a number of security techniques and means

-   strong compliance to standards

We shall pay a strong attention at standardization in the next paragraph.

---

[1] smart cards

4 Standardization issues

4.1 OSI security architecture

Standardization is the most important issue on any fieldbus – Internet application. In other words, the overall application development must comply existing standards. The basic and the fundamental one is the second part of the OSI Reference Model [1], which extends the Model by a definition of security architecture for open system interconnection. It specifies a number of *security services*, which could be provided optionally within the OSI Reference Model framework. However their implementation is necessary for any system, which is intended to be secure. The standard defines such services as authentication, access control, data confidentiality and integrity, and non-repudiation.

The standard also defines particular *security mechanisms*, which can be used to implement security services or combinations of them. It could be encipherment and digital signatures [3], access control and authentication mechanisms, notarization, and others.

4.2 Secure Internet management protocols

Any fieldbus-Internet application intends a usage of a certain communication protocol and specific data representation models. It is always rather difficult to decide to take either standardized or properiterian approach. However only the first way is beneficial. It is proofed by a great number of applications and large practice. Today, there are a number of protocols, which can be suitable for this purpose. Simple Network Management Protocol (SNMP) is regarded as one of the most promising protocols of this group. Since its first publication in 1988, it has become the most widely used network management tool for TCP/IP based networks. The Simple Network Management Protocol [5]:

- Defines a protocol for exchanging information between one or more management systems and a number of agents

- Provides a framework for formatting and storing management information

- Defines a number of general-purpose management information variables or objects.

Usually, the model of the network management used in SNMP deals with such key elements as management station, management agent, management information base (MIB) and network management protocol. This model perfectly fits to requirements to fieldbus Internet applications. It was already some work done, which approved this approach [6].

The most promising feature of the third version of SNMP is its security functions, which were significantly improved in comparison with the previous versions. Security functions of SNMPv3 are implemented in two subsystems called Security Subsystem and Access Control Subsystem. They seem to cover all main groups of threats [5] and provide a set of necessary security mechanisms satisfying requirements defined in the OSI Security Architecture. Therefore usage of SNMPv3 in fieldbus – Internet applications seems to be more than just reasonable.

4.3 Smart card technology

It is an axiom of the world of data security that all security mechanisms must be run on secure hardware, the hardware, which provides strong data protection. This concept can be implemented only on a smart card[2]. A smart card can be defined as a portable computer device with intelligence and provisions for identity and security [7]. Due to the highly protected architecture and implemented cryptographic mechanisms, the smart card is strongly protected against different kinds of security attacks.

Physically, a smart card is a plastic card having a number of semiconductor elements embedded in it (Fig. 2). Most of the smart cards are equipped additionally with a special dedicated cryptographic co-processor, which performs cryptographic operations. The microcontroller performs computational operations and decision making. It also handles any access to protected storage areas located on the card. The smart card microcontroller consists at a minimum of a central processing unit (CPU), Random Access Memory (RAM), Read Only Memory (ROM) and non-volatile memory.

Smart cards must be regarded as an inherit part of an architecture for any fieldbus - Internet application intended to be secure[8].

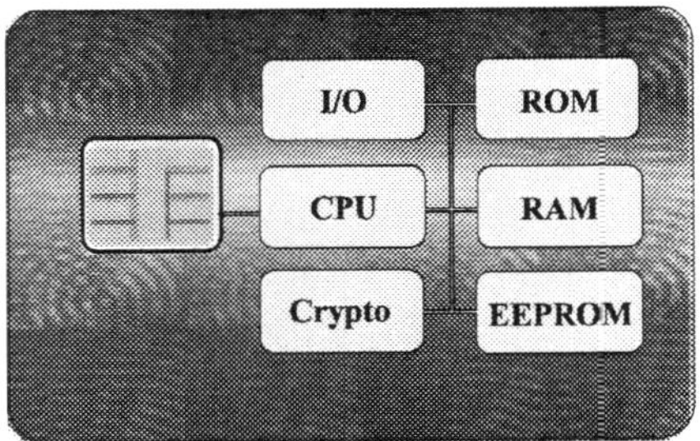

Fig. 2. General architecture of a smart card

5 Implementation issues

Taking all aspects described in this paper together, we shall focus on a general structure of the security architecture. Fig. 2 shows its graphical representation. Functionality of modules is intuitively clear. A main task of the fieldbus module is to provide an access to the fieldbus network. It acts as a bridge between the fieldbus and the SNMP entity.

---

[2] we omit some types of special dedicated secure hardware

There is another advantage of the SNMP-based approach. As we can see it in the figure, implementation of the client and the gateway is rather identical[3].

The SNMP entity interacts with the security module, which provides an access to the smart card security services, for instance encryption and digital signing. Their specification, as well as a specification of the SNMP security mechanisms, is directly obtained from the application security policy.

It is very important to pay reader's attention at one security aspect of such an architecture. Communication channel between the gateway (or a client) and a smart card can be a bottleneck of the overall application security. If it is not protected enough, it can be easily exploited by an attacker. One of the ways to solve this problem is to provide its security physically by integrating a smart card into the gateway.

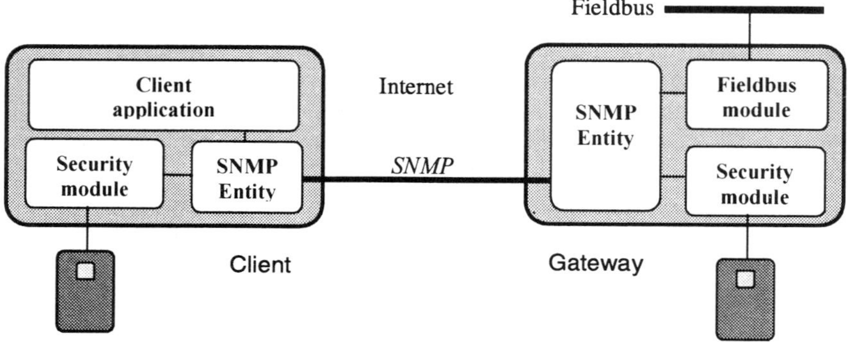

Fig. 3. General structure of the security architecture

This architecture can be implemented on various platforms using various software tools because SNMPv3, as well as smart card technology, is strongly supported by a big number of software manufactures, which offer a wide spectrum of different APIs. However Java seems to be one of the most promising approaches. Besides platform independence, it provides a number of additional security features. This is a significant benefit because Java security mechanisms can be effectively combined with the security mechanisms, which were presented in this paper. It contributes the requirements to the security architecture and leads to its more stronger security.

6 Conclusion

Role of the Internet is rapidly increasing in all areas of modern technology. It causes a strong need in efficient solution for data security. This objective factor directly

---

[3] till certain level of abstraction. Of course, SNMP agent and manager do not consist of exactly the same modules

concerns fieldbus technology. The major goal of this paper was to point it out and propose some possible ways towards secure fieldbus – Internet applications.

References

1. ISO 7498-2, "Information processing systems – Open system interconnection – Basic Reference Model – Part 2: Security architecture", ISO, 1989

2. Hedwig Brobeil, "Software-Angriffe auf PCs und Netzwerke", R. Oldenbourg Verlag, 1992

3. Bruce Schneider, "Applied Cryptography", John Wiley & Sons, 1996

4. K. Jackson, J. Hruska, editors, "Computer Security Reference Book", Butterworth-Heinemann Ltd., 1992

5. William Stallings, "SNMP, SNMPv2, SNMPv3, and RMON 1 and 2", Addison Wesley, 1999

6. M. Knizak, M. Kunes, M. Manninger and T. Sauter, "Applying Internet management standards to fieldbus systems", Proceedings of the IEEE International Workshop on Factory Communication Systems, WFCS´97, Barcelona, 1997

7. W. Rankl, W. Effing, "Handbuch der Chipkarten", Carl Hanser Verlag, 1999

8. M. Manninger, B. Burgstaller, H. Reiter, "Smart-Card-Security für Feldbussysteme", in: D. Dietrich, H. Schweinzer (Hrsg.), Feldbustechnik in Forschung, Entwicklung und Anwendung, Springer, Wien NewYork, 1997, S. 297-303.

P-NET-Management über das Internet

Melih Ayal, Martin Knizak, Thilo Sauter

Institut für Computertechnik, Technische Universität Wien,
Gußhausstraße 27-29, A-1040 Wien
T. ++43-1-58801-38430 F. ++43-1-58801-38499
E-mail: {knizak, sauter}@ict.tuwien.ac.at

Abstract. Bei der Anbindung von Feldbussen an IP-basierte LANs über SNMP ist es zweckmäßig, das dafür notwendige Gateway in zwei Teile aufzuspalten. Ein Teil übernimmt dabei die Abwicklung des SNMP-Protokolls, der andere die Kommunikation mit dem Feldbus. Dieser Artikel beschreibt die Implementierung des feldbusspezifischen Teils an Hand des in der Europäischen Norm EN50170 verankerten Systems P-NET. Als Treiber für den Feldbus wurde dabei die Windows-basierte Software VIGO verwendet. Ein Ansatz für die Konfiguration des Gateways wird ebenfalls präsentiert.

Abstract. For the interconnection of fieldbus systems and IP-based LANs, a gateway is required. A reasonable implementation approach is to split the gateway into two parts. One part then takes care of the SNMP protocol, while the other handles the actual communication with the fieldbus. This article describes the implementation of the fieldbus specific part for P-NET, a fieldbus specified in the European standard EN50170. The implementation uses the Windows-based software VIGO as a fieldbus driver. An approach for the configuration of the gateway is also presented.

1 Einleitung

Die Verbreitung von Feldbussen nimmt immer noch zu. Waren anfänglich die Einsparung des Verkabelungsaufwandes, bessere Wartbarkeit und Diagnose im Vordergrund, so legt man in letzter Zeit das Hauptaugenmerk auf die Vernetzung dieser jetzt vorhandenen Insellösungen. kein einheitlicher Feldbusstandard existiert, gilt es, Lösungen zu finden, die für alle (oder zumindest für einen Großteil) dieser Systeme verwendbar sind.

Prinzipiell sind zwei Wege möglich und werden den Erfordernissen entsprechend auch angewandt. Einerseits kann man zwischen zwei vorhandene Teilnetze ein Gateway setzen, in dem die PDUs (Protocol Data Units) auf das jeweilig andere spezielle System umgesetzt werden. Andererseits erscheint es manchmal effizienter, ein Koppelelement einzufügen, das keinem der beiden Systeme angehört und eine protokollunabhängige Darstellung der Daten liefert [1]. Der zweite Ansatz vergrößert zwar den Verarbeitungsaufwand, aber wenn der protokollunabhängige Teil einmal existiert und getestet ist, beschränkt sich die Arbeit für weitere Feldbussysteme auf die Implementierung eines für dieses System geeigneten Treibers.

Im folgenden wird das Konzept eines solchen speziellen Treibers für P-Net dargestellt, wobei für den eigentlichen Feldbuszugriff das System VIGO verwendet wird. Dieser Treiber dient dazu, einen modularen Agent (oder Gateway) für SNMP (Simple Network Management Protocol) an den Feldbus P-NET anzupassen.

2 Der Sub-Agent für P-NET

Im Projekt WEBFAN wird ein flexibles Netzmanagementsystem realisiert. Mit diesem auf SNMP [2] beruhenden System kann über das Internet auf verschiedene Feldbusse wie PROFIBUS oder P-NET zugegriffen werden.

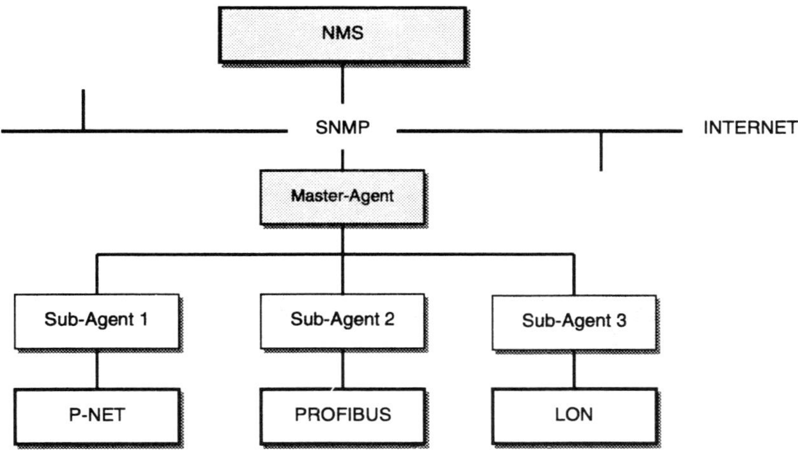

Abb. 1. Netzmanagementsystem WEBFAN

Um das Netzmanagementsystem möglichst modular und somit flexibler zu gestalten, ist der Agent in zwei Teile unterteilt, den Master-Agent (MA) und den Sub-Agent (SA). Der Master-Agent ist jener Teil, der über SNMP mit der Außenwelt in Verbindung tritt und eine beliebige Anzahl von Sub-Agents und damit auch verschiedene Feldbusse verwaltet. Er ist unabhängig vom verwendeten Feldbussystem aufgebaut. Anhand der Anforderungen der jeweiligen Network Management Station (NMS) trifft er die Entscheidung, wann und wie ein Wert eines Objekts gelesen oder aktualisiert wird. Der Sub-Agent ist für den eigentlichen Feldbus-Zugriff zuständig. Für jedes Feldbussystem muß daher ein eigener Sub-Agent implementiert werden. Die vorliegende Arbeit beschäftigt sich mit dem Sub-Agent für das Feldbussystem P-NET.

3 Zugriff auf den Feldbus

Für die Implementierung des Sub-Agents wurde ein PC verwendet, in dem sich eine P-NET-Karte und eine Ethernet-Karte befinden. Da mit Microsoft Visual C++ programmiert wurde, fiel die Wahl der Treibersoftware auf VIGO, ein Produkt, das speziell auf die Kommunikation mit Programmen unterMicrosoft Windows NT und Windows 9x abgestimmt ist.

Der Zugriff auf das Programmpaket erfolgt über eine OLE2-Verbindung [3]. Die einzelnen Werte des Feldbusses werden über virtuelle Objekte angesprochen (siehe Abb. 1). Die gesamte feldbusseitige Protokollimplementierung erfolgte innerhalb der verwendeten Treibersoftware.

Informationen können bei dieser Art der Implementierung nur über virtuelle Objekte des Treibers entnommen werden. Der direkte Zugriff auf den Feldbus ist nicht möglich. Es stehen jedoch genügend Statusmeldungen zur Verfügung, um sowohl den Zustand des realen Feldbussystems als auch den des Treiber feststellen zu können. Damit ist auch eine schnelle Lokalisierung einer Fehlfunktion möglich.

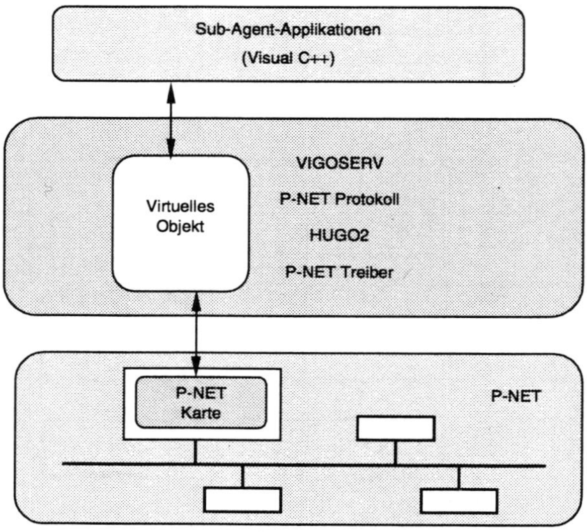

Abb. 2. Zugriff auf den Feldbus P-NET

Die Möglichkeiten des verwendeten Programmpakets werden bei weitem nicht ausgeschöpft. Der Grund hierfür ist, daß es sich bei diesem Programmpaket (VIGO, HUGO2, ...) um eine proprietäre Lösung handelt, die für P-NET optimiert wurde [4]. Das WEBFAN-Projekt hingegen hat eine vom Feldbus unabhängige Möglichkeit der Vernetzung zum Inhalt, bei der der Sub-Agent-Teil bei allen Feldbussen die gleichen Aufgaben hat und diese auf ein Minimum beschränkt sind.

Die in der gewählten Implementierung enge Verbindung zwischen Betriebsystem und verwendeter Software gewährleistet ein reibungsloses Zusammenspiel der einzelnen Komponenten. Eine Änderung des Betriebsystems, z. B. auf Linux, würde aber so viele Änderungen notwendig machen, daß ein Neuentwurf des gesamten Sub-Agents sinnvoller wäre. Die weite Verbreitung von MS-Windows macht diesen Nachteil jedoch akzeptabel.

4 Konfiguration des Sub-Agents

Bei Netzmanagementsystemen, die auf der Internet-Protokollfamilie basieren, wird angenommen, daß jeder verwaltete Knoten mehrere Objekte besitzt, auf die der Manager zugreifen kann [5]. Durch Lesen und Verändern dieser Objekte wird der Knoten beobachtet und kontrolliert. Nicht alle diese Objekte sind für Managementaufgaben von Interesse. Der Konfigurator des Sub-Agents ist ein Tool, das zum Auswählen der Objekte des Feldbussystems dient, die der Sub-Agent verwalten soll.

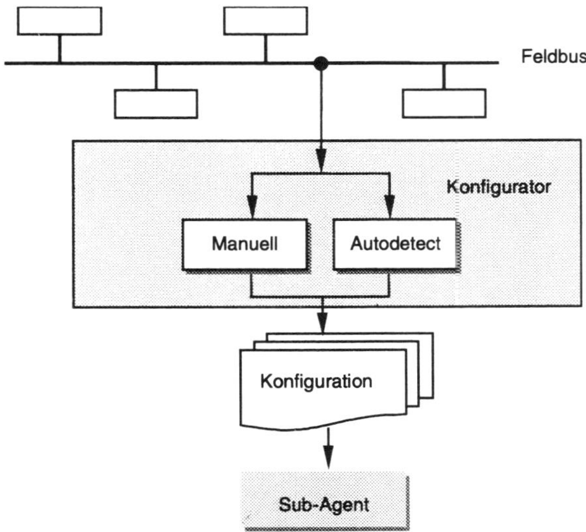

Abb. 3. Initialisierung des Sub-Agents

Er hat folgende Aufgaben (Abb. 3):

* automatisches Erkennen und Anzeigen aller verfügbaren Feldbus-Objekte,

* automatisches oder manuelles Auswählen der Objekte,

* manuelle Eingabe und Modifikation der Objekteigenschaften.

Nach der Auswahl werden alle Angaben in einer Konfigurationsdatei gespeichert. Bei der Initialisierung des Sub-Agents wird diese Konfigurationsdatei gelesen, und die in ihr gespeicherten Objekte werden in den Sub-Agent aufgenommen. Dem Sub-Agent kann über einen Parameter eine bestimmte Konfigurationsdatei angegeben werden, andernfalls wird ein Defaultwert verwendet.

5 Implementierung des Sub-Agents

Der Sub-Agent ist für den Zugriff auf das Feldbussystem verantwortlich. Dieser wird durch die in Punkt 3 vorgestellte Treibersoftware zu großen Teilen selbständig erledigt. Nach dem Start sucht der Sub-Agent als erstes nach einer Initialisierungsdatei. In dieser stehen der Name der Konfigurationsdatei und Portnummern, die bei der Kommunikation mit Master-Agents verwendet werden. Der Sub-Agent kann entweder mit einem oder ohne Parameter aufgerufen werden. Durch diesen Parameter wird eine Initialisierungsdatei angegeben. Damit wurde die Flexibilität des Programms erhöht, da man je nach Bedarf verschiedene Initialisierungsdateien benutzen kann.

Nachdem die Daten der Initialisierungsdatei verarbeitet worden sind, wird die Konfigurationsdatei, die mit Hilfe des Konfigurators erstellt wurde, gelesen. In der Konfigurationsdatei stehen die Objekte mit ihren Eigenschaften in einer vorgegebenen Reihenfolge. Diese Reihenfolge bestimmt auch die ID-Nummer der verwalteten Objekte, die vom Sub-Agent und den Master-Agents zur Identifizierung des jeweiligen Objekts benutzt werden. Der Master-Agent spricht diese Objekte nur über ihre ID-Nummer an.

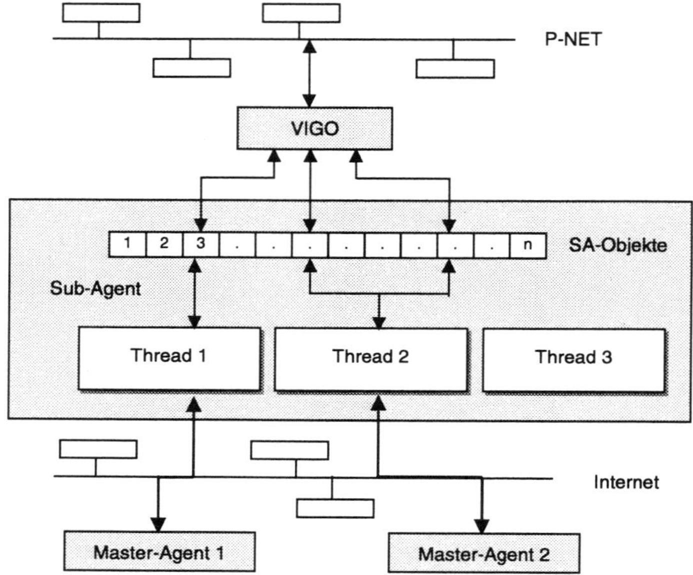

Abb. 4. Multithreading im Sub-Agent

Nachdem alle Daten aus der Konfigurationsdatei gelesen wurden, beginnt der Sub-Agent mit den Vorbereitungen für die Verbindung mit den Master-Agents. In der vorliegenden Implementierung ist der Sub-Agent multimasterfähig, das heißt, mehrere Master-Agents können vom Sub-Agent gleichzeitig betreut werden. Um eine möglichst parallele Verwaltung der Verbindungen mit den Master-Agents bewerkstelligen zu können, wurde der Sub-Agent als Applikation mit Multithreading-Fähigkeit implementiert. Ein einzelner Prozeß kann verschiedene Ausführungspfade (Threads) enthalten, die vom Betriebssystem verwaltet werden und über eigene Stacks verfügen, aber den gesamten Adreßraum mit Code- und Datenbereich benutzen dürfen. Der Sub-Agent verwendet für jede Verbindung mit einem Master-Agent einen separaten Thread, für den die globalen Daten der Sub-Agent-Applikation zugänglich sind (Abb. 4). Es existiert außerdem immer ein Thread, der auf eine neue Verbindung wartet.

Da auch die Treibersoftware selbst eine parallele Durchführung mehrerer voneinander unabhängiger Buszugriffe erlaubt, wird eine gleichzeitige Ausführung mehrerer

Threads, die gemäß der Kommunikation mit dem jeweiligen Master-Agent auf das Feldbussystem zugreifen, kein Problem verursachen. Nur ein gleichzeitiger Schreibzugriff auf ein Objekt durch mehrere Master-Agents sollte vermieden werden. Dies könnte Inkonsistenzen und andere Nebeneffekte mit sich bringen. Deshalb wird für den Schreibzugriff einem Thread, und dadurch einem Master-Agent, ein exklusives Zugriffsrecht auf ein bestimmtes Objekt erteilt. Solange dieser Thread dieses Recht besitzt, können andere Threads nicht darauf zugreifen.

6 Kommunikation mit dem Master-Agent

Nach der Initialisierung wird immer ein Thread gestartet, der auf eine Verbindung mit einem Master-Agent wartet. Wenn ein Master-Agent eine Verbindung mit dem Sub-Agent aufnehmen will, muß er zuerst eine Socket-Verbindung mit dem für diese Aktion vorgesehenen Port aufbauen. Ob beide Teile auf derselben Maschine laufen oder nicht, ist nicht von Bedeutung. Der Aufbau erfolgt nur über die Internetadresse und die Portnummer. An dieses Ziel schickt nun der Master-Agent eine `ConnectionRequest`-Meldung. Falls dieser Verbindungswunsch vom Sub-Agent akzeptiert wird, sendet er die Meldung `ConnectionResponse` mit einem Fehlercode Null zurück, und der Verbindungsaufbau ist erfolgreich abgeschlossen.

Über diese nun stehende Verbindung kann der Master-Agent mittels eines für diese Aufgabe entworfenen Protokolls mit dem Sub-Agent kommunizieren und somit auf das Feldbussystem zugreifen. Um dieses Protokoll einfach zu halten, wurde keines der bereits bekannten [6] verwendet, sondern ein auf die Minimalerfordernisse zugeschnittenes entworfen. In diesem Protokoll gibt es fünf Anfragen, die von einem Master-Agent zu einem Sub-Agent geschickt werden:

1. `ConnectRequest`: Diese Anforderung besagt, daß der Master-Agent eine Kommunikation mit dem Sub-Agent aufbauen will. Dabei wird auch die Anzahl der von diesem Master-Agent verwalteten Sub-Agents mitgeschickt.

2. `GetMaxNumberRequest`: Mit dieser Anforderung fragt der Master-Agent den Sub-Agent, wieviele Objekte er insgesamt zu verwalten hat. Diese Anfrage dient zur Initialisierung des Master-Agents.

3. `GetRequest`: Mit dieser Anforderung kann ein Master-Agent den Wert eines verwalteten Objekts vom Sub-Agent erhalten.

4. `GetDetailRequest`: Mit dieser Anfrage fragt der Master-Agent nach den Eigenschaften des verwalteten Objekts, wie die Beschreibung und Status des Objekts oder der ASN.1 Typ seines Wertes sind. Diese Daten werden dann in die MIB des Master-Agents eingetragen.

5. `SetRequest`: Mit Hilfe dieser Anfrage kann der Master-Agent gemäß den Anforderungen der NMS den Wert eines Objekts verändern. Dabei werden die ID-Nummer und der Wert des Objektes gesendet.

Für jede Anforderung vom Master-Agent gibt es auch eine entsprechende Antwort-Meldung (Response) des Sub-Agents, die auch einen Fehlerwert mitschickt. Dieser Wert ist Null, wenn kein Fehler aufgetreten ist, oder enthält einen feldbusabhängigen Fehlercode.

7 Zusammenfassung und Ausblick

Durch die Aufteilung des Netzüberganges in einen allgemeinen und einen feldbusspezifischen Teil wird der Arbeitsaufwand für die Erstellung eines Interfaces für ein neues Feldbussystem bedeutend verringert. Am Beispiel von P-NET sieht man deutlich, daß nicht alle vom Feldbus angebotenen Tools und Fähigkeiten genützt werden können, da sonst der Kreis der für dieses Konzept geeigneten Feldbusse extrem eingeschränkt wird. Bei der Auswahl für die ersten Implementierung hatten P-NET und PROFIBUS als in der EN 50170 genormte Feldbusse eindeutig die besseren Argumente. Es sind jedoch auch Sub-Agents für EIB und LON in Arbeit. Diese Bussysteme, die speziell für die Gebädeautomation verwendet werden, eignen sich besonders für den gewählten Ansatz, da in diesem Bereich die Anforderungen an die Reaktionszeiten nicht so hoch sind wie in der industriellen Automation. Der Einsatz dieses Modells, eventuell erweitert im Master-Agent um ein Html-Interface und Java-Elemente, für die Fernwartung von Gebäuden erscheint ebenfalls eine sinnvolle Anwendung. Weiters bildet dieses Konzept einen guten allgemeinen Weg, um Protokollierung, Fernwartung und Management von Feldbussen über andere Netzwerke wie etwa das Internet abzuwickeln.

Literatur

1. Sauter, T., Knizak, K, Manninger, M.: Einbettung von Feldbussen in das Internet-Management, in: Dietrich, D., Schweinzer, H. (Hrsg.), *Feldbustechnik in Forschung, Entwicklung und Anwendung*, Springer, Wien NewYork, 1997, S. 223-230.

2. Rose, M. T.: *Einführung in die Verwaltung von TCP-IP-Netzen*, Hanser Verlag, München, 1993.

3. Chappell, D.: *ActiveX und OLE verstehen*, Microsoft Press, München, 1996.

4. PROCES-DATA A/S: *VIGO Users Manual for Version 3.0. - 1997.*

5. Perkins, D. T.: *Understanding SNMP MIBs*, Revision 1.1.7, 1993.

6. Agent Extensibility (AgentX) Protocol: Request for Comments Dokument Nr. 2257.

Wireless Data Transfer System
for Oil Drawing Plants

Angel Alonso Perez, Vitalio Alfonso Reguera, Felix Alvarez Paliza

Telecommunications and Electronic Department.
Central University of Las Villas, Cuba

Abstract. The present work describes the design and building of a data transmission system for Collector Centres monitoring and automation in an Oil Drawing Enterprise in Cuba (**EPEP**). Packet-radio modems were used, working in the VHF bands at 1200 and 9600 bits per second, resulting in an economic solution. This data transmission system is a wider and modern version of the previously developed at 1994 by the authors. The system goal is automatic transfer of files and data collected by instruments, from several oil drawing centres, to the computer network located in the headquarters building. This task is done without interfering with other applications normally executing in these plants. The new system runs in multitasking environments such as Windows95 and WindowsNT.

1 Introduction

This Enterprise is located near Varadero, in a place where there are many Oil Platforms either alone or gathered in groups. In both cases there is a pumping unit that pull the oil into a tank from where it is driven toward the Collector Centre through pipelines. In these places gas and water are removed from the oil and its temperature is increased. There are about 10 centres like these, they are connected to two pumping stations that send the oil to the plant located in Cardenas, where oil is processed before shipment.

In every place, from drawing plants to Cardenas plant, there are control systems that supervise the local processes by measuring temperature and pressure, and also supply signalling and protection functions. Fieldbus networks interconnect the digital measurements systems in these plants. Computers collect all the information needed, allow process monitoring and control them.

The plants automation and updating project consist of settling a hierarchical digital communication system, working from the lower drawing stations level, up to the central plant level where an efficient and accurate drawing control and plant monitoring takes place.

To carry out communications radio transmission was chosen, according to the following facts:

- There is not any available structure for wired communications, moreover the enterprise has a long experience and the equipment needed for radio communications in VHF band.

- The environment is highly corrosive in the zone, so if wired connections were selected, specially protected wire would be needed, increasing the system cost.

- The amount of data to transmit is relatively small, due to the low speed nature of the processes to be supervised. This allows traffic to be supported by a single low-speed shared channel.

- The distances between drawing plants and collector centres are about 2 kilometres and about 15 from Cardenas plant to headquarters plant. These distances permit good quality and secure communications to be covered by VHF radio links.

2 System Description

The first step in the project was to settle the link between Cardenas and the EPEP. Packet radio modems types **KPC-9612** were used, they can work at 1200 or 9600 bits per second. Currently we are using the same type of modems but the programs are flexible enough to support almost any type of them. Formerly, 1200 bps was used and then the radio transceivers were adapted to work at 9600 bps.

The Enterprise's investment in field instrumentation is focusing in digital transmitters linked by field-bus networks, supervised by **AIMAX** software. As it works on Windows platforms it can communicate with other applications via **Dynamic Data Exchange** (DDE) and **Open Database Connectivity** (ODBC). The present system uses the same support and does not interfere with AIMAX.

The software for data transfer was designed taking into account the conditions related above. It can be managed from a communications server via remote commands using a master-slave arrangement.

The system consists of two programs, one for the master side and the other for the slave side of the link. Both perform similar functions such as modem control, data link protocol, error detection, and received and transmitted files handling. Communication between program and modem follows the KISS protocol, which is customary in amateur packet-radio modems, while the data link protocol was specially designed for this application to reduce protocol overheads.

The program for slave stations requires also internal communication with AIMAX system, this was done using Dynamic Data Exchange (DDE). The program includes a DDE Client whose field variables are user configurable. AIMAX DDE Server automatically updates these variables.

The program's master side polls each station cyclically according to the time established by the user. It also includes an electronic hard key to prevent illegal copies of the software.

3 Protocol Description

The purpose of the "KISS" protocol is to facilitate the use of amateur packet radio controllers (TNCs) with host computers, particularly in the development of experimental protocols. It is almost identical to SLIP, a popular method for sending IP datagrams across ordinary dialup modems.

The linklayer protocol is removed entirely from the TNC, as are all command interpreters and the like. The TNC simply converts between synchronous HDLC, spoken on the half-duplex radio channel, and a special asynchronous, full duplex frame format spoken on the host/TNC link. Every frame received on the HDLC link is passed intact to the host once it has been translated to the asynchronous format; likewise, asynchronous frames from the host are transmitted on the radio channel once they have been converted to HDLC format.

The "asynchronous packet protocol" spoken between the host and TNC is very simple, since its only function is delimiting frames. Each frame is both preceded and followed by special FEND (frame end) character, analogous to an HDLC flag. No CRC or checksum is provided.

FEND	00	Frame	FEND

Fig. 1. Asynchronous Packet Protocol

The reason for both preceding and ending frames with FENDs is to improve performance when there is noise on the asynch line. The FEND at the beginning of a frame serves to "flush out" any accumulated garbage into a separate frame (which will be discarded by the upper layer protocol) instead of prepending it to an otherwise good frame. As with back-to-back FLAGs in HDLC, two FEND characters in a row should not be interpreted as delimiting an empty frame.

Frames are sent in 8-bit binary; if an FEND ever appears in the data, it is translated into the two byte sequence FESC TFEND (frame escape, transposed frame end). Likewise, if the FESC character ever appears in the user data, it is replaced with the two character sequence FESC TFESC (frame escape, transposed frame escape).

4 Data Link Control Frame Format

We combine functions of the data link layer with some of the file transmission in the frames of this protocol. This is made with the objective of reducing the overload introduced by the protocol. The format of the frame includes the following fields:

Address	Type	Function	Data	Checksum

Fig. 2. Data Link Control Frame Format

The **Address** field is 8 bit long. It is used to identify the slave station in every frame.

Up to 254 networked stations is allowed.

The control field is also 8 bit long. Four of them (Type) are used by the data link control protocol and four (Function) by the file transmission protocol.

Type: Identifies the kind of frame, Information, Acknowledge, Connect Request, Disconnect, etc. This field also includes the sequence bits, used by information frames only. The first sequence bit is the next frame to send, and the second is the frame expected. This is a one bit sliding window protocol.

Function: This field contains the function code from the master station in order to request the desired file operations, and the response codes from the slave station. When the requested operation is performed sucessfully the response code is the same of that that of the request. These are some examples:

0001 **OpenFile**: The Data Field contains the full pathname of the file that will be opened in the slave station. The slave response contains the same code in the function field and the file size in the data field.

0100 **WriteFile**: The Data Field contains the data to append to the file. The response contains no data.

0011 **ReadFile**: The Data Field contains the amount of bytes to be read from file. The response includes the read data.

0010 **CloseFile**.

0101 **SeekFile**. We use this command when part of the file is already at the master station. The Data Field contains the offset of the first byte to be read.

0111 **GetItem** The Data Field of this frame includes the name of a **Topic** and the name of an **Item**, in order to request the current value from the **DDE Server** at the slave side. The response frame contains the current value of the item.

1001 **Access denied**.

1000 **Error**. The Data Field includes the reason of the error.

Checksum: This field is 16 bit long, it is the well known cyclic redundancy code, using CRC-CCITT as the generator polynomial.

5 Conclusion

This system is being used since December 1997. Currently there are one master station at headquarters and eight slave stations in the packet radio network. The use of this system in conjunction with the automation project provides a notable reduction in production and maintenance costs. It can be used in applications were wireless transmission of file, command and/or data is needed and also in process monitoring.

References

1. Alonso, A., A. Sarmientos. Curso por correo electrónico de Redes. Conferencia No. 7. Universidad Central de Las Villas. 1995.

2. Bdale Garbee,The KA9Q Internet Software Package, User Manual for 890421.1 Revision, May 1989.

3. Stallings, W. "Data and Computer Communications". Macmillan Publishing Company, NY, USA, 1985.

4. Tanenbaum, A. "Computer Networks". Third Edition. Prentice Hall, NJ, USA, 1996.

Chapter 3: Profiles

Profiles for Fieldbuses -
Scope and Description Technologies

Christian Diedrich

ifak Institut für Automation und Kommunikation e.V. Magdeburg
Steinfeldstr. 3, 39179 Barleben, Germany

Abstract. After years of distributed protocol specifications the added value for end users in terms of more functionality e.g. for diagnosis and maintenance becomes topic more and more. The question which functions of the field devices can be used in a similar standardized way as communication services is of interest instead of the question, how data are transferred. So the fieldbus organizations offer profiles on top of the communication protocols. The state of the art of specifying profiles is described. In the profile specification the field device is seen as a stand alone component not as part of a distributed control system mostly. This results in high efforts for the integration of field devices in the all over life cycle of the distributed system. Therefore the paper offers an approach, how field devices are reflected as system components during the specification work, and how implementations of devices and tools can be derived from the specification. These specification method is based on the Unified Modeling Language (UML).

1 Introduction

New work assignments within the life cycle of distributed control systems are coming up with the existence of smart field devices and digital industrial communication systems. Field device functionality has to be integrated in the distributed control system (DCS), i.e. library functions of the DCS are offered by field devices now. Standardization activities have to be launched to specify field device functions manufacturer independently and the necessary data exchange between field devices and DCS components.

According to the starting point of the description/standardization activity, the variables and/or functions of field device itself, the function distribution between PLC and field device or the all over system distribution including supervisory stations and its engineering are covered. These standards or activities are for instance IEC 61131 for the PLC programming, profiles (PROFIBUS, CAN), companion standards (WorldFIP) and function block specifications (PROFIBUS-PA, Fieldbus Foundation) from the field bus area, and ISO184 SC5 WG5 for an open framework of distributed systems (Fig. 1). However, the implementation of the functionality is in terms of program code and data. The engineering, commissioning and maintenance tools have to interact with the device functionality. This forces the tools to support the standard, which unfortunately are based on different models. The information flow between the tools needs interface adapters or repeated manual editing of information. Many times the models can't be compared to each other. The models are implemented by the tool manufacturers implicitly.

This paper offers a description technology which makes the model borders of different specifications visible and suggests a way to lower fences between the models step by step.

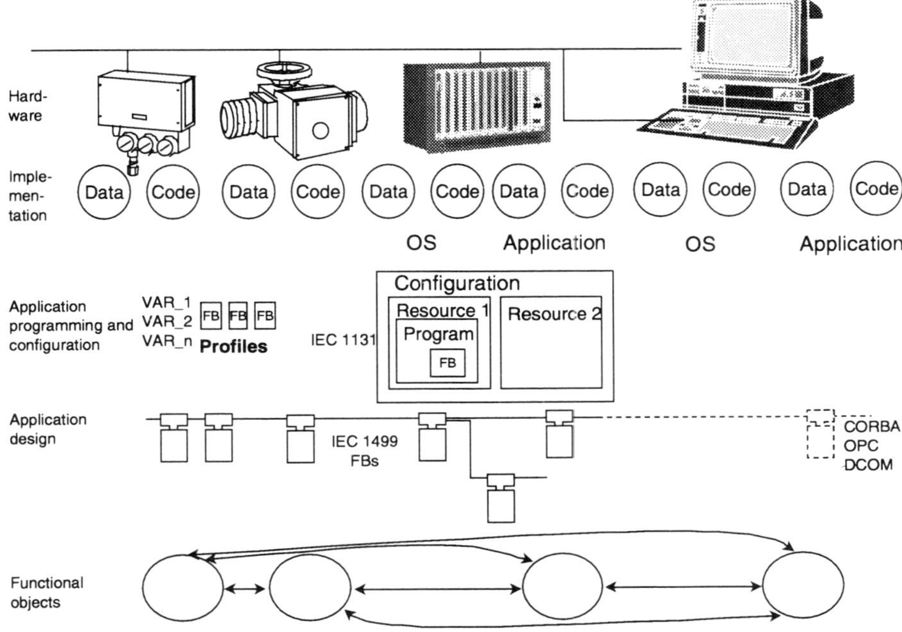

Fig. 1. Representation of automation functionality within different application scope

2 Scope of fieldbus profiles

Fieldbus profiles describe the field device itself and partly the integration of the field device functions in a distributed system. Therefore, the use of the ISO OSI Reference model is common for the specification of profiles. A profile is a choice of functions and parameters of existing definitions and standards. In the area of fieldbus there are Communication profile (Choice of services and protocol configuration parameters), Device profile (Application functions and parameters of devices) and Application profile (Co-operation of several standardized device functions within one control application, also known as profile of a branch). The following sections focus on device profiles only.

3 Steps for the definition of fieldbus profiles

Profiles should strongly distinguish between functionality in terms of the purpose of the profiles and the platform and model which the profiles are dedicated to. Therefore the following steps specifying a profile are suggested. This approach brings the knowledge of the field device experts and the fieldbus experts together.

3.1 First step: Device classification

Devices are classified using several aspects as e.g., the attachment type to the process (measurement, actuation devices), signal type (analogue, binary, discrete), measurement and actuation type (temperature, pressure, ... modulated actuation, variable speed drive, ...), measurement principles (e.g. resistor temperature measurement, thermocouple temperature measurement) and others more. There are German (DIN 19259) and international standardization (draftIEC 61987) activities to harmonize measurement and actuation terminology based on a device classification. A device profile has to be positioned in this classification precisely.

3.2 Second step: Device functions overview

The device is described in a top down approach, i.e. starting with the process input/output or connection (e.g. sensor, valve, motor) and the main control input/output (e.g. SP, measurement value). Depending from the device class there should be different degrees of detail so that a general overview and a detailed functional structure is visible. It is common to describe this functional overview in terms of functional diagrams.

3.3 Third step: Variable/Parameter description

The details of the Variable/Parameter have to be defined. There are variables and parameters which are related to functions. Additionally there are parameters, which are like containers in the device or describe device features (construction, device label, ...)

3.4 Fourth step: Selection or specification of the used device model

From an remote point of view device functionality can be seen as variables, commands, function blocks or objects. For each of these variants the relations between the device, its variables and functions have to be defined clearly. Most of all some basic architectural definitions are done (see [5]) additionally. The documentation of these definitions is done verbally, using tables and drawings.

3.5 Fifth step: Descriptions of the profile functions and variables in terms of the defined model

At least all functions and variables have to be transformed into the defined device model. This transformation doesn't extend the profile functionality because the semantic of the profile is already defined. Of course the semi-formal description of the profiles will make visible those definitions which are not unambiguously.

The presented steps doesn't reflect, that field devices are parts of an all over distributed systems, which has to be designed commissioned and maintained. The profile experts provide these knowledge as its background usually. The next section presents an approach, how the system aspect can be integrated in the profile specification more formally.

4 A common device model for field devices [5]

The device model contains the overall valid structure which combines the functional and the device view. It is fixed as shown here within the NOAH European project.

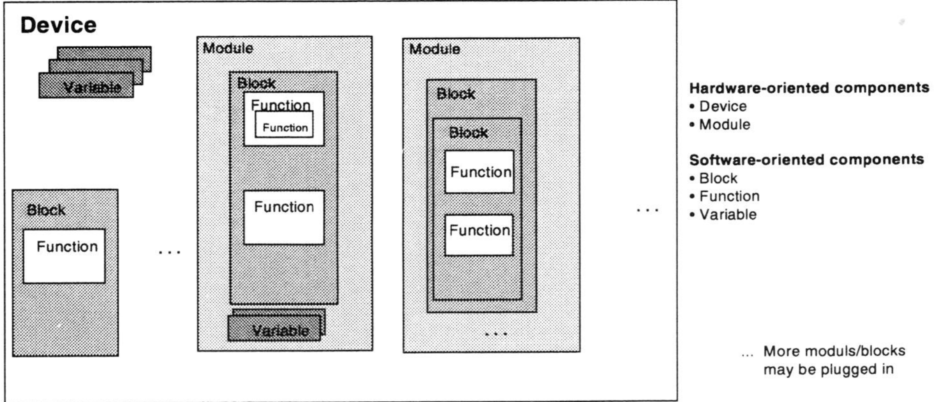

Fig. 2. Overview device components.

The more hardware oriented components are devices and the modules (e.g. plug in I/O modules). The software oriented components are variables, functions and the encapsulation of variables and functions, the blocks. Fig. 2 shows these components and some of their relations to each other. For example a device contains modules, variables and blocks. The blocks contain functions. Blocks may contain blocks, functions may contain functions. The graphical representation isn't precise enough to answer all questions of possible device configurations. Therefore a more formal specification is needed.

The UML (Unified Modeling Language) class specification gives the opportunity to specify components, their properties, their operation and their relation to each other unambiguously. The graphical representation has to be mapped to UML language elements using the following approach:

- Components will become classes (Device is CDevice class, Function is CFunction class, ...)
- The properties of the components become class attributes (e.g. Data Type of Variable)
- The relations between the components become relations between classes (e.g. functions are aggregated in blocks, ...)

The result is the class diagram shown in Fig. 3. The application related variable, function and block class (CVariable, CBlock, CFunction) are integrated in the device structure consisting of CDevice, CArray and CModule. The functions (i.e. the automation algorithms) are seen as name, i.e. they are references only. That's why the algorithm implementations are not standardized and hidden because they are specific know-how of the vendors.

The life cycle of devices in the distributed control system needs unambiguously identification of the devices. Therefore the according attributes are part of the CDevice class (Device_Vendor, Device_Model, Device_Revison und Device_Ser_No).

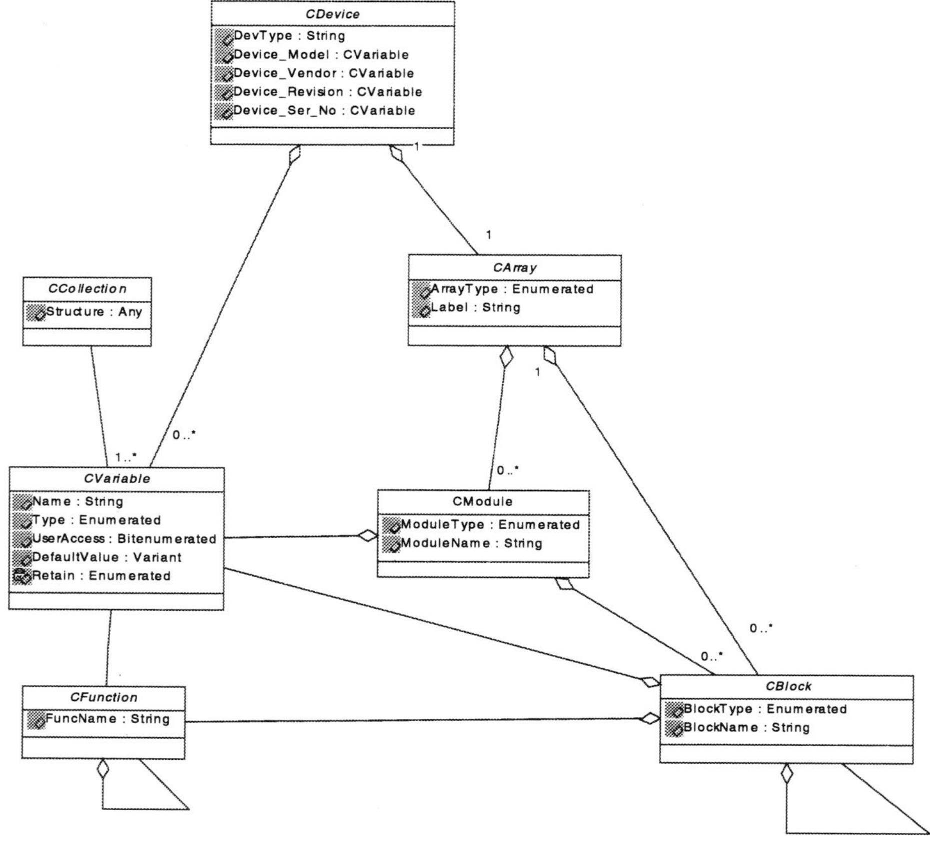

Fig. 3. Device model kernel class diagram

The device has the capability to contain a certain number of blocks and modules. The device has something like a rack for the plug in of the modules and the blocks. This is modeled with the CArray class. The CArray is part of the device and the blocks and modules can be plugged in the array. From a pure configuration point of view there is no difference between modules and blocks. In future the hardware configuration will become a software configuration.

5 Description technology for fieldbus profiles

Profiles, i.e. field device functions have a specific purpose from the user point of view. They are for control, diagnosis or maintenance reasons. Fulfilling these purpose there are interactions between so called actors performing a specific task. In the UML (Unified Modeling Language) description paradigm this is expressed in terms of so called use cases (for more details see [6]). The use cases give an overview about the actors in the systems and their interactions. In case of a DCS field devices and PLCs performing the control and field devices and operator performing diagnosis and maintenance. The control engineer, e.g. designs the system and commission the field devices. Doing this a data base system (or external system) is used.

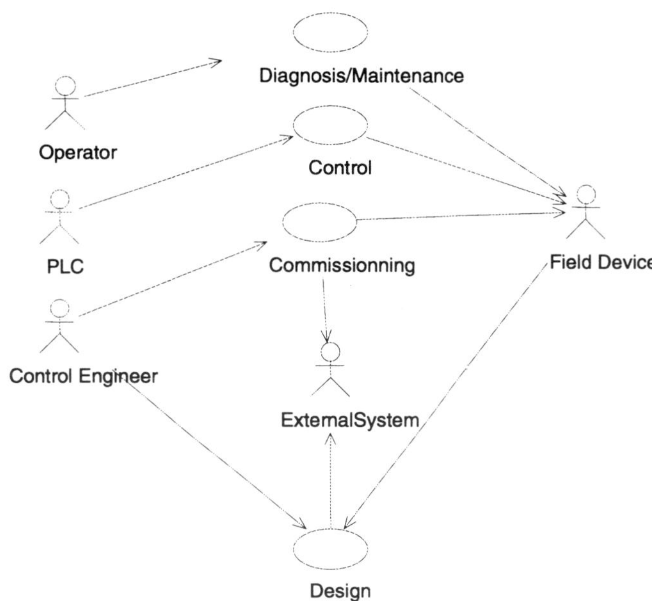

Fig. 4. Use case of distributed control system (partly)

Field devices, PLC, DCS (not shown in the figure) and tools are implementing the intended functionality in a distributed way. So each system component is a separated subsystem with fixed functionality. In terms of UML packages represent such subsystems. Because of the limits of pages granted for this paper, the transformation between the use case and the package structure isn't shown.

As subset of the all over distributed system, contained of field device implementations based on the profiles, the PLC representations of the field device functionality (profile proxy) and the description of the field devices in terms of the Electronic Device Description [7] is shown in Fig. 5. It is visible, that all components are based on the device model kernel. The kernel is a representation of the device model of step 4 (see 3.4) in terms of an UML class diagram (see also [5]).

The classes of the device model (e.g. Block) are specialized in the profile (Fig. 6) to e.g. PA-Function Block and Analog Input Function Block (this example see [4]). This specification is done in two separate packages. The aggregations of specific variables (e.g. measurement value, status, high-limit and low limit) to Function Blocks are part of an additional package. A modular structure of the specification can be achieved. The advantages is, that other profile specifications can be based on exactly the same device model and that the differences between the specifications are visible. If a specification can be started on a green field this approach is possible. It is also possible to reengineer existent solutions to compare the specifications using UML descriptions.

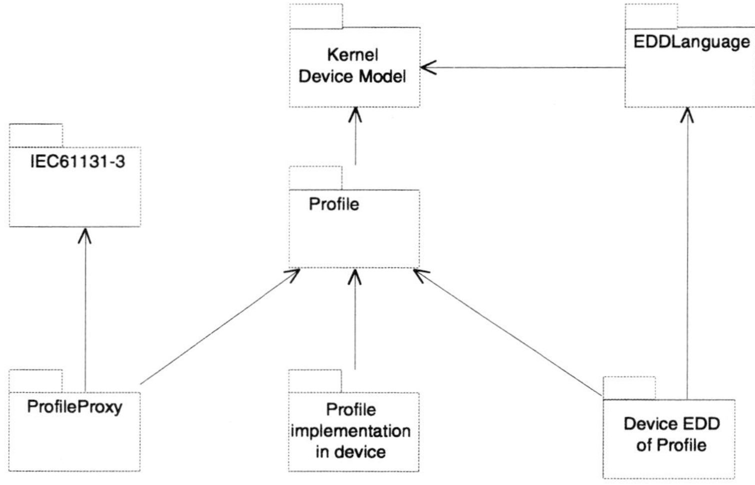

Fig. 5. Package overview of the considered system subset

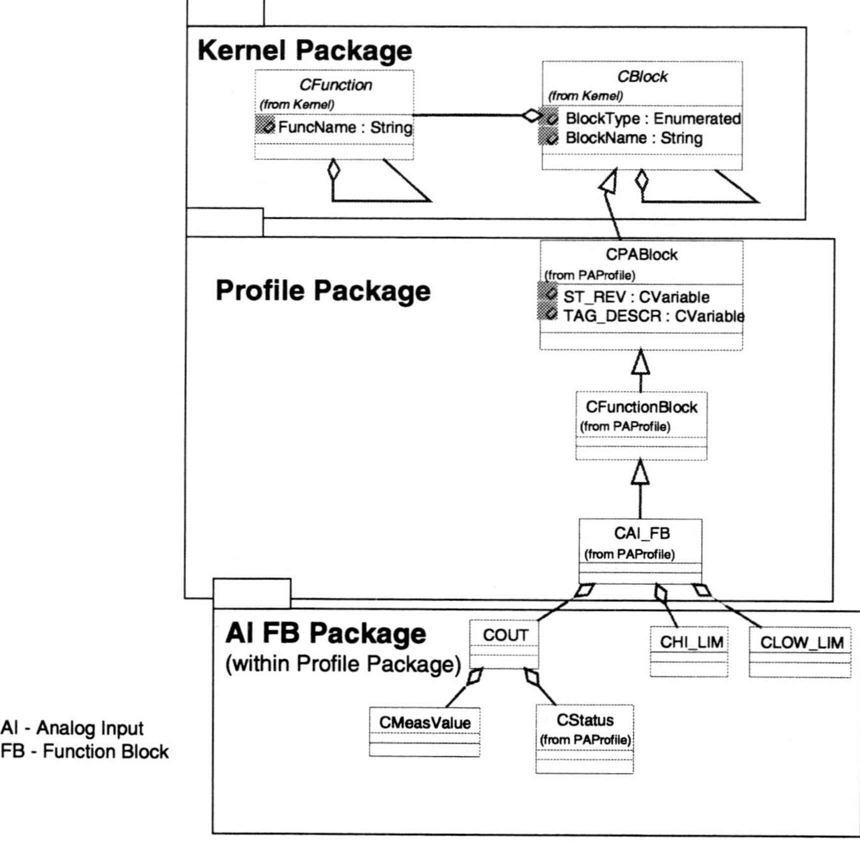

Fig. 6. Function Block/parameter relations of device model and profile specification (partly)

The example in Fig. 6 should make the general approach visible. In detail, the Electronic Device Description language, the PLC proxy, the device implementation can be derived from the device model and the profile specification. Relations, additions, modifications are clearly defined using associations, new classes, attributes and operations. Even the representation of the field device structure or off-line configuration parameters of the field devices in data bases can be derived from the device model and profile specification. Using a UML tool data type declarations for the development of tools and device software can be generated.

6 Summary

Functions on top of fieldbus protocols are becoming of broader interest. That's why end users are asking for added value of fieldbuses in terms of application functionality. The state of the art of the profile specification work in the fieldbus organizations are described.

Field devices are becoming components of distributed system and are not stand alone components anymore. This has to be considered during the profile specification work. Usually the device experts specifying the profiles provide these knowledge as background. The paper introduces a method, how field device functionality should be specified as part of a distributed control system. This is based on a so called device model which combines the functional view of an application and the devices view (i.e. the instrumentation). Some use cases, the packages structure and class diagrams using the UML paradigm are shown to make the principles of the method visible.

References

1. EN 50170: General Purpose Field Communication System, CENELEC 1996.

2. WorldFIP Companion Standards, WorldFIP, Paris 1999

3. P-NET Standardised general purpose channel types, October 1998, Process Data

4. PROFIBUS-PA Profile for Process Control Devices, PNO Karlsruhe 1999

5. Fieldbus Profile Harmonisation Approach of NOAH EP 26951, FET'99, conference proceedings

6. Terry Quatrani. Visual Modelling with Rational Rose and UML. Addison Wesley, 1998

7. Electronic Device Description Technology Approach of NOAH EP 26951, FET'99, conference proceedings

A Functional Profile for
Laboratory Measurement Equipment Based on
Measurement Bus and Profibus-DP/PA

Ulrich Wagner

Institut für Grundlagen der Elektrotechnik und Messtechnik, University of Hannover
Appelstr. 9A, D-30167 Hannover

Abstract. A new functional profile for labaroratory devices is the result of a working group „Laboratory Data Communication" in the Standardization Committee „Laboratory Devices and Laboratory Equipment" of DIN. Based on a very complex functional profile for process devices (Class A) of the Profibus User Organisation (PNO), a set of only seven simplified function blocks have been proposed, wich meets the main communication requirements of low-cost laboratory devices. Two fieldbus systems are elected for the necessary transport-oriented communication services: Measurement Bus („DIN-Messbus") and Profibus-DP/PA. All function blocks and its defined variables will be mapped on the specific application services of these two bus systems.

1 Laboratory Automation

Measurement and automation devices used in chemical, physical, biological, pharmaceutical and medical laboratories are provided mostly just with analog output or digital point-to-point-interfaces with varying protocols and communication skills. Consistent measured data acquisition, parameter entry and control of these appliances from a laboratory-guide-system, normally a PC, are therefore only possible with significant hardware- and programming efforts. The development and standardization of a consistent range of functions for the communication to the measurement and automation devices, based on selected communication systems (e.g. field bus systems) will open - comparable to other areas of automation - new cost-effective and timesaving possibilities for networking, program development and the actually laboratory working process.

2 Standardizing a Laboratory Communication Functional Profile

2.1 Communication Requirements

For the device specific communication in the area of laboratory measuring and control techniques the following requirements have been found:

- up to 30 participants with Master-Slave-operation

- broadcasting to all participants or groups of participants

- user data range of 1 to 50 byte

- reaction time of max. 100 ms

- implementation with standard microcontrollers and standard peripherals

- cost of one node max 50,- Euro

On the basis of these requirements all established and standardized field bus systems have been checked by the responsible working group in respect to their qualification for the intended laboratory communication system. The fieldbus systems *'Measurement Bus'* and *'Profibus-DP/PA'* have been selected as best qualified communication systems by the standard committee. One or both of these communication systems may be used at the proposed laboratory communication system with a standardized laboratory data functional profile (see Fig. 1.).

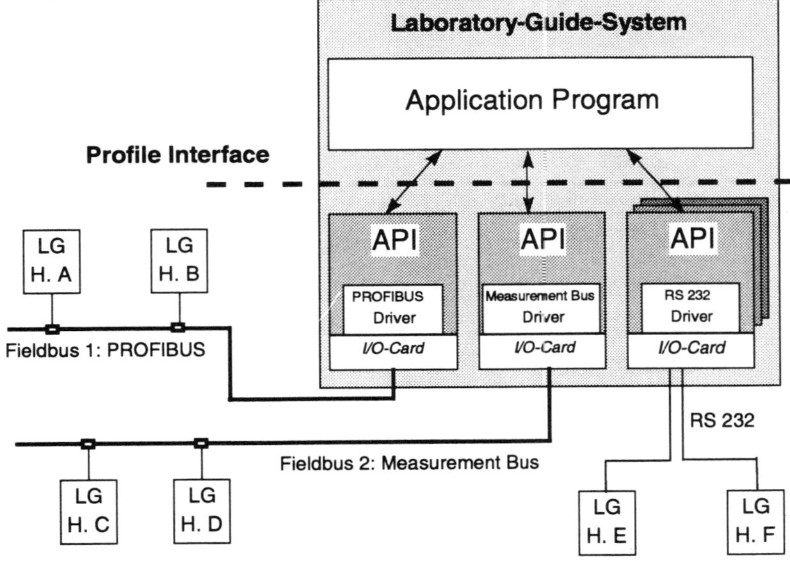

Fig.1. Architecture of a Laboratory Communication System: Functional Profile for different Measurement Devices and a Laboratory Guide System

2.2 Functional Requirements

According to *different technical functions* inside of the laboratory measurement and experimental equipment a set of function blocks and their parameters have to be defined. The result of this will be a *"Laboratory Functional Profile"*, usefull also as a basic principle for all similar measuring equipments and will allow a consistant method of application programming, e.g. assisted by use of commercial measuring and control operation programms.

For typical low-cost laboratory devices the following functions are most important for monitoring and control by a laboratory guide system:

- Analog measuring instruments, e.g. temperature or pressure measurement

- Analog sources or actuators, e.g. stirrer or heat source
- Binary sensor or actuator, e.g. pneumatic valves or switches
- Control units
- Time programmable control unit
- Terminal and display unit

2.3 Function Blocks

To handle these technical device functionality by a laboratory communication system the following (logical) function blocks have been provided:

- Binary-Input (binary sensor)
- Binary-Output (binary actuator, optionally with feedback message)
- Analog-Input (analog sensor, transducer, optionally with scaling)
- Analog-Output (analog actuator, transmitter of analog quantities, opt. with scaling)
- Control Function (laboratory device containing a control unit)
- Programmer (output of a time function for control of an analog output device or laboratory device containing control unit)
- Operator's Communication Unit (keyboard-display-unit).

Each realized specific measurement device profile of a laboratory device consists of some device independant *basic functions* (physical block and block list) and at a minimum one selectable function block (see Fig. 2.).

The *physical block* consists of 18 common used device specific variables or constants, e.g. identification data and software version-number. 9 of these parameters are mandatory. The *block list* contains a header and the list of implemented function blocks and their (device specific) absolute adress indizes.

A *function block* consists of 8 basic parameters (see table 1., only five of them are mandatory) and a variable number of block specific parameters. For each proposed function block the mandatory or optional parameters are standardized by identifier, name, data type, size, access mode, type of storage, update mode and default value.

The specification of the individual function block parameters was worked out - in opposition with other functional profiles of automation technology - not in the way of specific measuring variables (e.g. "rotational speed") instead of in the manifold applications of the measuring practice appropriating common method (e.g. for the function block analog-input in the sense of an "actual measured value"). In Fig. 2. an example of a measuring device with one analog measuring input and one (bus controlled) binary output is shown. Note that a function block not describes the complete hard- or software functionality of a specific device but only the part of functionality which is able to be controlled by the connected communication system, e.g. a fieldbus system.

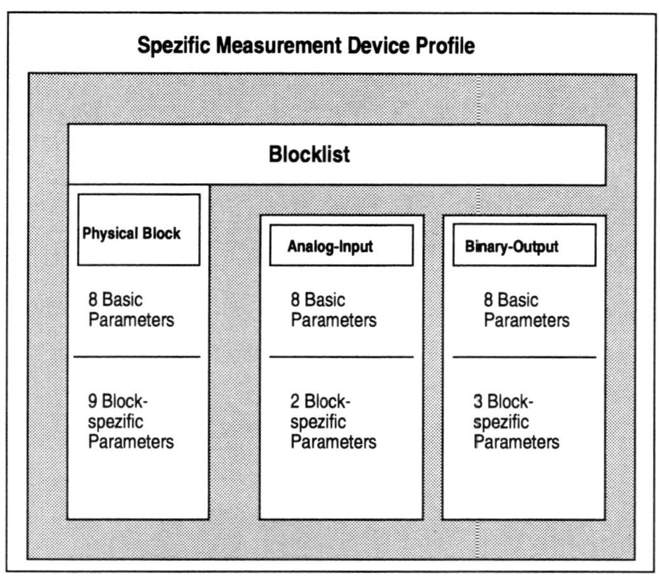

Fig. 2. Example of a spezific Laboratory Measurement Device with Function Blocks "Analog-Input" and "Binary-Output"

Table 1. Basic Parameters, necessary for each Function Block

Rel. Index	Parameter Name	Mand. Opt.	Objekt type	Data type	Size	Access	Store	Method of Updating	Default value
0	BLOCK_OBJECT	m	Record	DS-32	20	r	Cst	async	block-spec.
1	ST_REV	m	Simple	Unsigned 16	2	r	N	async	0
2	TAG_DESC	o	Simple	Octet-String	32	r,w	S	async	32xSP
3	STRATEGY	o	Simple	Unsigned 16	2	r,w	S	async	0
4	ALERT_KEY	o	Simple	Unsigned 8	1	r,w	S	async	0
5	TARGET_MODE	m	Simple	Unsigned 8	1	r,w	S	async	block-spec.
6	MODE_BLK actual permitted normal	m	Record	DS-37	3	r	N Cst Cst	async	block-spec.
7	ALARM_SUM	m	Record	DS-42	8	r	D	async	0,0,0,0

To give an impression of the parameter description, in the following tables all parameters, needed in the above example are listed. First the basic parameters are shown in Table 1. wich have to be implemented triple, specific for the physical block, the analog input function block and the binary output funktion block.

In Table 2. the - ownly two - parameters of an analog input function block are shown. The parameter OUT contains the actual measuring value and shall be updated synchronously.

Table 2. Parameters of the Function Block "Analog-Input"

Rel. Index	Parameter Name	Mand. Opt.	Objekt type	Data type	Size	Access	Store	Method of Updating	Default value
0 to 7: Basic Parameters (see Table 1)									
10	OUT	m	Record	DS-33	5	r	D	sync	-
12	OUT_SCALE	o	Record	DS-36	11	r,w	S	async	0 resp. 100 %

Table 3. shows the list of parameters of the binary output function block. Only the parameter SP_D is mandatory, who contains the - binary - value, which the laboratory guide system has sent, to set the specified output actuator, e.g. a switch. The - optional - parameters offer the possibility to reed the setted binary output value (OUT_D) and/or to reed the actual state of the actuator if there a binary state-sensor is implemented (REEDBACK_D).

Table 3. Parameters of the Function Block "Binary-Output"

Rel. Index	Parameter Name	Mand. Opt.	Objekt type	Data type	Size	Access	Store	Method of Updating	Default value
0 to 7: Basic Parameters (see Table 1)									
8	REEDBACK_D	o	Record	DS-34	2	r	D	async, sync	-
9	SP_D	m	Record	DS-34	2	r,w	D	async, sync	-
10	OUT_D	o	Record	DS-34	2	r	D	async	measured of the variable, state

Fig. 3. shows the state diagramm of a simple function block (note that this is not the state diagramm of the complete device). The Running-Mode must be implemented for each specified function block; this is the normal mode.

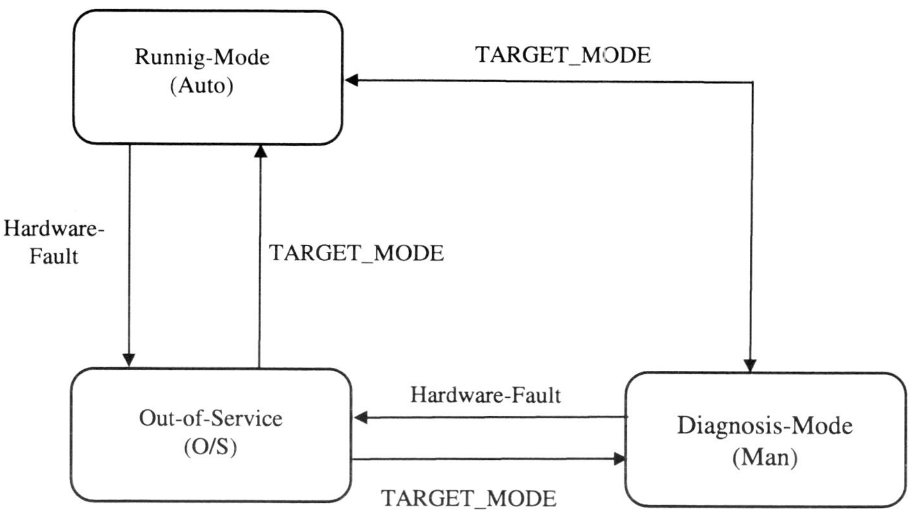

Fig. 3. State Diagram of the Function Block "Analog-Input"

Diagnosis-Mode and Out-of-Service-Mode are optional and offer the possibility for - manual operating - diagnosis mode and/or in case of a hardware defect a selfdiagnosis state. A change of state needs a 'write-of-variable' telegram to change the parameter TARGET_MODE.

The proposed *Control Function Block* consists of 8 basic + 19 specific parameters (6 of them are mandatory). This function block serves for control and monitoring of laboratory devices which contain an automatic control unit, e.g. a laboratory centrifuge. More complex application may be realized by using this function block multiple and or using in combination with other function blocks.

The *Programmer Function Block* has been defined to generate a time-dependent output function. The output parameter OUT must be the input value of another function block, e.g. an analog output function block of the same device. Thus it is possible to generate a linear spline profile beginning at a start-value and changing this value at every following time-step. The programmer function block contains (8+) 12 parameters, which all are mandatory.

The *Operators Communication Unit Function Block* is defined for use of a terminal-display-unit and consists of (8+) 5 mandatory parameters.

3 Realization

The proposal for a laboratory functional profile is nearly finished. Several companies have already begun to realize this functional profile for their measuring and laboratory devices based on Measurement Bus and Profibus-DP/PA. The necessary link between the outlined function blocks and the standardized fieldbus services of OSI layer 1 to 7 will be the next item of the working group. Two researching groups have begun with mapping the function blocks to either Measurement Bus or Profibus-DP/PA application layer services.

Thus using a standardized fieldbus system and the appropriate function blocks for his laboratory application programming the user is able to work independent from device manufacturers.

References

1. H. Cullmann: Normung von Datenschnittstellen an Laborgeräten. Vortrag auf der 25. ACHEMA '1997 in: Internationales Treffen für Chemische Technik, der Zukunft. Vortrag auf der 25. ACHEMA '1997 in: LaborPraxis 6 (1997) S. 414

2. U. Wagner: Measurement Bus Umweltschutz und Biotechnologie, Kurzfassungen der Vortragsgruppe Labor- und Analysentechnik, Frankfurt 1997

3. U. Höfling: Das Labornetzwerk - Characteristics and Areas of Application, Feldbus-Tutorial, Sensor '97 Kongreß, Nürnberg 1997

Mapping of Fieldbus Protocols to Standardised Field Level Objects

Peter Fischer

Fachhochschule Dortmund
Fachbereich Nachrichtentechnik
Sonnenstr. 171
D-44137 Dortmund

Abstract. The specification of the Field Level Object standard is finished and the corresponding European pre-standard will be published at the end of 1999. In this paper the mapping of a fieldbus communication system to the objects, functions and services defined in the pre-standard is presented. As an example the EIB system is used for which a detailed mapping guide as annex for a draft of an European pre-standard exists. The mapping is shown in detail for the analogue object explaining some of the mapping problems.

1 Standardised Field Level Objects

In the Task Group TG3|4 of CEN/TC247 "Controls for Mechanical Building Services" with members of Working Group WG3 "Building Management Products and Systems for HVAC Applications" and WG4 "System-neutral Data Transmission for HVAC Applications" the preliminary draft of ENV 13154-1 "Field Level Objects" [1] was specified. As shown in earlier presentations [2, 3] the scope of this specification is to make easier the communication between the four fieldbus communication systems BatiBUS, European Installation Bus EIB, European Home System Bus EHS and LonTalk Protocol, all specified in the ENV 13154-2 [6], and the communication to systems on the automation or management net level, like e. g. BACnet [7].

For implementing the Field Level Objects specification there is a common part for all communication systems and a specific part related to the corresponding one. This is a kind of an application layer interface realising the link between the application layer and the Field Level Objects (see Fig. 1).

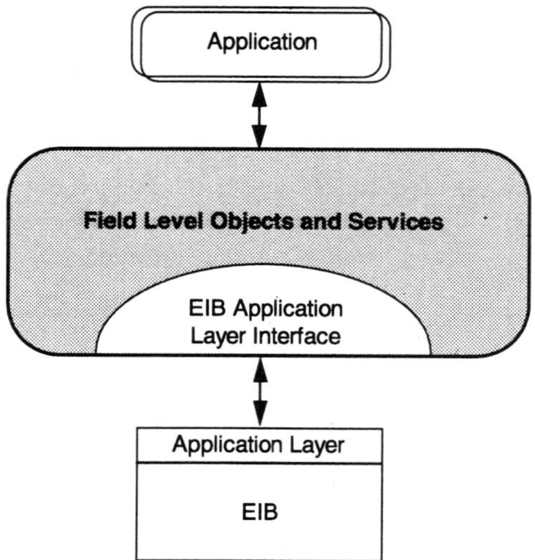

Fig. 1. Link between EIB and Field Level Objects

The preliminary draft of ENV 13154-1 is based on the functions specified in the BMS[1] Information List (see [5, 4]) which contains the standard input, output, communication, processing, management and operator functions for a BMS. It is subdivided into the four main sections for plant specific functions:

- Input and Output Functions (physical I/O, communications I/O)

- Processing Functions (monitoring, interlocks, closed loop control, calculation/optimisation)

- Management Functions (communication, historical data)

- Operator Functions (visual display, other messaging functions)

On the field net level the input/output functions and the processing functions are more important than the others. Therefore the Field Level Objects are sub-divided into two groups: the datapoint objects and the functional objects (see Fig. 2).

Datapoint Objects	**Functional Objects**
Analogue Object	Device Object
Binary Object	Alarm Notification Object
Counter Object	Event Notification Object
Multistate Object	
PID Loop Object	

Fig. 2. Datapoint and Functional Objects

---

[1] BMS: Building Management System

Every object type has mandatory, optional and engineerable properties. The mandatory properties are needed for every object (like e. g. the Object Identifier), so in every implementation this information has to be available. The optional properties can be used (like e. g. alarm limits for Analogue Objects), so if it is necessary this information can be available. The third group of the engineerable properties are part of the related object, they can't be changed and in most cases they are not visible from the outside (like e. g. default values).

2 Mapping of Field Level Objects

It seems to be reasonable to publish such a Field Level Object Standard when the different consortia of the above mentioned communication systems have proven the mapping of the Field Level Objects to the objects and services of the application layer protocol. The request was sent to the consortia and they responded as follows. The BatiBUS Club International BCI and the European Home System Bus Association EHSA didn't show any interest because of the convergence activities of the three systems BatiBUS, EHS and EIB. The German LON-Nutzerorganisation LNO and the European Installation Bus Association EIBA agreed on it as well as the PROFIBUS-Nutzerorganisation PNO and the BACnet Interest Group Europe B.I.G-EU.

The implementation of the mapping does not only mean the one-to-one mapping of the standardised field level objects and their properties to the corresponding objects of the application layer but also to the corresponding services which are provided or used by the application layer. In some cases not all information specified in the field level objects are part of the application protocol data unit (APDU) and therefore they are not accessible directly.

As an example the mapping of the Analogue Object FLN[2] to the EIB Analogue Object Types is shown in the following section.

---

[2] FLN: Field Level Network

3 Analogue Object FLN and EIB Analogue Object

Analogue Object FLN		
Object_ID_Number	m	R
Object_Type	m	R
Object_Name	m	R
Value_Presentation	m	R
Present_Value	m	W
Status	m	R
Units	m	R
Priority	o	W
Description	o	R
Commanded_Value	o	R
High_Alarm_Limit	o	W
High_Warning_Limit	o	W
Low_Warning_Limit	o	W
Low_Alarm_Limit	o	W
Maintenance/Out_Of_Service_Flag	o	W
COV_Enabled	o	W
Warning/Alarm_Enabled	o	W
Acknowledged_States	o	W
Alarm_Acknowledge	o	R

m: mandatory; o: optional;
R: read only; W: read and write

Fig. 3. Analogue Object FLN

The Analogue Object FLN can be used for the description of analogue inputs, outputs and values (see Fig. 3) as well as the EIB Analogue Object Type; the mapping of the object type properties is shown in Fig. 4.

EIB Object	PID_Object_Type	FLN Object	Object_Type
Analogue-Input	100	analogue-input	0
Analogue-Output	101	analogue-output	1
Analogue-Value	102	analogue-value	2

Fig. 4. Mapping of the EIB object types to the FLN object property Object_Type

Fig. 5 shows the Analogue Object Type EIB.

Property Name	Property ID	Type	O/M/W
Object Type	PID_OBJECT_TYPE	PT_UNSIGNED_INT	M
Object Name	PID_OBJECT_NAME	PT_UNSIGNED_CHAR[]	O
Description	PID_DESCRIPTION	PT_UNSIGNED_CHAR[]	O
Present Value	PID_PRESENT_VAL	PT_UNSIGNED_CHAR PT_UNSIGNED_INT PT_UNSIGNED_LONG PT_SIGNED_CHAR PT_SIGNED_INT PT_SIGNED_LONG PT_FLOAT PT_DOBBLE PT_EIB_FLOAT	M/W
Status	PID_STATUS	PT_UNSIGNED_CHAR	O
Units	PID_UNITS	PT_UNSIGNED_CHAR	O
Priority Array	PID_PRIORITY_ARRAY	same as Present Value	O
Relinquish Default	PID_DEFAULT	same as Present Value	O
Priority Value	PID_PRIORITY_VALUE	PT_UNSIGNED_CHAR[]	O
Commanded Value	PID_COMM_VALUE	same as Present Value	O
Event/Alarm Control	PID_EVT_ALARM_CTR	PT_UNSIGNED_CHAR	O
High Alarm Limit	PID_HIGH_ALARM_LIMIT	same as Present Value	O
High Warning Limit	PID_HIGH_WARNING_LIMIT	same as Present Value	O
Low Alarm Limit	PID_LOW_ALARM_LIMIT	same as Present Value	O
Low Warning Limit	PID_LOW_WARNING_LIMIT	same as Present Value	O

O: optional, M: mandatory, W: writeable

Fig. 5. EIB Object Type Analogue Input/Output/Value

The main difference is the object identifier. Each FLN object has the explicit property "Object_ID_Number" which does not exist for EIB objects. Every EIB object has an identifier only for addressing which contains the physical device address and the index of the object in the device (see Fig. 6). Each object identifier in EIB is unique for the EIB network. The FLN Object_ID_Number is a 24 bit long flat identifier and shall be unique in the whole project. It is mapped to the EIB identifiers.

Physical Address		Object	Object_ID_Number
High	**Low**	**Index**	
8 bit	8 bit	8 bit	24 bit

Fig. 6. EIB object identifier and FLN Object_ID_Number

4 Mapping of EIB Objects to FLN Objects

There are four different cases for mapping the properties of an EIB object to the properties of a FLN object:

- property mapping

- property mapping and type conversion

- property mapping and value mapping by a table

- complex conversion by an EIB application layer interface

Mandatory properties from the FLN object which are not existing in the EIB are set by the application layer interface program. Fig. 7 shows the mapping of the EIB Analogue Input/ Output/ Value Object to the Analogue Object FLN.

EIB Property	O / M / E		FLN Object Property
	EIB	FLN	
PID_OBJECT_TYPE	M	M	Object_Type
PID_OBJECT_NAME	O	M	Object_Name
PID_DESCRIPTION	O	O	Description
PID_PRESENT_VAL	M	M	Value_Presentation
		M	Present_Value
PID_STATUS	O	M	Status
PID_UNITS	O	M	Units
PID_PRIORITY_ARRAY	O	O	Priority
PID_DEFAULT	O	E	
PID_PRIORITY_VALUE	O	O	Priority
PID_COMM_VALUE	O	O	Commanded_Value
PID_EVT_ALARM_CTR	O	O	Warning/ Alarm_Enabled
PID_HIGH_ALARM_LIMIT	O	O	High_Alarm_Limit
PID_HIGH_WARNING_LIMIT	O	O	High_Warning_Limit
PID_LOW_ALARM_LIMIT	O	O	Low_Alarm_Limit
PID_LOW_WARNING_LIMIT	O	O	Low_Warning_Limit
		O	COV_Enabled
		O	Acknowledged_States
		O	Alarm_Acknowledge

O: optional; M: mandatory; E: engineerable

Fig. 7. Conversion of the EIB Analogue Object Type to the FLN Object

Regarding COV reporting and alarm handling there are some "white places" in the EIB column. The reason of it is the realisation in EIB. In an EIB system alarms and COV notifications are sent using group objects. There is no specific EIB object for alarm and COV notifications. Therefore the implementation of notification objects is part of the application layer interface. It has to add all mandatory parts to create an Alarm/COV-notification behaviour which is conform to the Field Level Object specification. The distribution of alarms and COV notifications is done via the EIB group mechanism.

5 Realisation in EIBnet

The as an example used mapping of the Field Level Objects specification to an existing EIB system is described as an informal annex [6] to a preliminary draft CEN ENV 13321-2 [7], called EIBnet. In this annex the mapping is described in detail for all objects, functions and services of the Field Level Object specification.

References

1   CEN Draft ENV 13154-1, Data Communication for HVAC Application Field Net - Part 1: Field Level Objects, September 1997

2   Fischer, P.: Ein universelles Profil für Feldbus-Protokolle in der Gebäudeautomation. In: Dietrich D., Neumann P. (Hrsg.): FeT '95. ÖVE 1995, 131-138

3   Fischer, P.: Ein Feldbus-Profil auf dem Weg zur europäischen Norm. In: Dietrich D., Schweinzer, H. (Hrsg.): Feldbustechnik in Forschung, Entwicklung und Anwendung. Wien: Springer, 1997, 116-122

4   CEN TC 247 WG3, Building management products and systems for HVAC application, Part 2: Equipment Functionality, Draft July 1994

5   VDI 3814, Blatt 1, Gebäudeleittechnik (GLT), Strukturen, Begriffe, Funktionen - Berlin: Beuth, Juni 1990

6   Field Level Objects mapping to EIB, Draft of an informal annex to the draft CEN ENV 13321-2, 1998

7   CEN Draft ENV 13321-2, Data Communication for HVAC Application Automation Net - Part 2: EIB, September 1997

Device Based Process Control in Foundation Fieldbus

Dirk Thiele, Terry Blevins, and Willy Wojsznis

Fisher-Rosemount Systems, Inc., 8627 Mopac Expressway North, Suite 400, Austin, Texas 78759, USA
dirk.thiele@frco.com

Abstract. Devices based on the Fieldbus Foundation's specification will, as an integral part of their design, allow process inputs and outputs to be precisely sampled by the device without aliasing or skewing. Fieldbus Foundation also standardizes a device description language (DDL) in which device manufacturers can present a description of their field devices. Given these standards it became possible to distribute function block control technology to fieldbus devices of different vendors. An overview of monitor, control, and diagnosis techniques used in Foundation Fieldbus is presented. Also, the manner in which fieldbus may influence advanced control implementations, such as process identification for controller self-tuning is reviewed.

1 Introduction

Often the primary means by which distributed control systems and digital controllers access or provide input to field devices is a 4-20 mA signal. The capability of smart transmitters to provide additional information is often limited by their low communication rate. An all-digital fieldbus is utilized by devices based on the Fieldbus Foundation Specification. Two levels of communication were defined: H1 which at 31.25K baud could be used with existing wiring and meet intrinsic safety requirements of the process industry and high speed Ethernet to meet requirements of manufacturing automation.

In addition to defining the physical layer for fieldbus the Fieldbus Foundation has standardized the device's communication stack and a user layer which models the application as function blocks. Such capability allows the base level process control and measurement done today in distributed control systems and single loop digital controllers to be implemented in fieldbus devices.

2 Infrastructure

To install Foundation fieldbus devices, the normal twisted pair with shield wiring used in your plant today may serve as the fieldbus. In an existing installation existing wiring to be reused when Foundation fieldbus devices are installed. A maximum of 32 field devices may be connected to a single fieldbus segment. Twelve of the devices on a fieldbus segment may powered from the fieldbus segment similar to the powering of tradition two-wire transmitters.

Most fieldbus installations will utilize a tree topology as shown in Fig. 1.

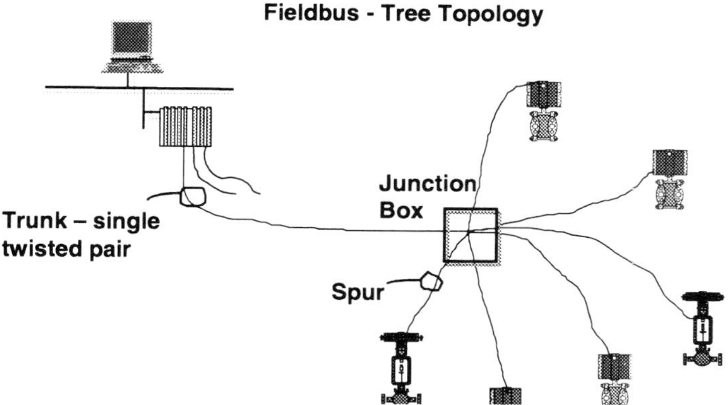

Fig. 1. Fieldbus segment utilizing topology

3 Function Blocks in Field Devices

In a fieldbus environment, the user application is defined through the configuration of function blocks. This approach is similar to the configuration of control and monitoring in DCS systems today. However, the fieldbus function blocks support applications distributed between fieldbus devices, which involve measurement and control. Such capability allows the base level process control and measurement done today in distributed control systems and single loop digital controllers to be implemented in fieldbus device.

Monitoring, calculation and control functions may be defined by configuring function blocks within a fieldbus device and configuring the connections between function block input and output parameters. For example, a flow measurement might be implemented in a transmitter using the analog input block, AI block. This transmitter publishes its measurement value and its status as shown in Fig. 2. A valve on the fieldbus segment

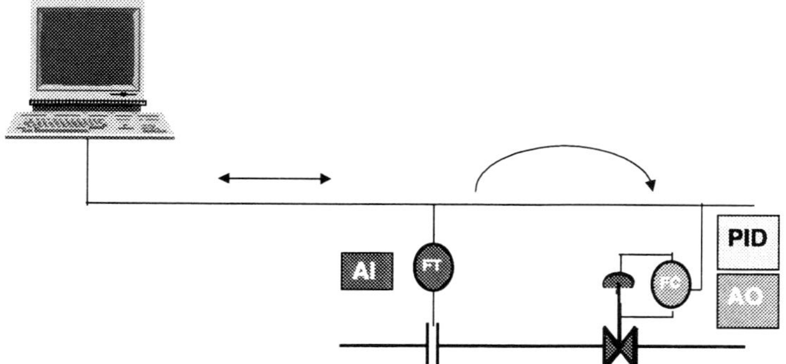

Fig. 2. Distributed control and calculations between fieldbus devices

may support a control block, PID block, which subscribes to the published measurement value. Based on this value and its target setpoint, the PID block may calculate an output required to maintain setpoint. Within the valve, this output might be used by an output block, AO block, to adjust the valve.

10 basic and 19 advanced function blocks have been defined by the function block specification. These blocks allow both analog and discrete monitoring, calculations, and control to be done in field devices. It is estimated that as much as 80% of the DCS and PLC controller functionality may be distributed to the fieldbus devices using this capability.

4 Scheduling

To successfully distribute control between fieldbus devices without degrading control by communication delay, the scheduling of function block execution and related communications is critical. The low processing power of intrinsically safe devices combined with the distribution of control between fieldbus devices imposes some constraints on the design of fieldbus devices. Each device maintains a schedule of which blocks within the device are to be executed. Based on this, it is possible to schedule communication of block output values between devices for calculations and control.

System management provides a mechanism responsible for synchronizing device clocks with a system master clock. The purpose of this capability is to provide a uniform time reference to all device applications on the network. This uniform time reference supports time-tagging of events and permits system schedules to be developed and coordinated. Through this scheduling capability, function block execution may be synchronized. Such capability allows the sample and communication delay introduced

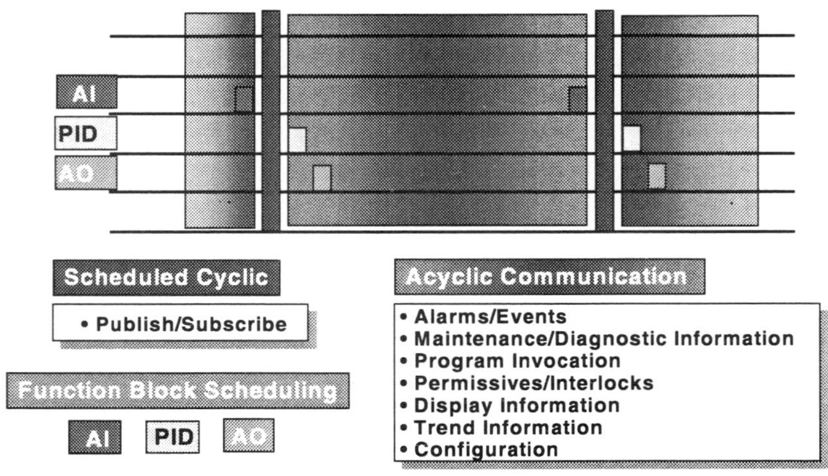

Fig. 3. Fieldbus Scheduling

into closed loop control to be minimized without the need to oversample inputs and outputs.

A typical schedule for the above example is illustrated in Fig. 3. When no control communications are scheduled, then other information for display, alarm notification, etc. may be communicated.

5 Interfacing process system

A new generation of control systems that is designed to effectively utilize fieldbus devices is available from major control system vendors. Such systems are smaller in size physically because of the reduction in the number of I/O card required to interface with field devices. This is especially true where multivariable fieldbus transmitters are installed.

To allow the configuration and calculated parameter values in a fieldbus device to be accessed, each manufacture has to provide a device description (DD) of his fieldbus devices written in the device description language (DDL) standardized by the Fieldbus Foundation. Configuration and operator interface stations, which have been designed to utilize device descriptions, are able to access device function blocks after the device description is loaded.

6 Diagnostic Support

Features are included in the Fieldbus Foundation Specification for the support of process analysis and troubleshooting field problems. Special consideration is given to signal filtering and sampling to insure that the resulting digital values adequately represent the true process measurement. Conditions detected by the field device that would impact the validity of the measurement are made available as the status of the digital value. To allow this information in a device to be used, an efficient mechanism using view objects is provided to access field device information without disrupting the control distributed between field devices.

As part of input block processing, checks on associated hardware and software are performed. Analog values will be transferred as floating point values in engineering units. The status of an output parameter is calculated by the block to give an explicit indication of the quality of the value; *good, uncertain, or bad* as shown Fig. 4. A sub-status attribute indicates the primary condition, which determines the quality. Also, there is an explicit indication in the status attribute of whether the value is limited high or low. When a block input or output parameter is access for viewing at an operator console, historian, or diagnostic tool, both the value and its associated status may be communicated. The quality information and its sub-status may be useful in diagnosing the cause of a Bad or Uncertain measurement.

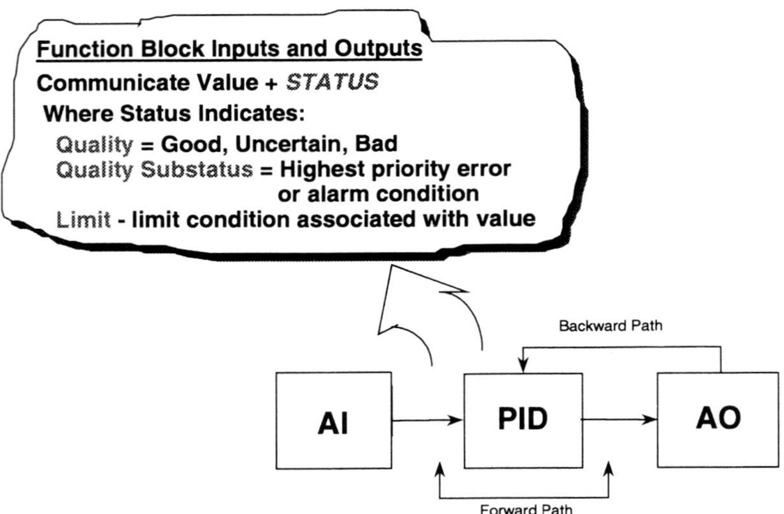

Fig. 4. I/O Status – Each function block input or output supports a value and status

In addition, auxiliary measurements such as stem position or limit switch status for an on-off valve will often be made available through fieldbus. A whole new generation of diagnostic and calibration tools will be accessible over fieldbus. These new features reduce the time required to diagnose and correct a field device problem. Asset management solutions[i] can be utilized to monitor device wear and schedule preventive maintenance.

7 High speed data collection

When accessing information for process analysis and diagnostics, it may not be practical or technically feasible to communicate individual parameter values as fast as the block executes rate. To address these issues, trend objects are defined in fieldbus devices as illustrated in Fig. 5.

At least one trend object is included in each field device. A trend object maintains a collection of 16 parameter samples. Through trend object configuration, it is possible to specify the input or output parameter to be collected at the collection rate. The block collects both the parameter value and status. Samples rates are an integer multiple of the parameter's block execution rate. Sample values may be averaged if the collection rate is slower than the block execution rate. When 16 new samples are collected, this information along with the time-stamp of the last value collected will be automatically reported to one or more devices.

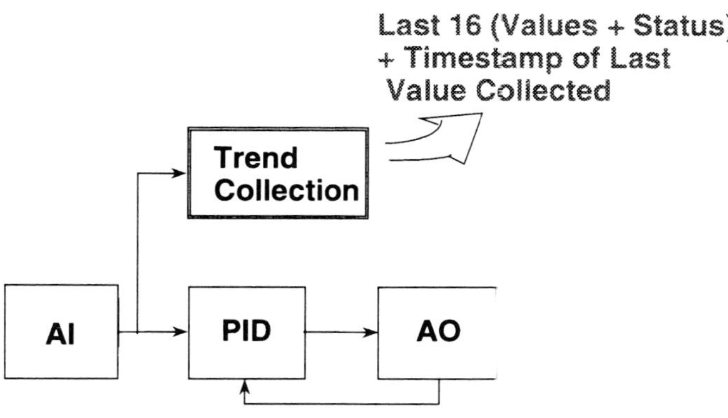

Fig. 5. Trend Object

The variable delay and jitter introduced by communications in many DCS systems has placed restrictions on the analysis of fast loops and the manner in which self-tuning is implemented. Trend objects allow analysis of fast processes and provide new flexibility in distributing features such as process identification for self-tuning. Other process identification applications which feature automated self-tuning on demand are implemented as function block in field devices. This reduces the bus communication to a minimum and guarantees process identification as close as possible.

References

1. Terry Blevins, "An overview of the ISA/IEC Fieldbus", 1998, Fisher-Rosemount, Austin, Texas

2. Terry Blevins, "Fieldbus ushers in a new era in process control", 1998, Fisher-Rosemount, Austin, Texas

3. Fieldbus Foundation, "Foundation™ Specification Function Block Application Process", 1994-1998

4. DeltaV[ii] website: www.easyDeltaV.com

---

[i] AMS – Software packet by Fisher-Rosemount Systems for remotely managing all plant equipment

[ii] DeltaV™ - Scalable process control system by Fisher-Rosemount Systems with field-based architecture (system extends to and includes field devices)

CANopen Device Profile for
Hydraulic Proportional Valves

Martin Wollschlaeger*, Erik Unger**, Heiko Witte**

*Institut für Prozeßmeßtechnik und Elektronik (IPE)
**Institut für Förder- und Baumaschinentechnik, Stahlbau, Logistik (IFSL)
Otto-von-Guericke-Universität Magdeburg
PO Box 4120, D-39016 Magdeburg

Abstract. The trend towards decentralized concepts in fluidic systems requires effective communication structures between the networked components. Field-busses are designed to meet these requirements. When using similar devices from different vendors, a harmonization concerning communication objects and devices' behavior has to be reached in order to minimize efforts in design and configuration of the whole system. Thus, a bus-system independent device profile for fluidic components has been defined. This profile has been mapped to CANopen. The article describes the profile and the CANopen mapping and the description technique used.

1 Motivation

The increasing functionality of electronic components enables the realization of valves with high complexity, suitable for fluidic systems. The introduction of decentralized fluidic components requires effective solutions for communication between such components. The most important solution for those requirements is an implementation of a fieldbus. In order to integrate devices from different vendors into a single project, and to implement user-friendly software tools, it is necessary to define the communication-relevant behavior of components with identically or similarly functionality. This can be done with device profiles.

2 Device Profile for fluidic components

In April, 1997, the "Fachgemeinschaft Fluidtechnik" of the Society "Verband Deutscher Maschinen- und Anlagenbau e.V. (VDMA)", established a group with focus on device profiles for fluidic systems. The main goal of this group was to define a fieldbus-independent device profile for hydraulic and pneumatic components used in stationary systems. Caused by the main application area of valves as integral parts of hydraulic drives, the group was merged with a similar group focused on profile developments for hydraulic drives. This was an important step in order to harmonize definitions and descriptions of sub-components and parameters of both device classes. The result of the standardization process was the definition of the device profile "Fluidtechnik", that was released in Version 1.0 in April 1999 [1]. This profile can be used for hydraulic proportional valves, as well as for hydraulic drives of high complexity. In principle, it can be used in pneumatic systems, as well. Figure 1 shows the principle environment for hydraulic valves, while Figure 2 illustrates the internal architecture of the devices.

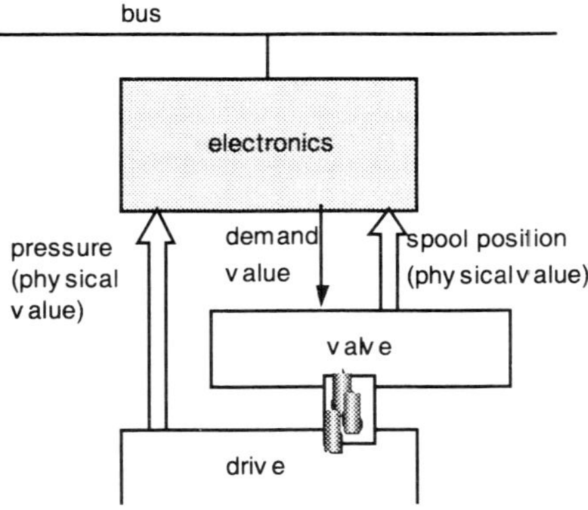

Fig. 1. Application environment for hydraulic valves.

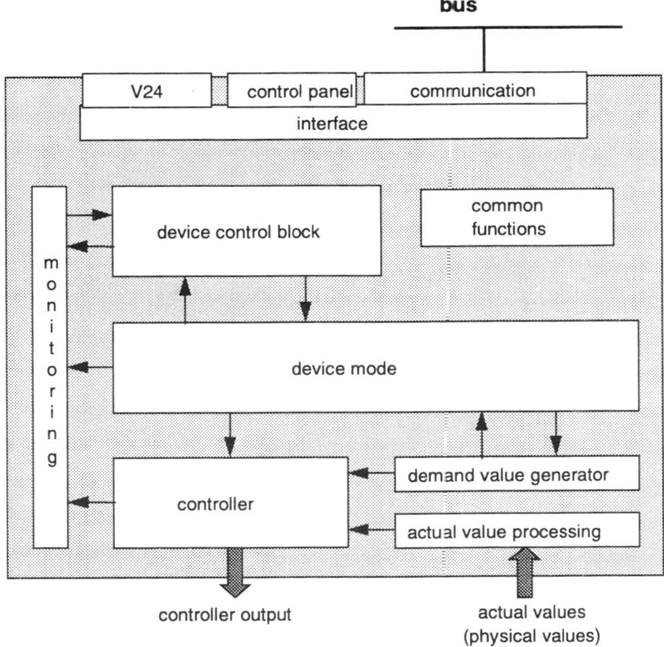

Fig. 2. Internal device architecture

The behavior of devices in both classes follow an unique architecture. Depending on the class, the defined function blocks are of different complexity and may be optional. In any case, the device control block is mandatory and has to follow the state machine shown in Figure 3. The transitions between the states are either performed automatically (startup, fault reaction), or can be controlled by a mandatory control word. The actual state is mapped to a mandatory status word.

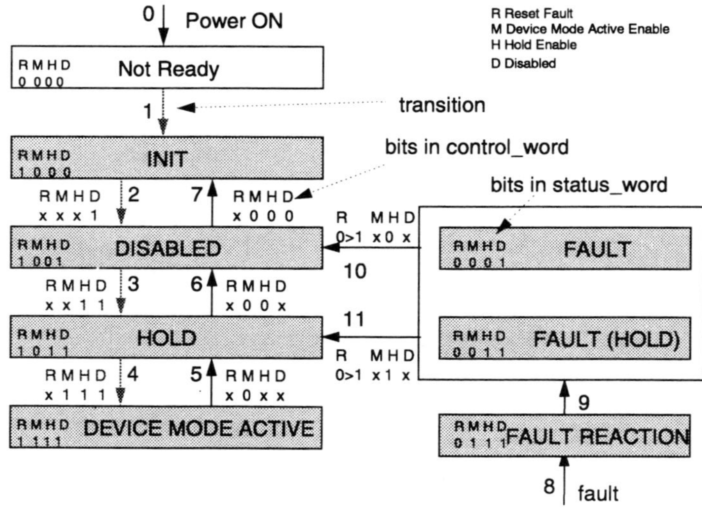

Fig. 3. State machine of the device

For hydraulic drives, the following controller modes are defined:

- spool position control (open loop and closed loop)
- pressure control (open loop and closed loop)
- pressure flow control closed loop (p/Q)

Depending on the application areas, devices can be designed supporting only a fixed controller mode, or having different modes implemented that can be switched by a parameter. Together with the controller modes, the relevant parameters have been defined. As an example, Figure 4 shows a valve in pressure control closed loop.

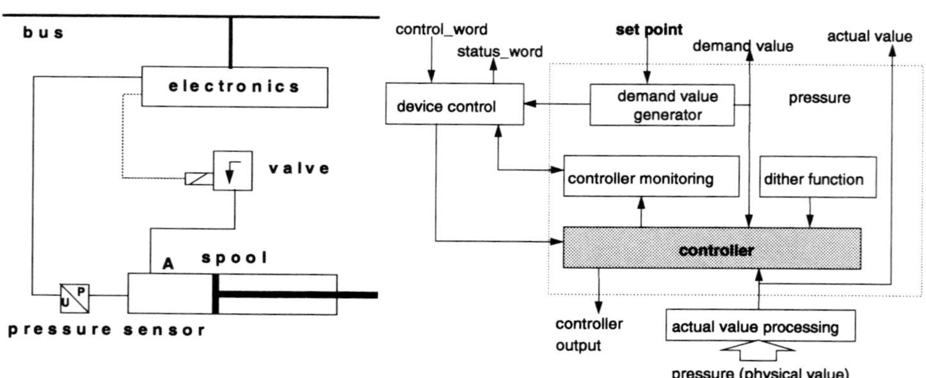

Fig. 4. A valve in pressure control closed loop (environment and internal structure)

Common functions, suitable for valves and drives, are described in the common part of the profile. These functions are for example actual value processing, components of the demand value generator (e.g. ramps, direction-depending gain, zero point offset), controller monitoring, error codes, and device information.

The profile in [1] was designed to be completely independent of the fieldbus system. Therefore it is necessary to map this profile to different fieldbusses. This shall be done by the user organizations. Depending on the supporting companies' strategies, INTER-BUS-S, CANopen, DeviceNet, and PROFIBUS-DP have been announced as target systems for the mapping. The mappings are currently under development. The actual status of the document [1] is a released version 1.0. This document has been passed to the user organizations of the fieldbusses named above. There is a continuing work on the bus-system independent profile to address further concepts, and to include the comments from bus-specific mappings and implementations into a next version.

3 Mapping to CANopen

The mapping of the profile in [1] to CANopen is done by a CANopen Special Interest Group (SIG) "Hydraulics", that has been established by CAN in Automation (CiA). Besides stationary systems, requirements from mobile hydraulic systems, e.g. construction machinery, have been considered. Thus, the device profile covers a broad application area for hydraulic components. The current result of the mapping process is described in the document CiA DSP-408 [2], that is released in version 1.0.

The mapping is based on the new version of the CANopen communication profile based on CAL, version 4.0 [3]. Besides the definition of device-depending information and objects in the CANopen standardized communication object area, the CANopen mapping covers the definition of communication objects accessible from Service Data Objects (SDO) communication, and the definition of mappings for Process Data Objects (PDO).

3.1 Device-depending definitions

Following the CANopen conventions, additional device-specific error codes have been defined. An example for device specific information in standard communication objects is the object "Device Type" (index 1000h) (Figure 5). Besides the device profile number in bits 0 to 15, this object specifies, whether the device is a hydraulic drive (bit16=1), or a proportional valve (bit24=1). Furthermore, the bits 25 to 29 indicate device capabilities by representing the valves' supported controller modes.

Objects can be implemented with SI units and prefixes. In this case, the coding for SI unit and prefix is included into parts of a record structure and has to follow that one specified in CiA DR-303-2 [4]. If SI unit and prefix are configurable, the associated sub-components have ReadWrite (rw) access, otherwise ReadOnly (ro).

OBJECT DESCRIPTION

INDEX	1000h
Name	Device_type
Object Code	VAR
Data Type	UNSIGNED32
Category	Mandatory

ENTRY DESCRIPTION

Access	ro
PDO Mapping	No
Value Range	UNSIGNED32
Default Value	No

Byte: MSB   LSB

Device Type																
device specific																Device Profile Number
31	30	29	28	27	26	25	24	23	22	21	20	19	18	17	16	408=198h

Bit16=1: hydraulic drive
Bit24=1: hydraulic proportional valve
Bit25=1: spool position control open loop supported
Bit26=1: spool position control closed loop supported
Bit27=1: pressure control open loop supported
Bit28=1: pressure control closed loop supported
Bit29=1: pressure flow control closed loop (p/Q) supported

Fig. 5. Object description and entry description of the object Device Type

3.2 SDO-Mapping

All parameters defined in the device profile are accessible trough SDO communication. The object dictionary was structured according to the functionality of the sub-components and function blocks used in the different controller modes (Figure 6). So, for example, the objects with indices from 6040h to 604Fh contain parameters independent of the controller mode, e.g. parameters of the device control block, controller mode and actual value processing functions. information o

object index range	meaning
1000h ... 1010h	standard communication object area according to CAL
....	
6040h ... 604Fh	device control, controller mode, actual value processing
6050h ... 606Fh	spool position control (open loop and closed loop)
6080h ... 609Fh	pressure control (open loop and closed loop)
60B0h ... 60B3h	additional parameters for pressure flow control (p/Q)
60C0h ... 60D0h	device information

Fig. 6. Object index ranges

The object index range from 6050h up to 606Fh contains those objects, that are relevant for controller modes spool position control. Analogously, the index range from 6080h up to 609Fh is used for objects relevant in controller modes pressure control. The objects in both areas are used in controller mode pressure flow control, too. Additional objects for that controller mode are defined ranging from 60B0h up to 60B3h. Finally, the index range from 60C0h to 60C6h is used for device information. Using an index scheme like this, readability and expandability of the object dictionary is enhanced.

3.3 PDO-Mapping

The association of communication objects to PDOs is specified by the PDO mapping. Depending on the device and certain device modes, default mappings have been defined. They guarantee fast access to the mapped communication objects by PDO transfer, that is faster than SDO communication. This means, that first of all objects containing control-related values (e.g. set points, actual values) should be mapped to a PDO. As CANopen Communication Profile version 4.0 allows to specify up to four default mappings, they are defined to support the different controller modes. Within the mappings, Transmit PDOs (TPDO1-4) and Receive PDOs (RPDO1-4) can be distinguished (Figure 7). The terms transmit and receive are related to the scope of the device. In addition to the object references, the transmission type has to be defined. It is in all cases asynchronous, initiated by a device specific event. The default mappings for hydraulic drives have to be defined in the same objects. In order to distinguish between drives and valves, the device type object (1000h) has to be considered. User-specific PDO mappings can be defined covering all PDO-mappable objects.

	object1	object2	object3	object4	transmission type
TPDO1	status_word (6041h)	-	-	-	255
TPDO2	status_word (6041h)	spool_position_actual_value (6051h)	-	-	255
TPDO3	status_word (6041h)	pressure_actual_value (6081h)	-	-	255
TPDO4	status_word (6041h)	spool_position_actual_value (6051h)	pressure_actual_value (6081h)	-	255
RPDO1	control_word (6040h)	-	-	-	255
RPDO2	control_word (6040h)	spool_position_setpoint (6050h)	-	-	255
RPDO3	control_word (6040h)	pressure_setpoint (6080h)	-	-	255
RPDO4	control_word (6040h)	spool_position_setpoint (6050h)	pressure_setpoint (6080h)	-	255

Fig. 7. Default PDO mappings for hydraulic valves

While Transmit PDOs always cover the status word, the control word is mapped to all Receive PDOs. Thus, a fast transmission of status information of the device, and of control commands issued to the device is guaranteed.

4 Profile description technique

Based on the device profile, device specific software tools, visualization components, and other configuration information files can be derived. All those information is related to the same input – the object descriptions. Therefore it is useful, to reduce errors and inconsistencies by relating to a single document. This single document should be the device profile.

In order to meet the challenge of having such a single reference, an appropriate description method has to be selected. Relating to the trends in Internet technology, the eXtensible Markup Language XML [5] is used to describe the profile. XML expands the well-known description language HTML with user-defined tags, data types and structures. There is a clear separation between the data descriptions, the data themselves, and the representation in a browser. This enables to use a single XML description file for different tasks. Based on an appropriate Document Type Definition (DTD), an XML description of a CANopen Device Profile was created (Fig. 8).

```xml
<?xml version="1.0"?>
<?xml:stylesheet type="text/xsl" href="profile.xsl"?>
<!DOCTYPE profile SYSTEM "CiA_profile.dtd">
<profile>
  <profileDescription>
      <profile_class>Device Profile</profile_class>...
  </profileDescription>
  <GeneralDescription>
    Here goes the general description of the profile.
</GeneralDescription>
    <ObjectDictionary>
        <ObjectDescriptions>
          <Object Category="mandatory">
            <ObjectDescription>
              <Index>1000</Index>
              <Name>DeviceMode</Name>
              <ObjectCode>ARRAY</ObjectCode>
              <DataType>UNSIGNED32</DataType>
              <Explanation>This Object holds ...</Explanation>
            </ObjectDescription>
            <EntryDescription>
              Description of the object's elements (access rights,..)
            </EntryDescription>
          </Object>
          ... other objects  ...
        </ObjectDescriptions>
    </ObjectDictionary>
    other description information...
</profile>
```

Fig. 8. XML description of a CANopen device profile (extract)

This description holds all information necessary to generate the associated device profile document in HTML format. The generation of the document is supported by an XSL style sheet with formatting and filtering instructions. This style sheet has to be developed once, and it can be re-used for other profile descriptions. In addition, more generalized DTD and style sheet allow an easy porting of the device profile to other fieldbusses, too. Using the same XML file, specific style sheets and scripts are implemented to create input files for other tasks. For example, Electronic Data Sheet (EDS), or Device Configuration File (DCF) file used in CANopen-specific tools can be created by a Script. The same is possible for generating header files for C++ development, or parameter information for Java classes. Furthermore, an ODL-file (Object Definition Language) can be created, that is used for building COM-supporting applications (e.g. DCOM-objects or ActiveX-controls).

5 Conclusion

The mapping of the fieldbus-independent device profile to CANopen has been completed for hydraulic proportional valves. The mapping of the drives-related part is expected to be completed until the conference. In parallel, hydraulic components' vendors supporting the profile definition process are implementing the mapping into their products. This will allow to receive comments and hints, that will be included into the next version of the profile. It can be expected, that the device profile will help to reduce the complexity of fluidic systems, especially during design and configuration.

References

1. Profil Fluidtechnik. Stetigventile und hydraulische Antriebe. V1.0. Verband Deutscher Maschinen- und Anlagenbau e.V. (VDMA), Frankfurt/Main, 28.04.1999.

2. CiA DSP-408 V1.0, CANopen Device Profile for Hydraulic Drives and Proportional Valves, Juni 1999.

3. CiA DS-301 V4.0, CANopen Communication Profile based on CAL, Juli 1999

4. CiA DR-303-2 V1.0, Representation of SI Units Draft Recommendation, Juli 1999

5. Bray, T.; Paoli, J.; Sperberg-McQueen, C. M.: Extensible Markup Language (XML) 1.0. 1998, http://www.w3.org/TR/REC-xml

Chapter 4: Validation

Papers presented in Session 1.5

Comparing the Networks CAN and ARINC 629 CP with Respect to the Quality of the Service Provided to an Automatic Control Application
Blum, I.; Juanole, G.

Modelling and Evaluation of Systems for the Interconnection of Industrial Communications Networks
Sempere Payá, V. M.; Mataix Oltra, J.; Utrilla Ginés, E.

Use of Formal Specification and Design Language for Protocol Description – Field Report
Prüßner, M.; Gemici, A.

Further Papers

CANopen Conformance Test
Schumann, T.

Formal Description Software for WorldFIP Industrial Fieldbus
Mariño, P.; Domínguez, M. A.; Poza, F.; Nogueira, J.

Experiences in Different Fieldbuses Used together with PC-Based Control Systems
Saarimäki, V.; Siltala, N.; Partanen, P.; Vihinen, J.; Tuokko, R.

Comparing the networks CAN and ARINC 629 CP with respect to the quality of the service provided to an automatic control application

I. Blum and G. Juanole

LAAS-CNRS, 7, Avenue du Colonel Roche, 31077 Toulouse Cedex 4. France

Abstract. The objective of this work is, at first, to formally model and evaluate the Quality of the Service (delay) provided by the MAC layer of the networks CAN and ARINC 629 CP with respect to a particular periodic message stream among a set of periodic message streams and, second, to link this Quality of Service to the stability performance of a closed loop control system which is based on this particular periodic message stream. At first, the message scheduling and delay evaluation are made by using Stochastic Time Petri Nets. We show the relation (message priority - message delay), and when either CAN or ARINC is better. Finally, by considering the closed loop control system as a continuous linear system and then using the concept of transfer function based on the Laplace Transform, we evaluate its phase margin in function of the delay.

1 Introduction

In the present technological context, industrial applications like, for example, closed loop control systems [1,2], are more and more implemented by using networks (fieldbuses for example). The timeliness of the scheduling of the message exchange through the transmission ressource is essential to garantee the performances of the applications. The networks CAN [3] and ARINC 629 CP [4] are networks used in on board systems and then it is very important to formally model the scheduling of these networks in order to evaluate the delay of the messages in different scenarios. Linking this delay to the performances of an automatic control application implemented on a network is also very important (particularly from a user point of view) as this allows the user to make an actual comparison. The goal of this paper is precisely to deal with these different points. This paper includes three parts. The first part concerns the presentation of the study context. The second part concerns the formal modelling and analysis of the message scheduling in CAN and ARINC 629 CP by using Stochastic Timed Petri Nets (STPN model) [6]. The third part presents the results.

2 Study context

2.1 Global system

The global system, in which we have the considered application, is a distributed system composed of several sites (computers) connected by a communication network on a bus (figure 1). The sites i and j are only concerned by the considered application (industrial process control where the controlled variable s(t) is captured in the site i and the returned variable r(t) is compared to the set point in the site j). The role of the sites i and j (with the associated computers) is very simple: they implement (by using the network) the feedback loop.

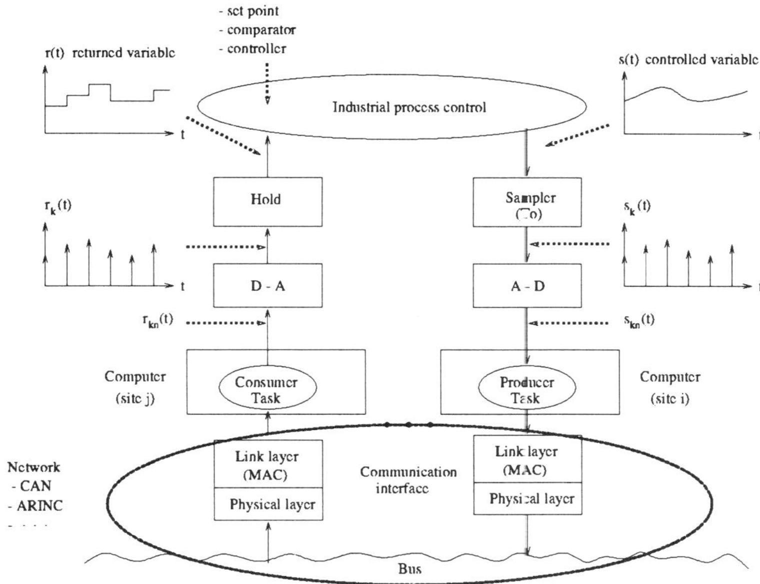

Fig. 1. Global system

In the site i, we have, at first, the sampling (period T_o) of the analog signal s(t) (analog samples $s_k(t)$) and the analog to digital conversion (A-D) which provides digital samples ($s_{kn}(t)$). The digital samples $s_{kn}(t)$ are considered in the computer of the site i by the task called the Producer task. Then the Producer task requires the service of the MAC entity for transmitting data. In the site j, the MAC entity transmits the data to the task called the Consumer task, which then sends the received digital sample values $(r_{kn}(t))$ to the digital analog computer (D-A). After the Hold, we have the returned variable r(t). On the figure 1, the dots between the computers of the sites i and j mean that there are other computers connected to the network and that the MAC entities of these computers are competing with the MAC entity of the site i for the use of the bus (the traffic managed by these MAC entities constitutes, with respect to the traffic from the site i to the site j, what we call the external environnement). We consider that we have no frames losses and then the quality of the service provided by the network to the automatic control application is expressed by

130

the delay τ_d (time elapsed between the generation of a sample $s_{kn}(t)$ and the reception of the corresponding sample $r_{kn}(t)$). This delay τ_d depends on the external environnement and on the access control technique in the MAC layer. These two points are studied in this paper. Furthermore, concerning the external environnement, we suppose only periodic traffics.

2.2 Scheduling mechanisms in the MAC layer of ARINC 629 CP and CAN

ARINC 629 CP is based on a CSMA/CA [4] mechanism which manages the exchanges of periodic and aperiodic traffics on the basis of the bus cycle concept. Here we only consider the periodic traffic (each station transmits one periodic message per cycle). The bus cycle is started with the emission of the periodic message by a station called leader which has a timer TI (Transmit Interval) which defines the cycle period. The other stations are called not leader and emit their periodic message on the basis of a timer TG (Terminal Gap) which has a different value for each station and which uses the events on the bus (bus becomes quiet: TG starting; bus becomes busy: TG resetting). The message priorities in the not leader stations are expressed through the TG values.

CAN is based on a CSMA/CD scheme with a deterministic collision resolution based on priorities associated to the identifiers (adresses) of the frames which carry the messages (smaller is the adress (i.e. it has many dominant bits (bits 0)), higher is the priority). A station, which does not recognize the adress of the frame it is emitting, detects a collision and stops. The only station which does not detect a collision (it is the station which (among the emitting stations) has the longer sequence of dominant bits) continues and is then the winner. The priorities associated to the frame identifiers reflect the priorities between the messages included in the frames.

2.3 Automatic control application

We consider that the system to control is a continuous linear system which can be modelled by using the concept of transfer function based on the Laplace Transform [1]. The model of the considered closed loop control system is represented on the figure 2. The transfer functions of the direct chain (integrator - first order) are functions which have been choosen arbitrarily for the study (as the numerical values) but they are functions which are found in many systems. The feedback loop integrates the parameter τ_d (delay of the network). Note that the transfer function of the zero order hold can still be approximated by a pure delay $e^{-\frac{T_0}{2}s}$ (the conditions are [1]: sampling frequency $(\frac{1}{T0}) \gg$ higher frequency of s(t), which is satisfied here with the direct chain transfer function which is considered). Then the feedback loop is characterized by a pure delay $e^{-(\frac{T_0}{2}+\tau_d)s}$. A fundamental performance is the stability which is characterized by the phase margin φ_m:

$$\varphi_m = 180^0 - arg\left(\frac{50\sqrt{2}}{j\omega} * \frac{e^{-j(\tau_d + \frac{T}{2})\omega}}{1 + j\frac{\omega}{50}}\right) \text{ (for } \omega \text{ such that } |\ \frac{50\sqrt{2}}{j\omega}\frac{1}{1+j\frac{\omega}{50}}\ | = 1). \ \varphi_m$$

depends on the parameter τ_d.

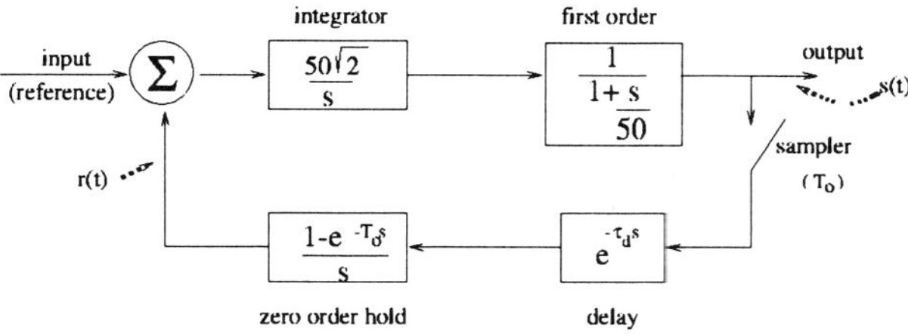

Fig. 2. Closed loop control system

3 Formal Modelling and Analysis

3.1 Producer and Consumer tasks

The Producer and Consumer tasks are represented on the figure 3.

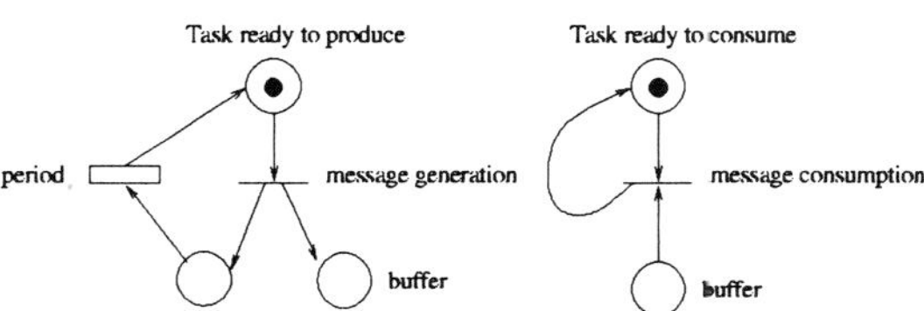

Fig. 3. Producer and Consumer tasks

3.2 Scheduling in CAN

We only visualize (figure 4) the transfert of three messages of increasing priority (priority x, priority (x+1) and priority y). The priority is modelled by using inhibitor arcs. The modelisation requires to model the bus quiet, the bus busy and the time between the frames carrying the messages (INT: INTerframe).

3.3 Scheduling in ARINC 629 CP

We only model here the scheduling of the stations (we consider one message stream per station and then there is not local scheduling; however, in the context of ARINC (like with networks based on the token mechanism) the local scheduling is an important problem on which we are presently working).
This scheduling is represented by the STPN model of the figure 5 which follows the scheme of the bus cycle (transitions with a bar are immediate transitions;

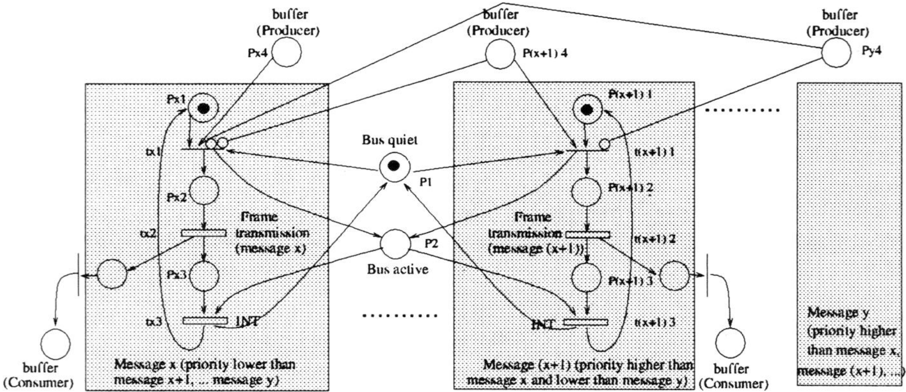

Fig. 4. Scheduling in CAN

transitions with rectangles represent durations). The message stream model in each station is not given here but it is like on the figure 3. It is important to note than the message stream to analyse can be, a priori, in any station (that will be considered in this analysis).

3.4   Analysis for the determination of the delay τ_d

A delay is obtained by considering quantitative abstract views of the randomized state graph [6] which represent the dynamic behaviour of the STPN model: we make a projection on the two states which represent respectively the message production and the message consumption. The duration associated to the transition between these two states gives the delay.

4   Results

4.1   Hypothesis

1. We have 35 periodic message streams (which are all in phasis and which have no jitter; concerning ARINC, the release occurs at the beginning of the cycle which is not the best configuration): 10 streams with a period $0.5\ 10^{-2}$s and 25 streams with a period of 10^{-2}s. The message stream to analyse (it represents the sample stream of the automatic control application given in section 2) is one of period 10^{-2}s. A message of each stream has 64 bits length.
2.   – CAN: the message length induces a frame of 127 bits length (we consider 19 stuffing bits which is the maximun). Between each frame, we have 3 bits of intertransmission (INT). The bit rate is 1 Mb/s (it is the maximum value permitted in CAN in our knowledge).

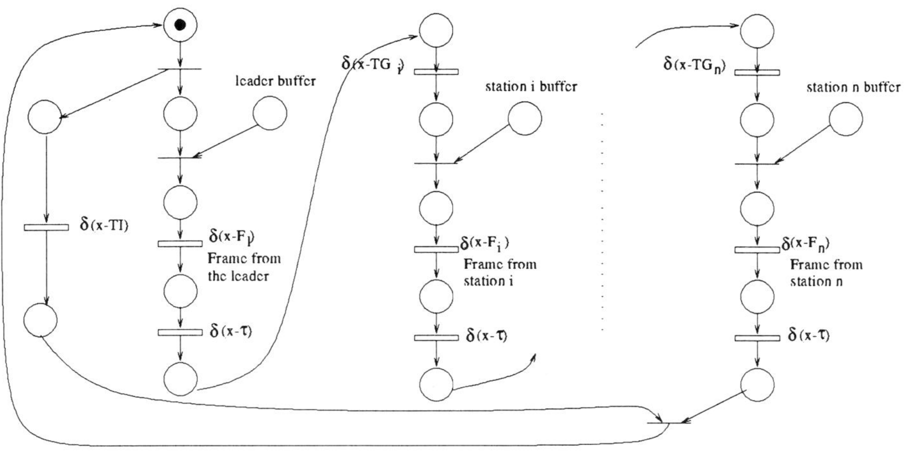

Fig. 5. Stations scheduling in ARINC

- ARINC: we have 35 stations. The timer TI in the leader station has a duration of $0.5\ 10^{-2}$s (the cycle duration is based on the smallest period of the streams). The timers TG in the non leader stations have values from $4.6875\mu s$ to $70.6875\mu s$ with jumps of $2\mu s$. Longer is the TG of a station, more distant of the beginning of the cycle is the transmission of this station and then bigger will be the delay with respect to a message produced at the beginning of the cycle (we still say that the station position in the cycle is larger). The bit rate is either 1Mb/s or 2Mb/s (in ARINC, the normal bit rate is 2Mb/s; however we consider too 1Mb/s in order to make a strict comparison with CAN).

3. Scheduling of the message streams: in CAN, we consider the Rate Monotonic Algorithm [5] (the message streams with a period of $0.5\ 10^{-2}$s have the highest priorities (1 to 10); the message streams with a period of 10^{-2}s have the lowest priorities (11 to 35)). In ARINC, as we have one stream per station, we specify a priority scheme based on the position of the station in the bus cycle: the 10 streams with the period of $0.5\ 10^{-2}$s are respectively in the leader (station 1) and the nine non leader stations which have the smallest TGs (positions 2 to 10); the 25 streams with the period of 10^{-2}s have the positions 11 to 35.

4.2 A comparison between CAN and ARINC

We consider the message stream to analyse (period 10^{-2}s) and we evaluate its delay τ_d with respect to the choice of his priority in CAN (in ARINC, the position of its station in the bus cycle) among the 25 streams which have the period of 10^{-2}s: the choice of this priority (position) can vary between 11 and 35. The results are given on the table 1 which shows how the priority in CAN and the position in the bus cycle in ARINC influence the values of the delay τ_d (in μs). It is still important to remark these following points (table 1). Concerning strictly CAN: the delay increases when the frame priority decreases. Concerning strictly ARINC: the delay increases with the position of the station in the bus cycle and decreases with the bit rate (all these results are coherent with the specification of ARINC). Concerning the comparison between CAN and ARINC (1 Mb/s): we can see that the delay in ARINC (till the position 25), which is smaller than the delay in CAN (till the priority 25) becomes higher from the position (priority) 30 in ARINC (CAN). This result can be precisely explained by giving the general expression of the delay experimented by the message of the stream to analyse. Call FL_{CAN}, FL_{ARINC} and X respectively the length of a frame in CAN, the length of a frame in ARINC and the number of the streams with priorities highest than the message stream to analyse (CAN formulation) or the number of the stations which have a better (smaller) position in the bus cycle than the station including the message stream to analyse (ARINC formulation):

- delay in CAN: $(FL_{CAN} + INT)X + FL_{CAN}$
- delay in ARINC: $(FL_{ARINC} + \tau)X + (FL_{ARINC} + \tau) + \sum_{i=1}^{X} TG_i$

The equality of the two delays is obtained for $X = 26$ (for a bit rate of 1Mb/s), which is coherent with the comparison given on the table 1. For a bit rate of 2Mb/s, we would obtain $X = 64$.
Note that with ARINC (2 Mb/s), we obtain the best performances (which is normal).

CAN	Priority	11	15	20	25	30	35
	τ_d	1430	1950	2600	3250	3900	4550
ARINC	Station position	11	15	20	25	30	35
	τ_d (1 Mb/s)	1246	1761	2449	3187	3974	4812
	τ_d (2 Mb/s)	696	1011	1449	1937	2474	3062

Table 1. Delay τ_d in CAN and ARINC

4.3 Performance of the automatic control application

Integrating the delay τ_d in the expression of the phase margin (section 2), we get the curves of the figure 6.

5 Conclusion

We want to underline three points. The first point concerns the interest of the Stochastic Timed Petri Nets for making the modelisation and analysis of the message scheduling in local area networks. The second point concerns the interest

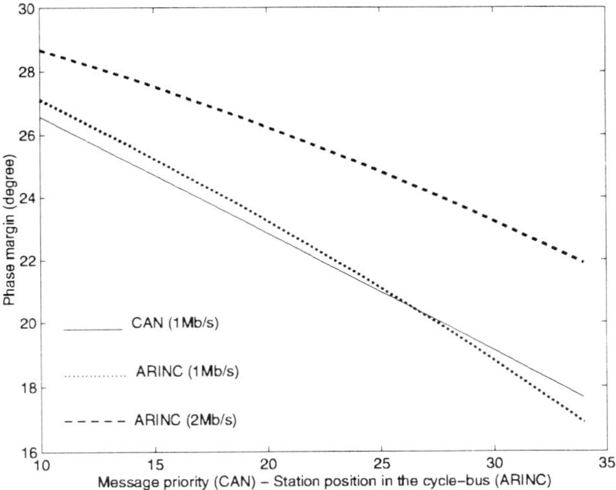

Fig. 6. Phase margin

of being able to compare two networks like CAN and ARINC, taking into account for the developpement, present and future (in our opinion), of these networks. Finally, we would like to emphasize the link which has been done on the quality of the service provided by a network (delay) and the performance of a closed loop control system realized by means of the network. This last point is an opening toward pluridisciplinary works which, in our opinion, are more and more necessary today taking into account the very sophisticated applications which can be implemented on the top of networks (we must be competent in networks and automatic control, for example).

References

1. Karl J. Aström and Björn Wittenmark. Computer controlled Systems: Theory and Design. Prentice Hall. Information and system Science Series. 1997.
2. Yoram Halevi, Asok Ray, Integrated communication and control systems. Journal of Dynamic systems, Measurement and Control. December 1988.
3. International Standard Organization (ISO), Road Vehicles - Interchange of Digital Information - Controller Area Network for high-speed Communication, ISO 11898,1994.
4. Airlines Electronic Engineering Comittee, Multi-Transmitter Data Bus. ARINC Specification 629-2: Part 1, Technical Description, Aeronautical Radio Inc. edition, october 1991.
5. C. Liu and J. Layland, Scheduling Algorithms for Multiprogramming in a Hard Real-Time Environnement. Journal of ACM, vol. 29, no. 1. 1973.
6. Y. Atamna, Réseaux de Petri Temporisés Stochastiques Classiques et Bien Formés: définition, analyse et applications aux systèmes distribués temps-réel. PhD thesis, Université Paul Sabatier, Toulouse, December 1994.

Modelling and Evaluation of Systems for the Interconnection of Industrial Communications Networks

Víctor M. Sempere Payá, Jorge Mataix Oltra (*), Estanislao Utrilla Ginés
(vsempere@dcom.upv.es) (jmataix@eupbl.upc.es) (esutgi@tlk.upv.es)

Universidad Politécnica de Valencia. Dpto. de Comunicaciones
(*) Universitat Politècnica de Catalunya

Abstract. The aim of this paper is to tackle issues regarding interconnection of Industrial Communications Networks. For this purpose, two interconnection strategies are presented and applied to a widely used fieldbus standard: Profibus. The adaptation of Profibus to the so called Extended Fieldbus is investigated and an extended address scheme for Profibus is proposed. A brief description of the developed models, focusing in the interworking unit, is given together with the verification and evaluation methodology followed. Finally, some performance results are presented which are used to compare the two interconnection strategies.

1 Introduction

Technological progress in electronics, computer science and communications has made a significant impact in the field of industrial communications networks (ICN). This, when added to the great demand for distributed control systems experimented by many sectors, such as industrial processes, automobile, buildings control, environment, etc., makes ICN one of the most dynamic and developed fields.

ICN's are used for the monitoring and control of manufacturing and process control applications. Considerable efforts have been made over the last years to integrate the three levels of factory automation hierarchy: plant level, cell level and process level [1]. At the lowest level of this hierarchy, field devices such as sensors, actuators, or PLC´s, exchange messages through a single communication medium -- the fieldbus [2,3]. The introduction of fieldbuses has produced great benefits in ICN process level communications by greatly improving performance and lowering maintenance costs.

Fieldbuses have to fulfil critical time constraints, as well as security and reliability requirements because of the nature of the environments where they are installed -- which are often harsh and dangerous [4]. These time constraints are imposed by the kind of traffic involved. The traffic profile is usually cyclical traffic generated by message transfer among field devices and process controllers in polling cycles; but unexpected bursts of messages in response to alarms or abnormal operational situations must be handled within specific deadlines.

Several fieldbus proposals have been implemented in different industrial environments. Among them, PROFIBUS stands out as the most important European standard for fieldbuses. It is an open, vendor independent and widely established network with a three-layer OSI based protocol architecture designed for time critical communications at lower factory automation levels. It is based in hybrid protocol between token-passing bus and master-slave polling, offering functionality for client-server communications between controllers (masters) and field devices (slaves).

Nowadays, there exists an increasing trend towards the effective centralization of huge amounts of often very distributed information. This centralization can enable better management as well as a more rational use of resources. ICN has not been designed to completely cover those issues. This has important implications in manufacturing and quality aspects of industrial activity. Many National and International research proposals have focused on topics related to the interconnection of these networks. This paper is devoted to the investigation of the problems associated with the interconnection of ICN´s, and concretely of PROFIBUS fieldbuses.

2 Extended Fieldbus

Extended Fieldbus concept [5] answers the needs of distributed and very closely related applications. It enables several segments to work together by means of gateway nodes interconnected with permanent ATM connections. Extended Fieldbus can be used by other current architectures of Fieldbus such as FIP, LON, CAN, etc. However, the adequate protocols for running Extended Fieldbus must be defined.

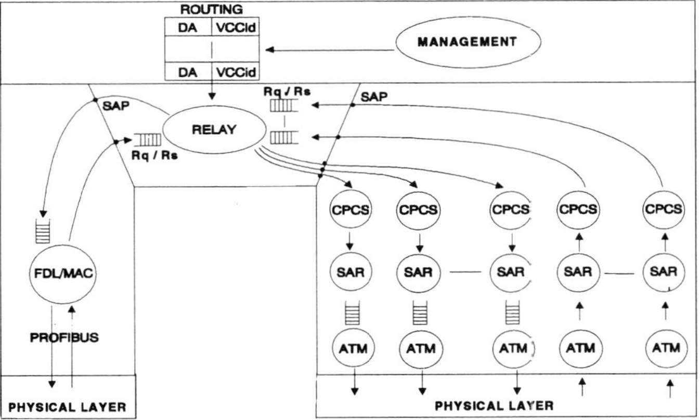

Fig. 1. Gateway model in EF

Interworking Units, or gateways, are interconnected with ATM permanent connections. The ATM backbone is just modeled as a fixed delay. The service class is Constant Bit Rate (CBR) for its suitability for allocating real-time traffic [6,7]. Peak Cell Rate (PCR) parameter has been over-sized assuming the worst-case traffic profile. Neither losses, nor jitter effects have been modeled. The functions performed by the gateway are basically : routing according to the address scheme performed by the EF; fragmentation

138

and assembling of Profibus messages; packing and unpacking of ATM cells; multiplexing and demultiplexing of ATM connections in message queues; and, service scheduling for pending requests and responses. In figure 1 can be seen how the Profibus interface and the ATM interface exchange job Requests / Responses through the Rq / Rs queues. The relay process performs routing and service scheduling functions depending on job priorities. ATM adaptation is performed in CPCS and SAP layers.

There are two strategies to interconnect Profibus segments via a WAN in the Extended Fieldbus framework : Virtual Fieldbus (VF) and Independent Fieldbus (IF).

2.1 Virtual Fieldbus

There exists only one token that circulates around the logical ring formed by the different segments. From the point of view of the token, the gateway just carries out packing and unpacking functions without a holding time. The token in a segment jumps from one station to the next according to the address policy. Once the token has passed through all the stations in the segment, it is forwarded to the gateway, where it is encapsulated in ATM cells and transmitted to another segment. It always retains a high priority. In the remote segment, the gateway receives the token from the ATM network, decapsulates it and sends it on to the first station in its segment. So the token passes through each gateway twice per cycle.

Two cases can be clearly distinguished in the operation of the segments – depending on whether they have the token or not. In the first case, the master with the token has permission to access the medium in order to send requests to the local segment or to a remote one. Requests to stations belonging to remote segments are received by the gateway, which packs them and puts them in an output queue to be issued. No limitations, other than the imposed by the ATM network, are applied to the messages in these output queues.

The gateway behaves like a master in the case of a segment without the token, and the rest of the stations in the segment behave like slaves allowed to access the medium only when requested by the gateway. The gateway schedules a FIFO policy where responses to previous requests have priority over newer requests.

2.2 Independent Fieldbus

The Extended Fieldbus is now formed of independently operating segments – with each having its own token. When a master station has the token, it is allowed to make local or remote requests. The latter being processed by the local gateway in the same way as in the other solution. However, the gateway is now a token contender like the other master stations and, therefore, requests coming from remote segments have to wait for the gateway to hold the token. When the responses generated to those requests are received by the gateway, it encapsulates them in ATM cells and forwards them to the appropriate remote gateway.

Figure 2 shows the topology of the model proposed. We can see that solid line represent token circulation in VF, whereas dotted line represent the token path for IF.

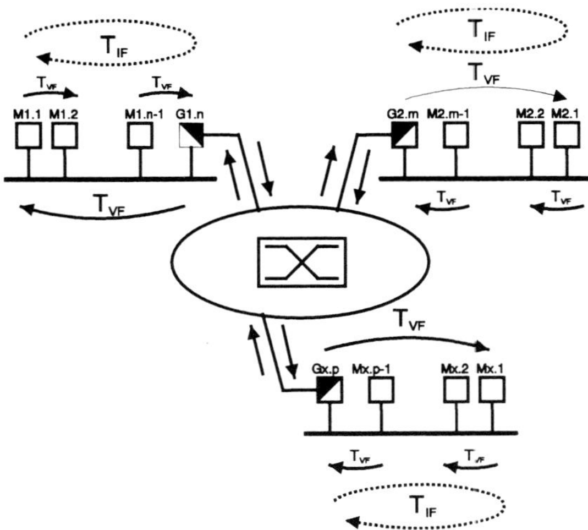

Fig. 2. VF versus IF

3 Modeling issues

In order to tackle the evaluation of the two proposed strategies, models are obtained in a way that are general enough for studying their main characteristics and for comparing them. Therefore, we have looked at a network with two Profibus segments. Each segment has three master stations and an undetermined number of slaves, in addition to the gateway.

The models have been specified by means of a *Petri nets* based language, SAN.

4 Petri Nets and Stochastic Activity Networks (SAN)

Petri net (PN) is a graphical and mathematical tool for modeling many complex systems like ICN. It is a tool for describing and studying systems which are concurrent, asynchronous, distributed, parallel and/or non-deterministic. As a graphical tool, PN can be used in visual communication aids like flow charts and block diagrams. As a mathematical tool, it is possible to set up state equations, algebraic equations, an other mathematical models describing the behaviour of systems. The graph of a PN is directed, weighted and bipartite, with two kind of nodes, *places* and *transitions*. Arcs are either from place to transition or transition to place. The former are called *input arcs* and the latter *output arcs*. Arcs are labelled with positive integers, where a k-weighted arc can be interpreted as the set of k parallel arcs. A *marking* (or state) is a vector, which assigns to each place a non-negative integer. If a marking assigns to place p an integer k, we say that p is marked with k *tokens* (k black dots in place p). The marking in a PN is changed by the following transitions :

A transition t is said to be *enabled* if each input place p of t is marked with at least $w(p,t)$ tokens. An enabled transition may or may not fire. Firing an enabled transition t

140

removes *w(p,t)* tokens from each input place *p* of *t* and adds *w(q,t)* tokens to each output place *q* of *t*, where *w(q,t)* is the weight of the arc from *t* to *q*.

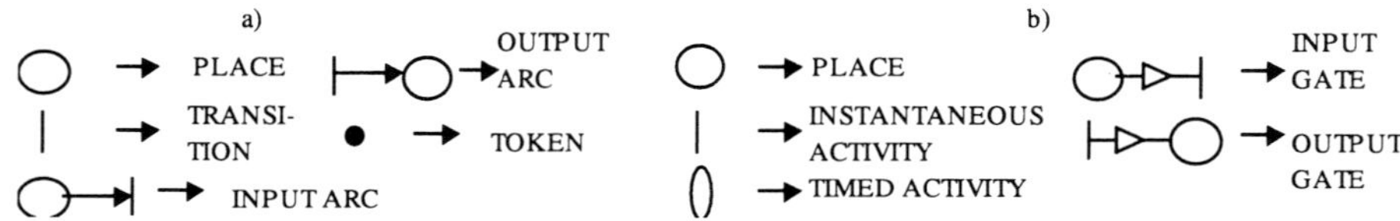

Fig. 3. Graphical representation of components of PN (a) and SAN (b)

In modeling, input places represent conditions; transitions represent events; and initial marking indicates the initial state. *Stochastic activity networks* (SAN) are stochastic extensions of petri nets (PNs). This high level modeling formalism has four primitive objects:

-   *Places.-* Represent the state of the modeled system as basic petri nets. Each place contains a certain number of tokens, which represent the marking of the place. The set of all place markings represents the marking of the network.

-   *Activities.-* Represent actions in the modeled system as transitions. There are two types: timed and instantaneous. The former has a duration, it takes an amount of time to complete, and the activity time distribution function associated with its duration is generally made of distributed random variables. Instantaneous activities represent actions which are completed in a negligible amount of time compared to other activities in the system. Activities have *case probabilities*, which conform to a probability distribution, that models the uncertainty associated with the completion of an activity.

-   *Input gates.-* Control the enabling of activities with predicates of place markings and define the marking changes that will occur when an activity is completed. They are represented graphically as triangles with their crown connected to the activity they control.

-   *Output gates.-* Define the marking changes which will occur when activities are completed, but only with a single case. They are represented graphically as triangles with their base connected to an activity or a case.

The UltraSAN tool has been used to obtain these results [8]. UltraSAN has a graphic editor for SAN models, it allows the use of global variables and it gives solution techniques which include many analytic and simulation based approaches.

5 Modeling of the EF Gateway

The parts that comprise the modeled gateway are different depending on its type. For the VF gateway, these are : access control (figure 4, part 1); timer management (figure 4, part 2); local medium-backbone stage (figure 4, part 3); and, backbone-local medium stage (figure 4, part 4). For the IF gateway, there is also a timer management (figure 4, part 2) which controls the token holding time of the station. Another difference between both types of gateways is the token management performed by the VF gateway in order to transfer the token through the backbone. This obviously does not exist in IF gateways. All these parts can be seen in figure 4.

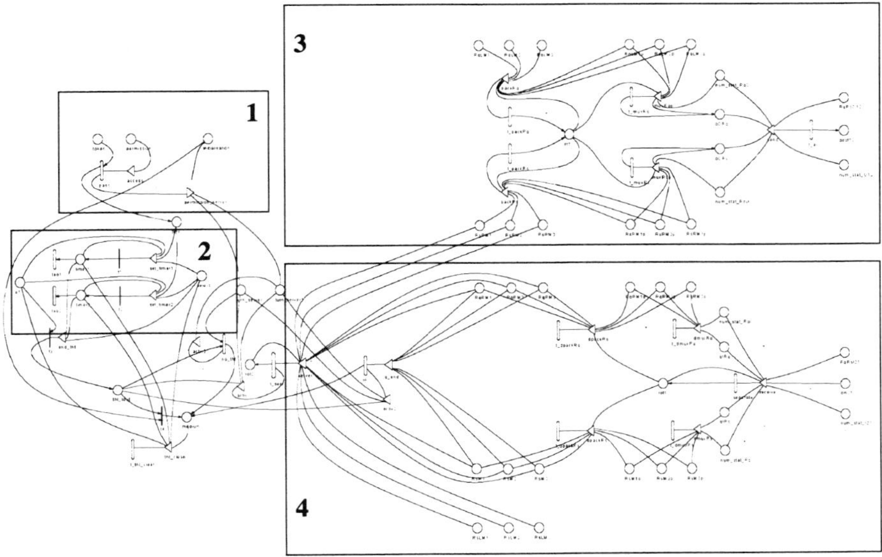

Fig. 4. Model of the IF gateway

5.1 Management of the access control

This part of the gateway, logically placed above the local medium physical layer is, basically, responsible for address identification and token reception.

5.2 Management of the local medium-backbone stage

This part is responsible for sending messages to the backbone. It has one interface with the local shared medium and another with the backbone. These messages can be different depending on the type of gateway. For the IF gateway, they can be requests or responses from a local master to a remote master or slave, whereas for the VF gateway, the token can also be sent to the backbone.

Each job request message is placed in RqLMx places, where x stands for one of the three local master stations which issued the request – service initiator master. Likewise, places RsRMx contain the job response messages replied by master x. Both requests and responses are immediately packed into ATM cells. This is modeled by time activities t_paqRq and t_paqRs, which upon completion move the corresponding mark, which represents the job, either from RqLMx to RqLMxp place or from RsRMx to RsRMxp place. Activities t_muxRq and t_muxRs model the multiplexing of queues represented by RqLMxp and RsRMxp places into the corresponding ATM connection – just one considered connection simplifies routing in this stage. Finally, the request or response is sent to the backbone with a PCR rate given by activity t_acc. Three places, namely RqRsO12, qout12 and num_stat_O12, contain all the information exchanged between gateways such as the service initiator master or the type of message – request, response or even the token for the VF case. Due to the incoming traffic rate and to the allocated PCR, gateway queues along this traffic direction are always empty.

5.3 Management of the backbone-local medium stage

This part is in charge of remote messages addressed to local stations. It receives messages, which can be requests, responses or the token for the VF case, through its interface with the backbone, and forwards them to the master or slave recipient through its interface with the local medium.

In a similar way as the local-medium-backbone stage, incoming messages are immediately demultiplexed over the corresponding RsRMxp place or RsMxp place according to the type and destination of the message. Then, unpacking is performed by means of activities t_dpaqRq or t_dpaqRs and marks representing jobs are then queued in places RqRMx and RsMx waiting to be served. In the case of requests, the generated responses are, once serviced, sent backwards to the service initiator master (see arcs going from the backbone-local medium stage to the local medium-backbone stage in figure 4). In the case of responses, the corresponding service initiator master simply marks the service as completed.

Since the access to the local medium is shared among all the masters within it and the gateway, is not guaranteed that the capacity of the gateway be enough to serve the possible traffic rate. Thus occupancy of gateway queues along this traffic direction (RqRMx and RsMx) is determinant for the time response of the system, being even possible achieving congestion situations.

Finally, say that as long as no bounds in the number of marks of places RqRMx and RsMx were imposed, lost of messages can not occur. Additionally, a priority scheduling policy has been implemented in every queue of the gateway.

5.4 Management of the timers

This part is only present in IF gateways since they can hold the token and therefore have to handle protocol timers, i.e. TRT (Token Rotation Timer) and variables: THT (Token Holding Time) and TTRT (Target Token Rotation Time). The time management procedure is exactly the same that the one performed in every master station and follows the algorithm stated by the Profibus norm. In figure 4, it can be seen how the calculation of the THT is implemented. It is performed by two timers, each one formed by one activity that periodically removes one mark from one place according the time base of the system. The numbers of marks within the timer's place represent its state in such a way that it expires when its number of marks reaches the value zero.

6 Performance Evaluation and Results

The performance used are: *utilization*; *throughput*; and *mean response time*. The *utilization* is the ratio of time a station holds the token over the simulation time. The *throughput* equals the ratio of job responses arriving at the station over completed job requests. And the *mean response time* represents the average time taken for serving a job. Studies show all the distributions of a payload in the system – from the case in which just one station serves local traffic within its own segment – to the case in which every station in the system has remote traffic to serve. The results obtained by simulations show several performance characteristics of the system like: congestion; effectiveness; or the most appropriate values for protocol parameters such as the Target Token Rotation Time. These results also show differences between the two strategies as well as the appropriate situations for each one. In general, they show a better behavior of IF over VF. The main results of this comparison are:

- IF shows more independence among segments. This is a function of the amount and distribution of remote traffic in the system. For example, when one segment, with low remote traffic, is connected to the backbone, the overall performance of the system deteriorates more in VF than in IF.

- The maximum traffic that can be introduced in a VF network is 1 Erlang because there is only one token. The amount of traffic can be greater in IF – depending on the number of segments. However, the amount of traffic is also limited by the capacity of the gateways, so more than 1 Erlang of traffic cannot always be introduced in the IF network without causing congestion.

- Generally, IF master stations show more token holding time than VF master stations. This means more traffic input rates in the overall performance of the system. This is because VF gateways spend more medium possession time than required for serving traffic– if we assume no congestion. This is because they depend on the time the token is outside the segment. Yet IF stations make use of their medium possession rights only when necessary, passing the token if there is no job demand. However, maximizing this traffic injection in the system, means fuller queues in gateways and consequently longer response times. In every case, only IF permits some kind of trade-off between maximum volumes of traffic and minimum response times. Figure 5 shows the situation when master stations inject as much traffic as allowed into the network. It compares VF and IF performance for three different studies. Results show that IF maximizes the utilization, and the number of jobs serviced, and so produces worse time responses.

- Other characteristics inherent in the strategies are: for example; the differing effects of failures in the network. In this sense, FI seems to have more failure tolerance since, for example, a token lost only affects one segment of the network and not the others.

- From these results, an understanding of the behavior of more generic systems can be achieved. For example, one situation in which the superiority of FI over FV is quite evident is when there are multiple segments generating remote traffic in such a way that traffic in the network is quite uniform, i.e. approximately the same traffic passes through every gateway in the system.

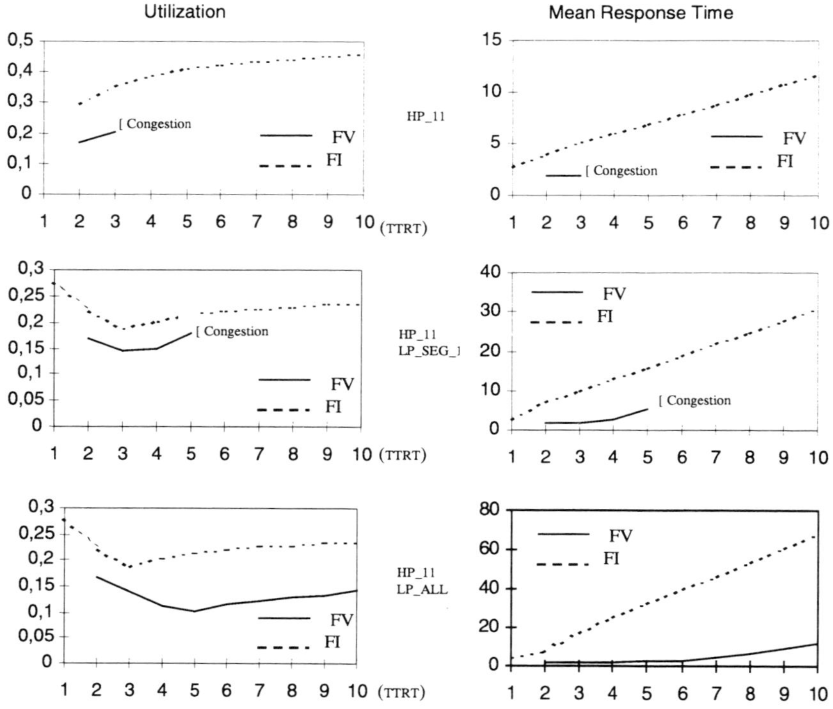

Fig. 5. Utilization and Mean Response Time for Master 1.1

7 Future directions

- Determination of the repercussion in the models of other traffic sources: bursty; cyclic; stochastic; etc: and with, for example, more remote traffic priorities.

- Study of real-time behavior of systems through the analysis of deadlines and measures like access delay or response time.

- Design of mechanisms for congestion avoidance and resource management: adaptive THT algorithms; double token VF protocols.

- Examination of more generic and real situations: finite queues; transmission errors; message discard; etc.

- Modeling of other ATM services like rt-VBR, as well as non-permanent connections with the corresponding connection establishment procedures.

- Study of other fieldbuses like FIP, CAN, LON, etc. and other interconnection mediums like Internet, mobile networks, Gigabit Ethernet, etc.

Acknowledgments

This work has been financed by *Comisión Interministerial de Ciencia y Tecnología* (CICYT) under project TIC98-0495-C02-02.

References

1. Ruiz (L), Decotignie (J.D), "Fieldbus: a Network for Process Level Communication", Ind. Electronics Society Newsletter, Sept (1994)

2. Plainevaux (P.), Decotignie (J.D.), "Time Critical Communication Networks: Field Buses", IEEE Network, vol. 2 May (1998)

3. Pimentel (J.R.), "Communication Networks for Manufacturing", Englewood Cliffs, NJ. Prentice Hall, 1990, pp. 418-446

4. Decotignie (J.D.), Pleinevaux (P.) "A survey on industrial communications networks", Annals of Telecommunications, N° 9-10 (1993), pp 435-448

5. Sempere (V.), Mataix (J.), Utrilla (E.), "Simulating a Wide Fieldbus Interconnected through ATM for its Real-time Performance Evaluation", European Conference on Networks & Optical Communications. ISBN-90-51994001 (IOS Press) June 23-25, Manchester (UK).

6. Mammeri (Z.), Haouam (K.D.), "Connection Allocation Schemes for Guaranteeing Hard Real-time Communications with ATM Networks", Proc. IEEE International Workshop on Factory Communication Systems, Barcelona, Oct. 1-3 (1997).

7. Mammeri (Z.), "Framework for Mapping Real-time Communications onto ATM Services". Int. Conf. for Computer Communications, Cannes France, Nov. 19-21 (1997)

8. Sanders (W.H.), Obal (W.D.), Qureshi (M.A.), Widjanarko (F.K.), "The UltraSAN modeling environment Performance Evaluation", N.H. Elsevier (1995)

Use of Formal Specification and Design Language for Protocol Description
- Field Report -

Matthias Prüßner, Aykut Gemici

Concept Development
Phoenix Contact GmbH & Co., Blomberg, Germany
Email: mpruessner@phoenixcontact.com, agemici@phoenixcontact.com

Abstract. Due to the growing demands placed on fieldbus systems, continuous development of protocols is required. Since it is becoming difficult to ensure consistency because of the increasing complexity of specifications based on informal text, the formal specification language SDL has been used in the current project. The experiences gained in relation to changes in the development cycle through the use of a formal specification language are described. The link between SDL and MSC, and its strengths and weaknesses are illustrated. The options for verification through simulation are introduced.

1 Introduction

The use of fieldbuses is state of the art. In Europe, the most widely available fieldbus systems are standardized in EN 50170 (includes PROFIBUS) and EN 50254 (includes INTERBUS). But in order to meet the latest demands in automation technology, such as remote diagnostics over the Internet or closed-loop applications, continuous development of protocols is required. In the past, specifications were created as informal text supported by state diagrams, if necessary. Due to increasing complexity - EN 50254, Volume 2, is approximately 400 pages long - the consistency of this type of specification can no longer be checked manually. It is also difficult to gain an overview of the effects of modifications/additions. Side effects are only discovered during implementation. The use of formal specification languages such as SDL (Specification and Description Language - ITU Z.100), ESTELLE (Extended Finite State Machine Specification Language - ISO 9074) or LOTUS (Language of Temporal Ordering Specification - ISO 8807) enables computer-supported processing. In addition to the option of machine processing, good legibility is particularly important for a specification language, and this must be possible without the time-consuming learning of language constructs. All three languages mentioned achieve this through the option of graphical representation. To make use of advantages such as verification, validation and testing a specification, appropriate tool support is required for the formal language. SDL is probably the most popular standardized specification and description language. It is used primarily for designing protocol architectures and reactive process systems. In the field of telecommunications, protocols such as GSM, DECT and TETRA are specified using SDL. But SDL is also used for standardization in other fields, such as IEEE 802.11 Draft For Wireless LAN. In addition to SDL, MSC (Message Sequence Charts - ITU Z. 120) makes possible interactions between communication partners in the form of partially arranged message sequences. Telelogic is one of the companies offering products such as SDT [1], a comprehensive tool package for SDL/MSC language which was used in this project.

2 Specification Using SDL and MSC

SDL is a formal specification and description language which can be used to design a system only by describing its behavior. Since this can be entirely in the form of a graphical representation, it is possible to read an SDL specification after a very short training period. SDL offers tools both for structuring the system and for specifying the dynamic behavior.

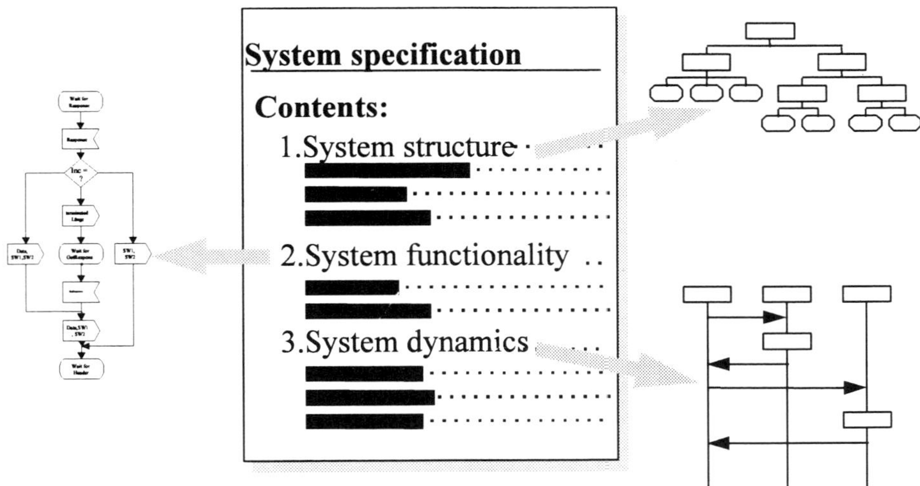

Fig. 1 System specification technology SDL/MSC

Using MSCs (Fig. 1 bottom right) the interactions on set interfaces can be specified. Furthermore, the tool offers MSC as an output option for the simulation, so it is easy to directly compare the previously determined and the current behavior of the specification. In this process it should be noted that SDL works on the model of finite automaton (SDL processes) which communicate with one another asynchronously over signal paths which may be delayed and buffers which may be unlimited. This means that only the sequence and parameters of the signals can be compared. The time intervals cannot be evaluated.

2.1 Structuring

A hierarchically structured working method is required for the development of complex systems. The requirements of the system are thus mapped in steps through repeated division into subfunctions which can then be converted more easily. The structuring is described using a top-down design. The aim of this is to reduce the complexity of the overall function by refining it step by step. The top-down description emphasizes a hierarchical method of viewing the system.

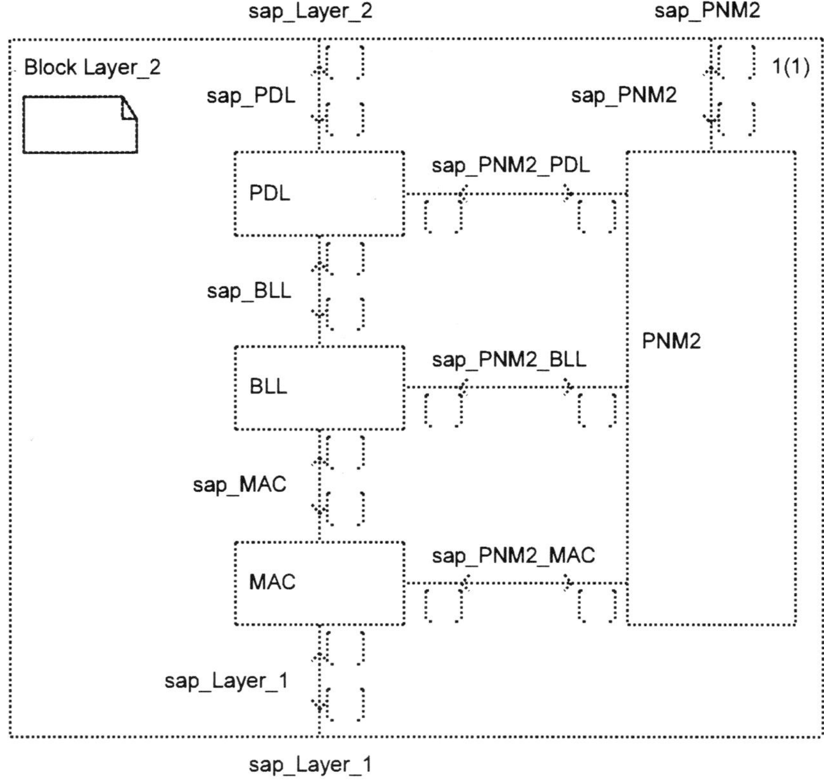

Fig. 2 Structure of Layer 2 of Interbus

SDL provides comprehensive linguistic and graphic tools for a top-down design. These are divided into:

- System level
- Block level
- Process level

Using the top-down design, the structuring process for the specification of a protocol stack is applied in the layer model. For example, Layer 2 of the INTERBUS protocol is structured as an SDL block. This contains further SDL subblocks which map the sublayers Peripherals Data Link (PDL), Basic Link Layer (BLL), Medium Access Control (MAC) and Peripherals Network Management of the Layer 2 (PNM2) (Fig. 2).

The lowest block level is used to arrange the processes determining the dynamic behavior of the blocks. Signals are used for communication between the processes. The exchange of signals is only possible over specified routes or channels. The valid signals are assigned for each path (route or channel). The interface between two SDL blocks is thus specified independently of the internal structure of the components. Other developers can specify subcomponents (blocks or subblocks) independently in parallel.

In the protocol specification, one channel between two layers provides what is known as the Service Access Point (SAP). SDL also ensures that each level only uses the allocated services of the adjacent layers for the execution of a task.

2.2 Behavior

The dynamic behavior is determined by process graphs (Fig. 1, left). Beginning with the START state, processes can enter several states. A change in state is always triggered by the receipt of a signal. It should be noted that each process has only one input queue and the signals are processed in the order in which they arrive. If a signal is not defined in the current state, it is rejected or buffered, if required. During a change in state, various actions can be executed, such as assigning variables, preparing and sending signals. The graphical representation offers the opportunity of providing an overview of the communication for a process: beginning with the state symbol, all valid changes are introduced by the receipt of a signal. The necessary internal actions and transmission of signals then follow in the process graph. The graph ends with the subsequent state.

If more complex tasks are to be executed during a change of state, one option is enclosure in an SDL procedure or C function to ensure that the graphical representation of the process remains clearly arranged. Since the integration of C functions has been resolved, their use is especially appropriate for more complex algorithms.

For more detailed information, see [2] and [3].

3 Verification Through Simulation

The aim of verification is to check whether the specification is internally consistent (free from contradictions), whether certain characteristics can be guaranteed and whether the requirements have been properly designed and implemented. Examples of the characteristics of a specification which can be verified include:

- Type check
- Deadlocks
- Attainability of states

In our project, simulation was used to test operational correctness. This meant testing not only the behavior of an individual protocol stack, but also the interaction of several devices. As a result, a test bench was set up in SDL comprising a master and 4 slaves. Each of the devices had one instance of the specified protocol stack. The devices were connected using a simulation of the medium specified in SDL (Fig. 3).

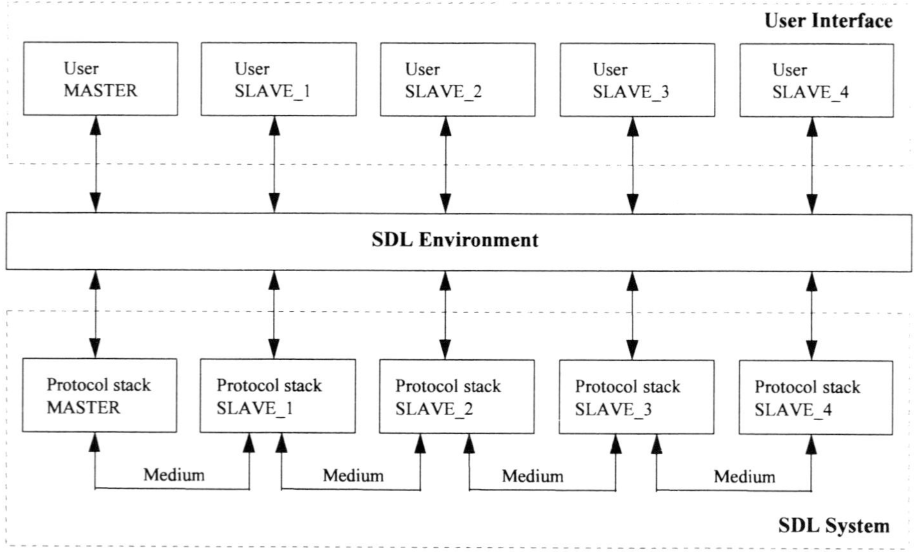

Fig. 3 Test environment for the Interbus protocol stack

While the visualization options (text trace and generation of MSC) in the simulator linked to the tool are well suited to detailed viewing, the simulator displayed weaknesses when viewing the overall system. While the text trace is primarily designed for testing sequences within a process, the option of generating MSC is well suited to representing interactions between two layers. Filters can be used to adjust the generation and, for MSCs, the subsequent representation. It is also possible to represent entire blocks as one communication partner. The subsequent processing of the representation of MSCs would actually be preferable because it is possible to change between various detail views. But recording MSCs with large numbers of processes proved too time-consuming.

In all, the simulation time was the main problem. One remedy was changing from the SDL simulation environment to an SDL application with its own monitor program which is based on the environment interface of the SDL system. In this way the simulation time for one data cycle was reduced from 10 minutes (not including generation of MSCs) to less than one second. At the same time, opportunities were created for clearly representing the data for the individual devices and operating the system with ease.

The effort involved in this changeover was quite considerable, but enabled us to carry out detailed tests of the protocol stack. By extending the mapping of the medium in

SDL it was possible to simulate the effects of transmission errors on the specified protocol stack.

4 Summary

The use of formal description techniques can be seen as positive for the area of fieldbus protocols. Advantages such as reducing the development cycle through the early identification of errors in the specification and above all the clear interpretation of the specification outweigh the training effort required for the methods. However, it does not completely replace informal text. To make it easier for others to understand the specification, each block and process is accompanied by a short informal description of its tasks and functions which is given on the first page of each one as a comment.

Experience was also gained of how to set up test conditions and test parameters for later implementation. For a hardware and software implementation, this interface should also be in the SDL system so that it is possible to compare the behavior of the specification and the implementation.

Further advantages are offered for the disclosure or standardization of a protocol: the formal specification can be used as a basis.

However, when using formal description techniques, the tool support is the most important aspect.

References

1. Tau, Version 3.4, Telelogic

2. Hogrefe: SDL, Estelle und Lotus, Springer Verlag, Germany

3. Ellsberger, Hogrefe, Sarma: SDL – Formal object-oriented Language for Communicating Systems
   ISBN 0-13-632886-5 Prentice Hall Europe 1997

CANopen Conformance Test

Thilo Schumann

CAN in Automation (CiA), Am Weichselgarten 26, D-91058 Erlangen, Tel. +49-9131-69086-0 (Fax –79), Email: headquarters@can-cia.de, URL: www.can-cia.de

Abstract. Conformance tests are used to guarantee that different products can work together. For that reason there was founded a group which defines a test procedure for CANopen devices based on the CANopen Communication Profile DS301 V3.0. This test procedure will guarantee that CANopen devices can communicate on the bus, the default values are set correctly to the standard and the network state machine is correctly implemented.

1 Conformance Testing

Within most fieldbusses the application layer is bound up with the protocol on layer 2. The CAN-bus is the only important exception of this system as it knows several different application layers. This situation is caused by the wide variety of applications, e.g. in the automotive industry, the biotechnology and automated manufacturing. Beyond that there are lots of autonomous systems with CAN applications that are hardly noticeable, e.g. in wheelchairs. At least in the European hemisphere the application layer CANopen won recognition in wide areas of the automation market by convincing more than 200 device manufacturers and their customers.

All the existing fieldbusses were specified and implemented in that way, that demands made by users and applications drove the process of collecting needed functions which had to be adjusted to the technical possible solutions and laid down in the specifications. Lots of problems are arising for system integrators out of this fact because they have to find a solution that fits best for the realisation of an application. The wide variety of offered applications for the CAN-bus might be severely diminished by the fact that they cannot cooperate problem-free with those of other manufacturers.

That's why conformance tests for fieldbus systems help to increase the interoperability of devices of different manufacturers noticeably if they are carefully implemented.

2 CANopen Conformance Testing

The test procedure for the CANopen Conformance Test was defined by a group of CiA members which where also involved in the specification of the CANopen Communication Profile. The specification for the test includes a static test where no timing requirements are taken. This is made because the timing depends on the application and must be defined by the system integrator and can not specified application independent. There is also no special device profile tested.

The CANopen Conformance Test is divided in different test sections. These are:

- Electronic Data Sheet

- Conformance to the CANopen protocol

- Correct behaviour in different network states and state transitions.

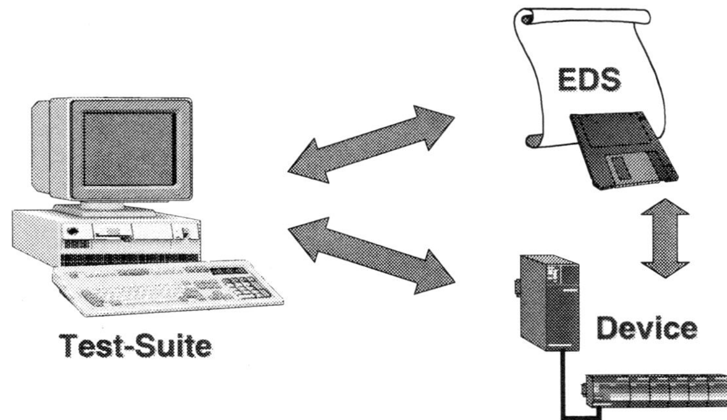

Fig. 1. CANopen test architecture

2.1 Electronic Data Sheet

The verification of the Electronic Data Sheet (EDS) is the first part of the CANopen conformance test because the EDS describes a device and his functionality in a electronically form. This is needed for all configuration tools to give help to a system integrator of a CANopen network. It is therefore necessary to avoid incomplete or erroneous data sheets.

The EDS is an ASCII coded file divided into different sections with a number of entries in every section. The size of the EDS is between 10 Kbytes for a small device with very restricted functionality up to more than 300 Kbytes for very complex devices. The test show that the following requirements are fulfilled by the contents of the EDS:

- value ranges should not be exceeded,

- mandatory entries must be supported,

- references must be point to existing entries,

- the EDS must be consistent.

The test steps of the EDS test can be seen below. As you can see there are not only errors reported but also warnings. E.g. these warnings results of not clearly defined PDO mapping because there exist the possibility to define more precisely which PDO is a consumer or a producer. This information is normally not necessary but it makes it easier for an automated configuration tool.

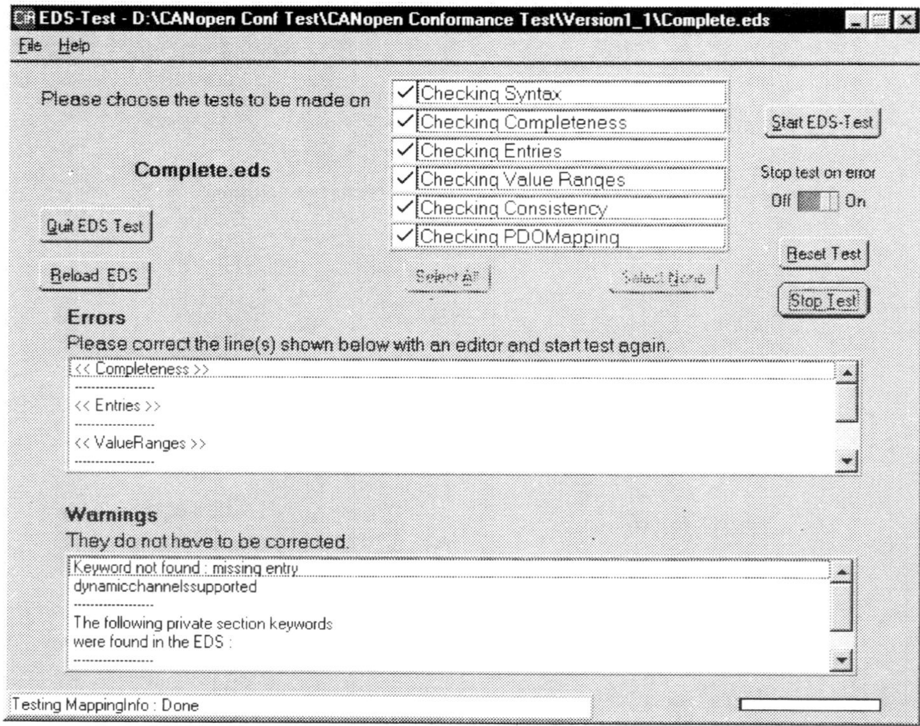

Fig. 2. EDS test

2.2 Conformance to the CANopen protocol

The test of the conformance to the CANopen protocol, named Protocol verification, is a static test where the error-free functionality and correct reaction to master commands is assured. All entries of the object dictionary will be double-checked against the EDS. It is the predefined connection set used to test:

- the SDO protocol,
- the PDO protocol,
- the object dictionary against the EDS,
- the emergency protocol,
- the Node guarding and Life guarding protocol,
- the Synchronisation protocol.

In this test all default values specified in the standards has to be set.

2.3 Correct behaviour in different network states and state transitions

The test of the correct behaviour in different network states and state transitions is another essential part of the device test because it guarantee that a device can be controlled in every situation and can be definably set in a defined network state.

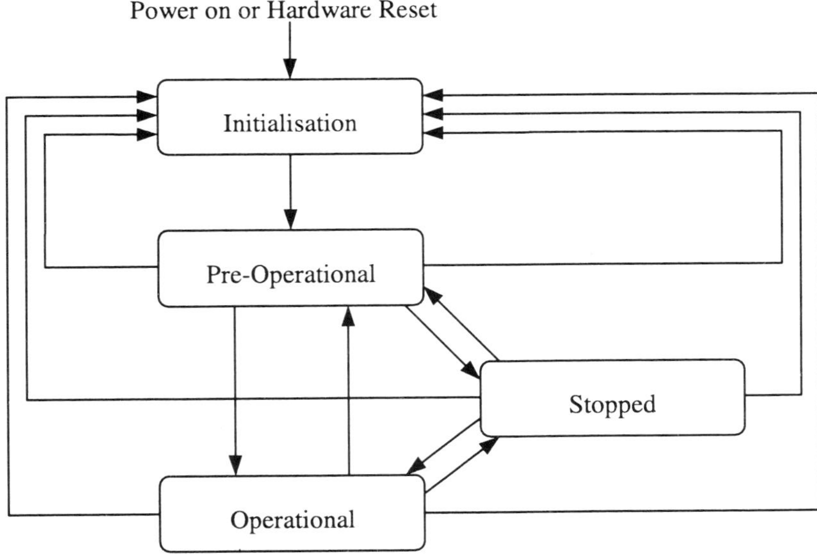

Fig. 3. Network states

2.4 Test results

During the test session the CAN telegrams can be viewed on the CAN telegram window. If it is necessary the object dictionary can be read or write. Every passed check will be instantly indicated by a check mark.

3 First experiences

There some first experiences collected with conformance test. That are:

- the default values are not set right,

- the EDS do not correspond to the device,

- the access rights are not correctly set including for PDO mapping

3.1 Default values

The conformance test does not test a configured device, it test a unconfigured device and for such devices the right default values must be set. The right default values are the values which are described in the standard.

3.2 EDS

The EDS describes a device with all the functionality. It is recommended that all objects of the device are described in the EDS including the default values, the right access type and the possibility for PDO mapping. All these is checked by the test and will be recognised.

3.3 PDO mapping

It is known that there is no defined procedure for PDO mapping described in the standard. It is allowed to write to a mapping entry at any time independent of the content of sub-index 0.

The mapping entries are the exception in the object dictionary. At any object in the object dictionary sub-index 0 describes the number of the follow sub-indices except the mapping objects. For these objects it is possible any time to write to a sub-index which is higher than defined in sub-index 0.

References

1. ISO 11898 (1993): Road vehicles – Interchange of digital information – Controller area network (CAN) for high-speed communication

2. CiA DS-301 (1996): CANopen communication profile

Formal Description Software for WorldFIP Industrial Fieldbus

Perfecto Mariño, Miguel Angel Domínguez, Francisco Poza, and Juan Nogueira

Electronic Technology Department and Applied Electronic Inst., Univ. Vigo (Spain)
E.T.S. Ingenieros Industriales, apartado Oficial, 36200 Vigo, Spain.
Fax: +34-986-469547, Tel: +34-986-812162,

Abstract. In this work the protocol WorldFIP for industrial communication networks, is specified with the formal specification language LOTOS. This specification is the foundation of new work lines such as efficiency analysis of WorldFIP performance with simulations in critical applications, the generation of benchmark patterns for conformance tests of WorldFIP communications equipment, and the proposal of new advances in the fieldbus area. The paper introduces the general characteristics of LOTOS, and the main features of WorldFIP particularly the related with its data link layer. Also the methodology and used tools are explained, and results and new guidelines of this work are quoted in last paragraph devoted to conclusions and future works. LOTOS specification of WorldFIP protocol provides powerful CAE tools for advanced designs in the framework of industrial communications.

1 Introduction

The development of industrial communications systems based on formal languages [1] [2] is the aim of this work. In the factory automation there is a great interest in network protocols able to assume time demanding requirements at shop floor level. Also the development of software tools to support all design stages as specification, checking and testing of these protocols, allows a quick prototyping and successive refinements. Information technology agents are involved in the task of create new communications standards [3] that can satisfy all diverse interests. Final agreements in the development of a standard have a long way to reach a full accuracy in definitive specifications.

2 Characteristics of LOTOS

The "Language Of Temporal Ordering Specification" (LOTOS) is a formal description technique (FDT) [4] standardized by the International Standardization Organization (ISO) [5], for designing services and protocols used in the communications of open systems (OSI: Open Systems Interconnection).

With LOTOS, the behaviour of a system can be specified as a sequence of events or actions that happen in an orderly way along the time. These actions are stated in

158

LOTOS by gates. In order to represent the temporary sequence of these events (gates) there is a set of operators with which behaviour expressions can be built [6].

The system is specified in LOTOS by a behaviour expression that states all possible sequences of actions (gates) happened as result of its real performance. For the best framework of specification, LOTOS allows the declaration of processes that describe behaviour expressions and that can be invoked in diverse specification sites.

A specification has two parts, the control part (behaviour) and the data part. In LOTOS does not exist any type of predefined data. All essentials data types, sorts, constants, operations, etc., must be specified, having thereby the advantage of create data suited to design needs, without to fit predetermined rules as is usual in conventional programming languages. For the design and specification of data are used "Abstract Data Types" (ADTs) based on the ACT ONE algebraic specification language [7].

3 Characteristics of WorldFIP

The fieldbus FIP (Factory Instrumentation Protocol) is based on the OSI model from ISO and its characteristics are recorded in the French standard C46-601 to 605 [8].

Afterwards this standard was described by CENELEC (Comité Européen de Normalisation ELECtrotechnique) into its standard EN 50170 Vol.3, with the name WorldFIP [9], being the same than former French standard but with the adoption of the international standard IEC 1158-2 [10] for its physical layer. Related with OSI model the WorldFIP standard only uses three layers of possible seven (Fig. 1).

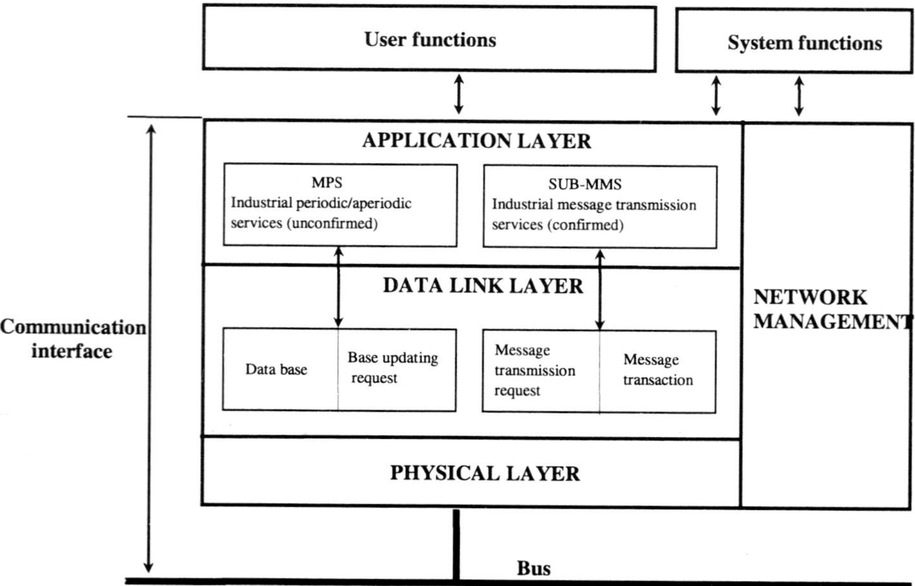

Fig. 1. Control equipment structure and services offered in WorldFIP

WorldFIP protocol allows interchange of information among different equipment connected to an industrial network such as sensors, actuators, Programmable Controllers (PLCs), Numeric Control Machines (CNCs), and so on. In a WorldFIP network the stations can have two types of functions: Bus Arbitrator (for managing medium access control) and Consumer/Producer of variables.

In a network only one station can be Bus Arbitrator. All of information interchanges are managed for the Bus Arbitrator, which is in charge of controlling the medium access from connected stations. There are three ways for giving access permission (Fig. 2): 1) Periodic polling of variables, messages and requests, 2) Aperiodic polling of variables, 3) Aperiodic polling of messages.

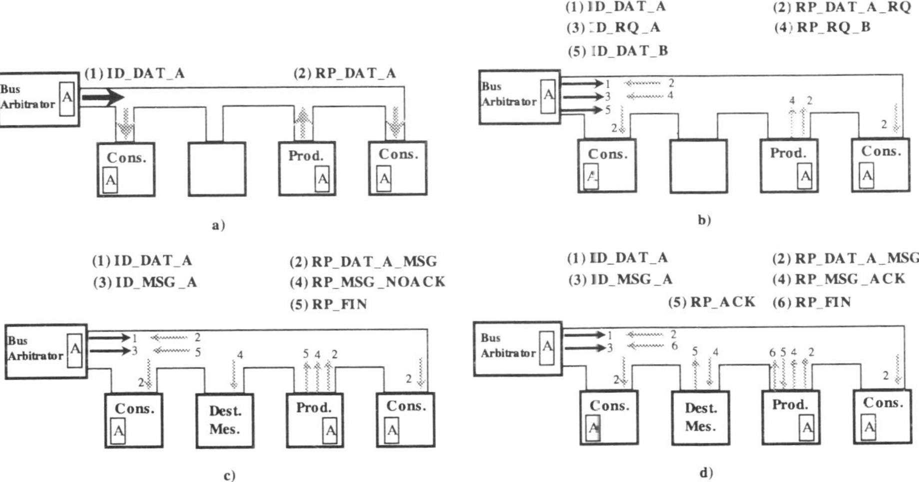

Fig. 2. Medium allocation mechanism: a) Periodic transfer, b) Request for aperiodic transfer, c) y d) Unacknowledge and acknowledged aperiodic message transfer request

Each kind of polling can be made in a "periodic window", an "aperiodic window for variables", and an "aperiodic window for messages". Also there is a "synchronization window" for padding each elementary cycle, and to comply so with periodicity imposed by periodic polling that are set by Bus Arbitrator in its initiation (Fig. 3).

Data link layer has two kinds of services for information transfer (Fig. 1): periodic/aperiodic services that manage variable interchange, and message transmission services that manage message transfers.

4 Specification of WorldFIP with LOTOS

The aim of this work is to specify in a formal way the data link layer (Fig. 1) of WorldFIP French standard for fieldbuses. For developing the LOTOS specification and having the following verifications, debuggings, simulations and validations [11], the

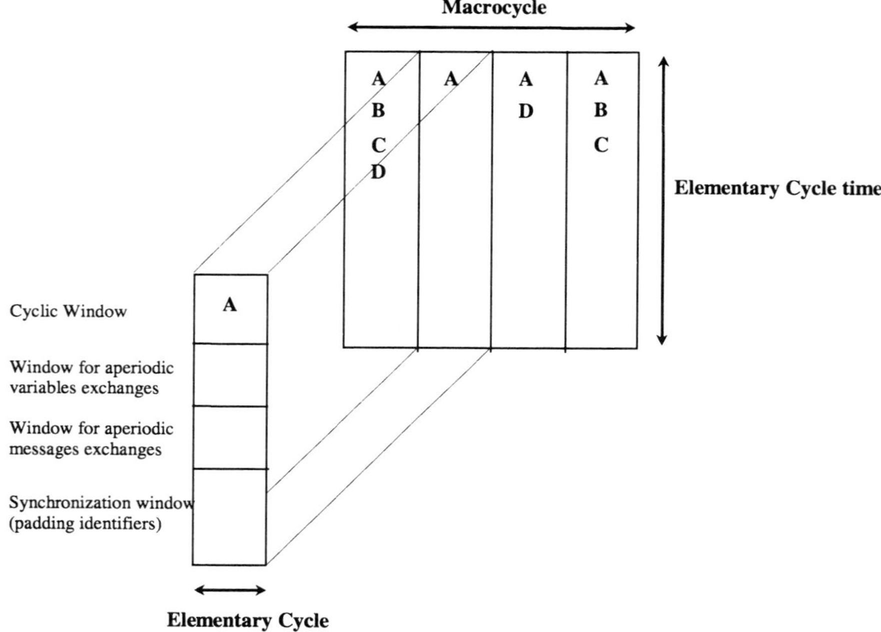

Fig. 3. Split of Elementary Cucles in 4 transactions windows

LOLA software tool is used [12]. Also TOPO tool [13] has been used for giving a program in C code from LOTOS specification, that related system events with visible happenings in program interfaces.

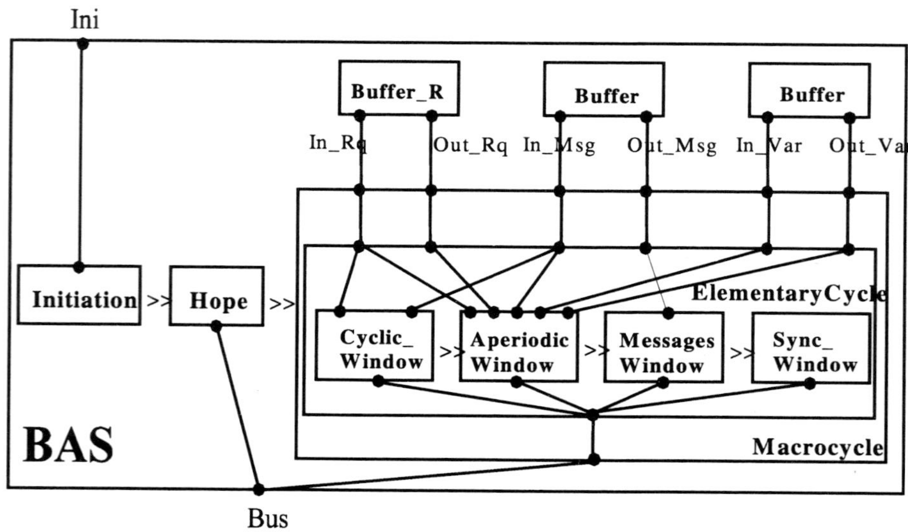

Fig. 4. Scheme of LOTOS specification for Bus Arbitrator Stations (BAS)

Given that in a WorldFIP network each station has two possible functions as Bus Arbitrator and as Consumer/Producer of information, in the first part of this work both functions are specified separately, implementing for each a succession of tests, simulations and verifications. For the Bus Arbitrator function is made a specification called BAS, which has two externally visible gates: Ini (allows user system configuration) and Bus (gives access to network physical layer for sending and receiving frames). In Fig. 4 all process existing in this kind of station are represented.

For the Consumer/Producer function a specification called CPS is made, with three externally visible gates (Fig. 5): Ini, AP (for communicating application and data link levels), and Bus.

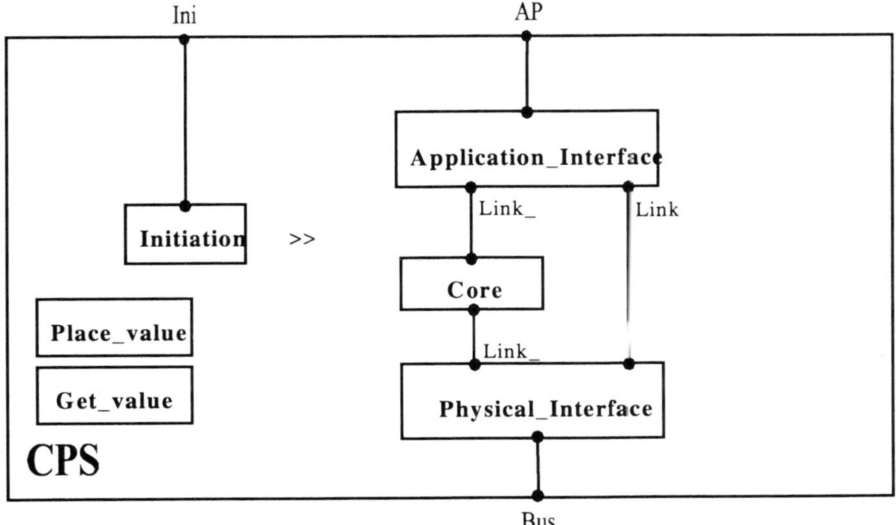

Fig. 5. Scheme of LOTOS specification for Consumer/Producer Stations (CPS)

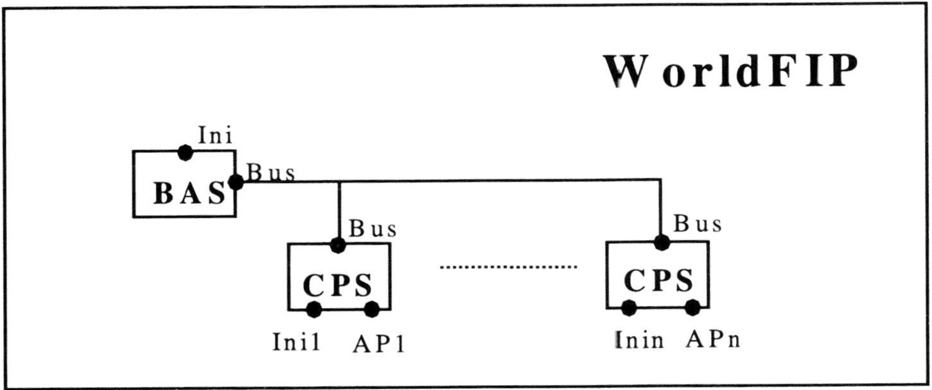

Fig. 6. LOTOS specification of WorldFIP network

162

After former functions have been specified separately and made simulations and verifications of their accurate behaviour, a whole specification of a WorldFIP network is implemented with several stations. In this new design the BAS and CPS specifications are defined as processes, and in each WorldFIP specification only one station will have a BAS function whereas the remaining ones will have CPS functions, working in parallel all stations and synchronizing in the "Bus" gate (Fig. 6). Then a set of simulations and acceptance and rejection tests are made, until validate the interaction and communication among stations as if were working in a real bus.

5 Conclusions and future works

With LOTOS specification of WorldFIP protocol comes up the possibility and benefit of using formal descriptions techniques for designs in the field of industrial communications. The potential of specifying the WorldFIP protocol in LOTOS allows to check the standard correctness, at the same time as certain ambiguity is detected given that the standard is implemented and explained in a natural (informal) language (French or English). This is a very important advantage of formal description languages in the design and definition of new standards so do not allow misunderstanding. In the development of this project about the formal definition of WorldFIP standard, currently are implemented in the stage of verification and validation, through step by step simulations and acceptance and rejection tests, the specifications "Bus Arbitrator" and "Consumer/Producer", and also the specification that defines a general WorldFIP network. For future works the proposed guidelines are:

- Implement a simulation enviroment for Consumer/Producer and Bus Arbitrator stations.

- Implement a design and testing enviroment for WorldFIP networks. This will be very useful for testing new designs of WorldFIP equipment. The designed new equipment would firstly be specified in LOTOS and its performance would be checked in the WorldFIP network simulator. In this way performance failures can be detected in the design stage, without waiting its physical implementation, which reduces the development costs of new equipment [14].

- Once implemented all this over the WorldFIP protocol, this design methodology could be extended to other industrial field bus protocols (as P-NET, InterBus-S, CAN, etc.) and to other formal description techniques as SDL and ESTELLE.

6 Acknowledgements

This work was made from three R&D projects sponsored by the following entities: R&D NATIONAL SECRETARY AND CICYT, Ref. TIC97-0414, CENTRAL GOVERNMENT (MADRID, SPAIN); UNIVERSITIES GENERAL OFFICE, Ref. 645021802, AUTONOMOUS GOVERNMENT (GALICIA, SPAIN); and RESEARCH VICECHANCELLORSHIP, Ref. 641021710, UNIVERSITY OF VIGO (SPAIN).

References

1. Mariño, P., Poza, F., Domínguez, M.A. & Nogueira, J.: "Formal specification applied to industrial LAN's design". Proc. of EUROMICRO 23rd Conference, IEEE Computer Society, pp. 215-221, Budapest (Hungary), 1-4 September, 1997.

2. Mariño, P., Poza, F., Domínguez, M.A. & Nogueira, J.: "Specification and verification of a real-time field bus with formal description languages". From book Transformation Based Reactive Systems Development, ed. Bertran, M. & Rus, T., Springer Verlag, Lect Notes in Computer Science 1231, pp. 415-429, Berlin, 1997.

3. Mariño, P.: Enterprise communications: Standards, networks and services. Ed. RA-MA, Madrid, 1995.

4. Turner, K. J.: Using Formal Description Techniques. John Wiley & Sons, 1993.

5. ISO: LOTOS – A formal description technique based on the temporal ordering of observational behaviour. Standard ISO 8807, 1989.

6. Bolognesi, T. & Brinksma, E.: "Introduction to the ISO Specification Language LOTOS". Computer Networks and ISDN Systems 14, North-Holland, pp. 25-59, 1987.

7. Meer, J., Roth, R. & Vuong, S.: "Introduction to algebraic specifications based on the language ACT ONE". Computer Networks and ISDN Systems 23: North-Holland, pp. 363-392, 1992.

8. AFNOR: Bus FIP pour échange d'information entre transmetteur, actionneur et automate. Standards NF C 46-601 to 605, 1990.

9. CENELEC: General purpose field communication system. Standard EN 50170, Vol. 3/3 (WorldFIP), 1996.

10. IEC: Fieldbus standard for use in industrial control systems – part 2. Standard IEC 1158-2, 1993.

11. Quemada, J., Azcorra, A. & Pavón, S.: Design with LOTOS (Chapter 4: Validation and Verification). Universidad Politécnica de Madrid (Spain), 1993.

12. Pavón, S. & Larrabeiti, D.: LOLA User Manual (Version 2.3). Universidad Politécnica de Madrid (Spain), 1995.

13. Mañas, J., De Miguel, T., Robles, T., Salvachua, J., Huescas, G.: TOPO: Quick Reference Front-End - Version 3R6. Univ. Pol. de Madrid (Spain), 1995.

14. Poza, F.: Contribution to research of industrial communications systems by formal description techniques. Ph. D. Thesis. Univ. of Vigo (Spain). December, 1997.

Experiences in Different Fieldbuses Used together with PC-Based Control Systems

Ville Saarimäki, Niko Siltala, Petri Partanen, Jorma Vihinen,
Reijo Tuokko

Tampere University of Technology, Institute of Production Engineering,
P.O. Box 589, 33101 Tampere, Finland, Tel +358-3-365 2111, Fax +358 3 365 2753,
Email: Ville.Saarimaki@pe.tut.fi, Niko.Siltala@cc.tut.fi, Petri.Partanen@pe.tut.fi,
Jorma.Vihinen@pe.tut.fi, tuokko@cc.tut.fi

Abstract. This paper deals with experiences about three different fieldbuses based on a comprehensive research study in the field of PC-based device and system control (soft PLC). The research has been made in Tampere University of Technology at the Institute of Production Engineering. The overall reaction time of PC-based control systems has been tested so that the delay caused by fieldbus is included. For the end user this is the reaction time that counts. The paper discusses also some topics concerning motion control and a case study of an application controlled by soft PLC.

1 Introduction

This paper deals with experiences about three different fieldbuses based on a comprehensive research study in the field of PC-based device and system control (soft PLC). The research has been made in Tampere University of Technology at the Institute of Production Engineering. Nine different industrial companies have participated in the project. The research has given the research team and participants versatile knowledge in this new technology area.

The goal of the project has been to compare traditional PLCs and PC-based control systems named soft PLC. Now the state of the technology of PC's and operating systems and possibility to integrate all kind of control applications on one computer makes it justifiable to think that PLCs will be replaced totally or partially by more open control systems based on PC-computers.

2 Reaction time measurements

The overall reaction time of PC-based control systems has been tested so that the delay caused by the fieldbus is included. For an end user this is the reaction time that counts. From the application point of view it's not so interesting to know the time that the computer processes the required task. More important is to know the overall reaction time between the input command signal and output response.

The built test environment (Fig. 1) for measuring the overall reaction time consists of digital I/Os, Fieldbus card and PC. A pulse signal was fed into the input module letting the PLC or Soft PLC to put the output on.

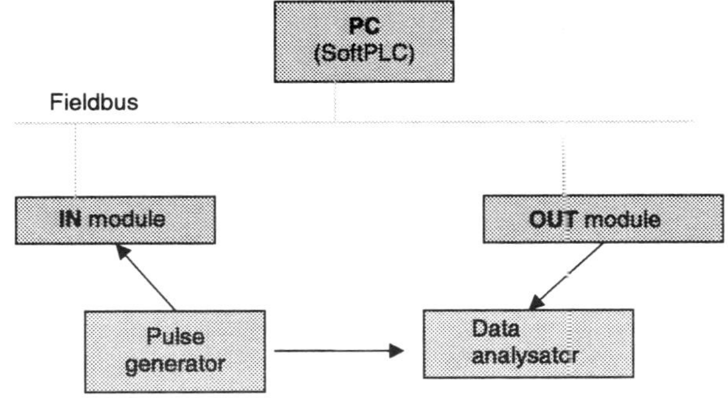

Fig. 1. Principle of reaction time measurements

The length of the pulse was 10 milliseconds and it was sent in every 50 milliseconds (Fig. 2). The time difference between input and output was then measured. Both the input and output modules were connected with a fieldbus to the controller.

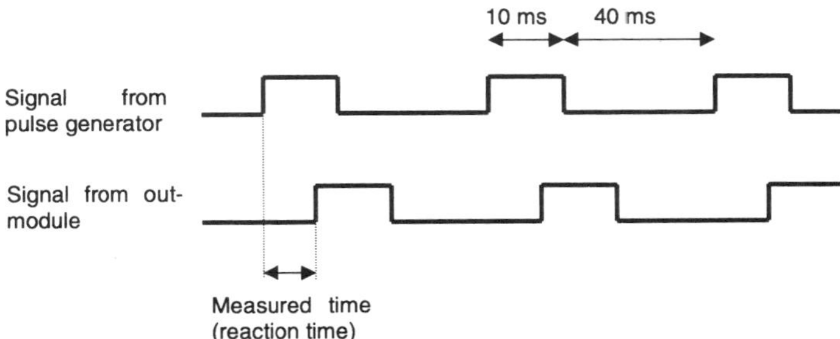

Fig. 2. Pulse signals in reaction time measurements

2.1 Measurements with Soft PLC and Profibus-DP and DeviceNet

A traditional PLC with Profibus-DP (Siemens S7 series with CPU 315-2 DP) was used to get a reference value. All of the distributed I/O equipment was the same than in other measurements. Average reaction time was 5,2 ms and minimum 4,1 ms and maximum 6,4 ms with standard deviation of 0,6.

The measurements have been made as shown in Figure 1 and Figure 2. In the tests each soft PLC product was measured with both Profibus-DP and DeviceNet. A fieldbus speed of 1,5 Mbits/s was used in the case of Profibus-DP, because in industry you often have implementations that need more than 100 meters of fieldbus cable. In case of DeviceNet the maximum fieldbus speed (500 kbits/s) was used. The devices used were as follows:

CPU   350 MHz
Memory   128 Mbytes
I/O modules   Siemens ET200B and Entrelec Schiele PMI
Fieldbus speed   1,5 Mbits/s (Profibus-DP) and 500 kbits/s (DeviceNet)
Fieldbus cable   Shielded copper wire

The results are an average of three different soft PLC products. Measuring frequency was 100 kHz. The amount of measurements was two thousand pulses with every product and every fieldbus.

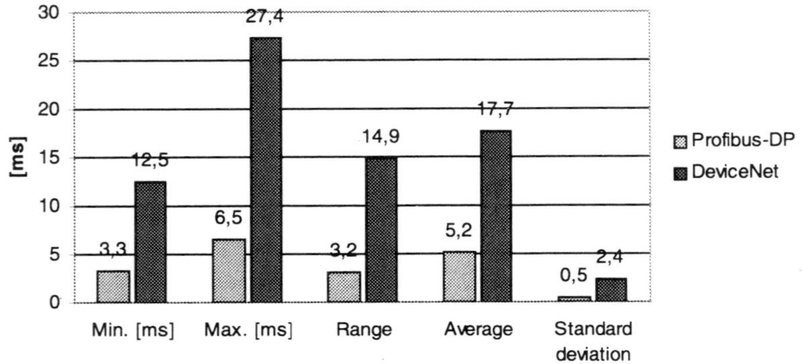

Fig. 3. Average reaction time results when using Profibus-DP and DeviceNet

The system with DeviceNet didn't notice all the pulses fed into the system. It lost 46 pulses from 2000 as an average. This has to be taken into account when reading the results.

2.2 Measurements with Soft PLC and Profibus-DP and Interbus-S

In the measurements the maximum speed of Interbus-S (500 kbits/s) was used. The same reference values are valid in this measurement also. The soft PLC product was tested with both Profibus-DP and Interbus-S. The equipment used in this test were as follows:

CPU   350 MHz
Memory   128 Mbytes
I/O modules   Siemens ET200B and Phoenix Contact IB ST 24 DIO
Fieldbus speed   1,5 Mbits/s (Profibus-DP) and 500 kbits/s (Interbus-S)
Fieldbus cable   Shielded copper wire

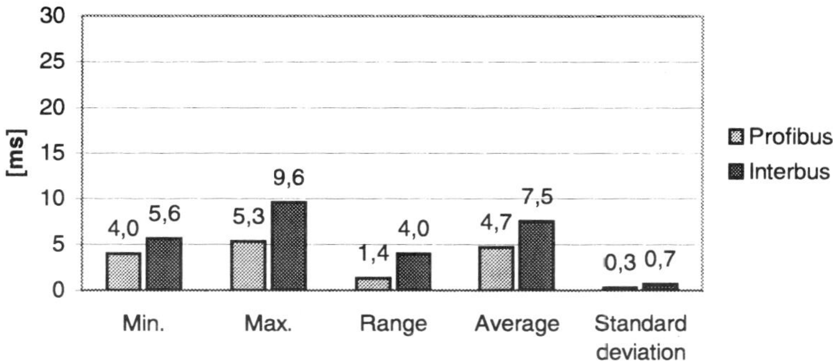

Fig. 4. Reaction times when using Profibus-DP and Interbus-S

3 Motion control

There are three main ways to implement motion control with soft PLC's (Figure 5).

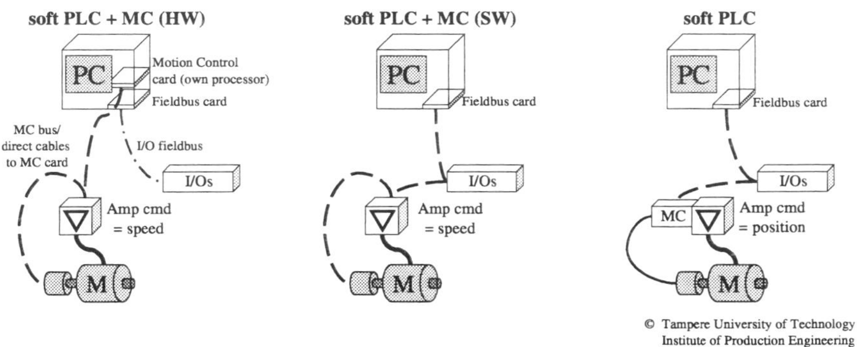

Fig. 5. Different ways to implement motion control in soft PLC solutions

At the left side in figure 5 the motion controller (MC) is right beside the PLC. MC is e.g. a PC card and it has its own processor. The Soft PLC program is running in the PC's processor. PLC and MC have shared memory and so the communication between them is very easy to be carried out. In the construction there is two different buses: one for MC and another for I/Os. This system is suitable for applications that are very fast or that need interaction of several axes. Disadvantage is need of two different fieldbus systems.

Second possibility (Fig. 5 middle) is to use the soft PLC program with an integrated MC-part. Both are software based and are running in the PC's processor. In this case we can manage with one fieldbus for whole application. The fieldbus speed will set limits for the behavior and accuracy of the application.

Because the feedback loop is closed over the fieldbus, the speed of fieldbus has an important role in both first two cases. Feedback information is transferred from encoders to controller via fieldbus and after calculations controller is commanding amplifier with e.g. a speed value and again via the fieldbus. Besides the speed of fieldbus poorly designed drivers for the adapter cards will increase a lot the cycle time of the control loop.

A third possibility (Fig. 5 right) is to distribute motion control in the field. In this case the speed of fieldbus is not as critical as it was in first two cases, because fieldbus is not closing the control loop. The soft PLC is giving only positions and parameters to MC via fieldbus. MC can be e.g. integrated into the amplifier.

4 Case study

This case study was concerning a motion control application of two axes of a machine. Both axes are driven with an AC-motor and frequency inverter. The position feedback was gotten from an absolute encoder located at the end of motor shaft. In this way the backlash caused by gears and mechanics was avoided.

In the application a 1,5Mbits Profibus-DP was used as the fieldbus. It was chosen from the set of fieldbuses given from the company. The motion control loop was closed at the PC with Profibus. The same fieldbus served other application IOs at the same time.

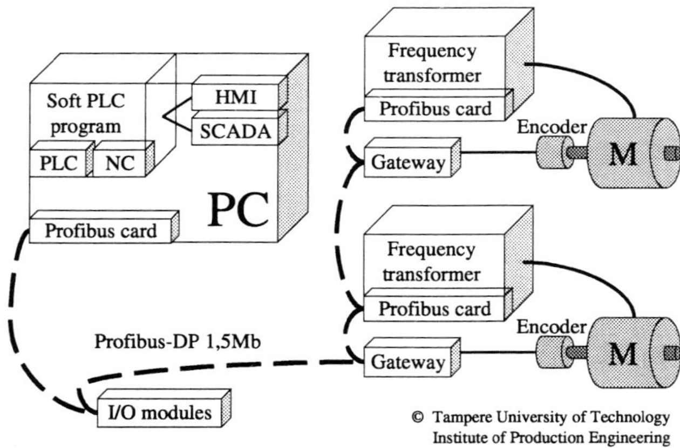

Fig. 6. Structure of the case system

The control loop started from the encoder that was directly connected via a supplier-made gateway to Profibus. The position information was transferred to Soft PLC and to its motion control part. Depending on the error between the target path and actual position a speed setup value was calculated for correcting this error. This calculated speed output was transferred into the frequency inverter, which was connected directly to the fieldbus with a supplier-made Profibus card. The frequency inverter commanded the motor without any speed or position feedback from the motor.

4.1 Some results

In the test the position accuracy aims were achieved. On the other hand that positioning accuracy was not reached with the wanted drive speed and without a too big overshoot. One reason for this can be the fieldbus speed. One of system component limited the baud rate to 1.5Mbits/s.

In the reaction tests it was noticed almost 1ms difference in averages of system reaction times between the fieldbus baud rates 12Mbits/s and 1.5Mbits/s. A delay of 1 ms may affect significantly functionality of the control loop. If we want to avoid these fieldbus delays we should validate components which are capable to maximum data transfer speed. Also with right PID parameter settings these delays should be possible to be minimized if there are component limitations. A second possible fail source was the system hardware combination. It was necessary to make some additional user specified program parts between the PLC and NC part to transfer a correct control word structure including the speed value for inverters. A solution to this problem is to use components recommended and specified by the software vendor.

5 Conclusions

The tests show that there are big differences between both soft PLC products and fieldbuses. Fieldbuses are developed for different purposes and they have different needs. /1./ Apparently in this case with devices and softwares used DeviceNet was not so suitable for this kind of high speed data transfer, or the companies who have developed these softwares have not put very much effort on developing drivers and interfaces for DeviceNet.

Every system, including the soft PLC and fieldbus is as weak as its weakest point. Either it is the fieldbus or driver or the fieldbus card or something else. In every case these points have to be taken into account when choosing the fieldbus. If you make a bad choice it becomes the weakest point of your system.

As to the tested three different fieldbuses the fastest seems to be Profibus-DP. Reaction time with Profibus-DP is even shorter when the bus speed is 12 Mbits/s. The second fastest is Interbus-S. There isn't that big difference between Profibus-DP and Interbus-S. The tests of Interbus-S were made using the maximum available bus speed, so that it is not possible to shorten the reaction time.

References

1. A. Steinhoff: Classification of serial bus systems, International Symposium on Open Control Systems, 30 May - 1 June 1999, Tampere, Finland

Chapter 5: Management

Mapping of Fieldbus Components to WWW-Based Management Solutions

Martin Wollschlaeger

Institute for Measurement Technology and Electronics (IPE)
Otto-von-Guericke-University Magdeburg
PO Box 4120, D-39016 Magdeburg
mw@ipe.et.uni-magdeburg.de

Abstract. The seamless integration of fieldbusses into enterprise wide communication systems enables a network centric approach to fieldbus components' management. This requires the mapping of fieldbus components and their functionality to WWW-related management solutions. Therefore different strategies are suitable. Developments in software technology allow the introduction of new concepts for management tasks of fieldbus based components. Besides the software oriented tasks, there are new description techniques that can be used in fieldbus systems. WWW-related management solutions for different systems are described in this paper.

1 Motivation

One of the most outstanding features of modern process control solutions is the integration of data acquisition systems into hierarchically organized process information systems. These systems are implemented using different layers for data exchange. Typically, fieldbusses represent the data transport solution in the lower layers, while common network implementations based on Local Area Networks can be found in the upper layers. The single transport systems are connected together via gateways, that are implemented in PLCs, Controllers, or PCs.

The impact of the Internet in the area of automation and control can be recognized by its ongoing introduction into appropriate software solutions. A process information system supports the adoption of internet technology and offers new opportunities for seamlessly integrated management solutions. In order to provide easy access to distributed plant and business data from any point within the technical and economical management of the enterprise, an integration of fieldbusses and LANs has to be performed.

This development is supported by the increasing use of PC-based system concepts. Modern software systems, designed to fit into networked environments, form another prerequisite for the integration of complex network systems. This is proved by the increasing use of concepts like Distributed Component Object Model (DCOM) and Java [1, 2] in the measurement and automation area. Concerning fieldbusses, integrated data exchange methods, like OLE for Process Control (OPC) [3], enable an easy, vendor-independent access from nearly any application platform to the data objects and to the functionality of fieldbus systems and their components. Figure 1 shows a system concept based on DCOM for integrating a fieldbus into a LAN.

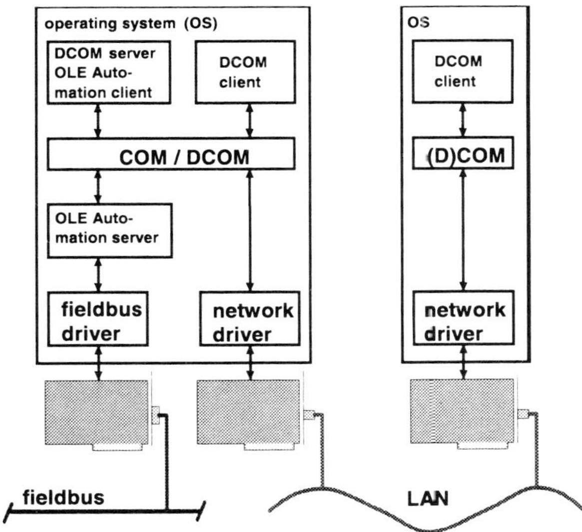

Fig. 1. Integration of a fieldbus into a LAN

2 WWW-based management framework

The implementation of a structure shown above is a prerequisite for the introduction of a WWW-based management framework. This framework is implemented on top of an enterprise network. It uses the Hypertext Transport Protocol (HTTP) as a network-wide unique transport layer for data exchange between different networked components. Data sources are implemented in Web-servers, while Web-browsers represent the user interface. An integrative software instance in the server coordinates data exchange between the fieldbus interface and the hypertext documents provided by the server, as well as the integration of software components into that documents [4]. The documents displayed in the browsers act as containers for data and for interactive components of the user interface. The principle structure of the WWW-based management concept is shown in Figure 2.

Since Web-browsers support modern software concepts, the interactive components can be identically to those used in any COM- or Java-based standard software tool provided by the vendors of automation and control software, or by vendors of fieldbus systems and components. Furthermore, there is an increasing number of general-purpose software components or even office-oriented solutions, that can be used within the same environment. In general, the benefits of using a Web-browser as a user interface are originated by its acceptance by the users. Web-based solutions have become state-of-the-art software systems. In addition, as WWW is heterogeneous by nature, management solutions for heterogeneous systems like fieldbusses within an enterprise network can easily be integrated.

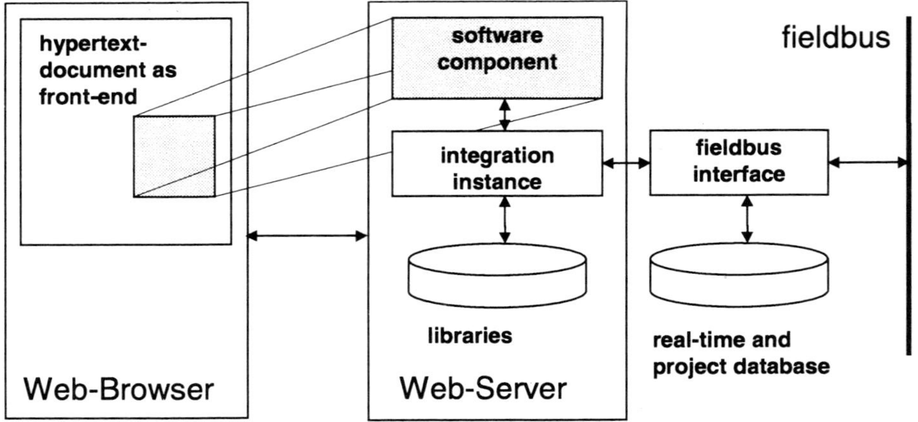

Fig. 2. WWW-based management framework

3 Mapping of fieldbus components

The mapping of the management functions of a fieldbus component requires an appropriate description of the communication objects and of the functionality the component provides. Both types of information are used to design the user interface, that will be embedded into hypertext pages. There are different approaches for that mapping.

3.1 Server oriented mapping using scripts

The access of the browser to fieldbus objects is performed using server based scripts, that can be implemented in any standard scripting language supported by the used WWW-server. Usually, this server is implemented in the system acting as a gateway to the fieldbus. Communication objects are mapped to variables, while management functions are mapped to procedures and functions of the script. When using the script based implementation, a request from a browser initiates the execution of a script at the web server. This script starts, for example, an instance of an OLE Automation client, that accesses the COM objects of the fieldbus mapped to the OLE Automation server. In addition, server related functions or access methods to other interfaces (e.g. data base systems) can be invoked by the script. However, the generation of component specific scripts is necessary.

3.2 Mapping to software components

This mapping concept uses DCOM technology or Java. ActiveX-controls or Java applets are instanciated on the client systems. The Web-server implemented in the gateway system provides the control or applet for download. After downloading it gets startup-information from the framework (usually the browser) and then uses the built-in interfaces (DCOM- or Java-based) to access the fieldbus objects over the network. The complete management functionality is implemented in the control or applet. The concept requires the development of fieldbus-component-specific controls or applets.

Compared with scripts, these software components can be re-used in any other DCOM- or Java-based environments (frameworks), for example in fieldbus-specific software tools. In addition, the combination of DCOM- and Java-technology offers a starting point for powerful future management solutions.

Both concepts are suitable for using a web browser as an unique interface widely accepted by the users. It provides access to fieldbus data as well as to any other in-formation provided by the networked environment. Especially the direct access to data in fieldbus related design and con-figuration tools, and to database driven applications will support future complex, but scalable software tools for fieldbus handling.

3.3 Description of fieldbus components in XML

The eXtensible Markup Language XML [5] expands the description language HTML with user-defined data types and structures. There is a clear separation between the data descriptions, the data themselves, and the representation in a browser. This enables to use a single XML description file for different tasks. The description can be hierarchically organised, using the capabilities of XML. Depending on the functions to perform, the XML data are filtered and associated to software components (controls, Java beans, etc.). The selection of the necessary information and the definition of their presentation details can be performed by means of style sheets (Figure 3). These style sheets are part of the development of XML. The XML file and the different style sheets are used as input files for HTML generators, that produce the HTML pages necessary for the certain management functions

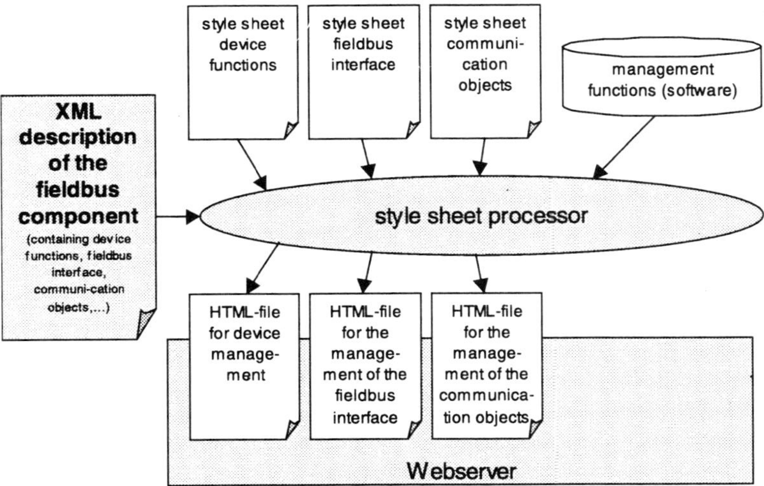

Fig. 3. Generation of HTML pages using an XML description and style sheets

The distribution of the generated HTML pages and associated software components is done following the concepts described above. The major benefit of this solution is a unique, reusable description with an excellent consistency and reduced efforts of the description process.

176

4 Exemplary solutions

4.1 Management of a P-NET system

The P-NET management solution is based on VIGO [6]. VIGO provides a COM-based interface to the project-related data stored in a object-oriented database (topology, addresses, parameter of application functions). In addition, it provides an interface to the data in the modules. The user interface uses a combination of Scripts, ActiveX-controls and Java applets [7]. While applets perform application function related tasks, existing ActiveX-controls are used to browse the project database. Scripts coordinate the data exchange between the different components. Fig. 4 gives an example for a hypertext document used in the P-NET solution.

```
<HTML><BODY>
<SCRIPT LANGUAGE=VBScript RUNAT=Server></SCRIPT>
<H1>P-NET access via VIGO and INTERNET</H1><HR>
Select VIGO variable from MibOCX Control:
<object
    classid="clsid:A436C660-1DAB-11CF-B597-0020AF38E834"
    ID=mib>
<PARAM NAME="PhysID"></object>
<HR><INPUT TYPE=BUTTON VALUE="Show Types" NAME="BtnShowTypes">
<HR>
Documentation of "UPI" in PDF Format:
<body leftmargin=0 topmargin=0 scroll=no> <embed width=100%
height=50% fullscreen=no src=docsrv\504008.pdf><HR>
<% set ad = createObject("VIGO")
ad.PhysID = P-NETsrc
a=ad.ExFloat %>
       .
</BODY></HTML>
```


Fig. 4. HTML document using P-NET related controls and COM-objects

4.2 Management of a CAN system

The CANopen application layer protocol and CANopen Device Profiles define the communication objects implemented in the networked modules. These objects have been mapped to variables in scripts. The scripts are used as control instances for ActiveX-objects, that the management functionality of the modules is mapped to. The interconnection to the CAN network is performed by an OLE Automation driver for CAN [8]. The scripts pass data from this driver object to the controls in a web page. In addition, DCOM-enabled software components have been implemented for a mapping of the management functions. These DCOM components can be embedded into or accessed from hypertext documents. They perform a transparent data exchange with a DCOM server across the LAN. The DCOM server is implemented in the gateway. Fig. 5 illustrates the topological structure of the CAN implementation.

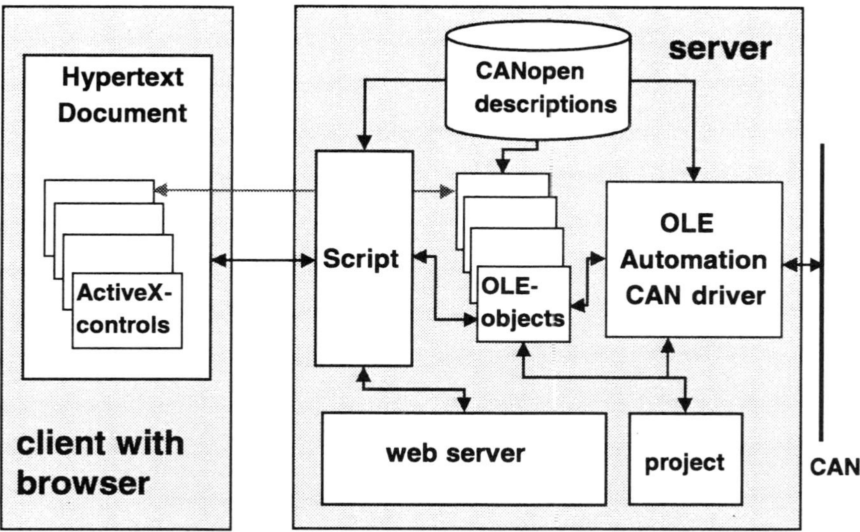

Fig. 5. Topological structure of the CAN management implementation

There are no major differences between the P-NET and CAN-related management solutions. Of course, the ActiveX-controls, Java applets, and scripts are different in detail, but the concept behind is the same. A test environment was set up, with both CAN and P-NET modules in a heterogeneous project. Both systems can be managed starting from a unique web-page. It has to be expected, that a generalized approach using, for example function blocks, will continue to reduce the differences between systems.

4.3 Implementation based on XML

The problem of re-using description information in different context (e.g. definition, management, visualization, application development) was addressed by creating XML descriptions for fieldbus components. Based on an appropriate Document Type Definition (DTD), an XML description of a CANopen Device Profile was created (Fig. 6). This description holds all information necessary to generate the associated document in HTML format (Fig. 7). The generation of the document is supported by an XSL style sheet with formatting and filtering instructions. This style sheet has to be developed once, and it can be re-used for other profile descriptions. In addition, more generalized DTD and style sheet allow an easy porting of the device profile to other fieldbusses, too.

Based on the same XML file, specific style sheets and scripts are implemented to create input files for other tasks. For example, the CANopen-specific Electronic Data Sheet (EDS), or Device Configuration File (DCF) files can be created by a Script. The same is possible for generating header files for C++ development, or parameter information for Java classes. Furthermore, an ODL-file (Object Definition Language) can be created, that is used for building COM-supporting applications (e.g. DCOM-objects or ActiveX-controls).

```
<?xml version="1.0"?>
<?xml:stylesheet type="text/xsl" href="profile.xsl"?>
<!DOCTYPE profile SYSTEM "CiA_profile.dtd">
<profile>
  <profileDescription>
      <profile_class>Device Profile</profile_class>
      ...
  </profileDescription>
  <GeneralDescription>
     Here goes the general description of the profile.
</GeneralDescription>
    <ObjectDictionary>
       <ObjectDescriptions>
         <Object Category="mandatory">
           <ObjectDescription>
             <Index>1000</Index>
             <Name>DeviceMode</Name>
             <ObjectCode>ARRAY</ObjectCode>
             <DataType>UNSIGNED32</DataType>
             <Explanation>This Object holds ...</Explanation>
           </ObjectDescription>
           <EntryDescription>
             ...
             Description of the object's elements (access rights,..)
           </EntryDescription>
         </Object>
         ...  other objects  ...
       </ObjectDescriptions>
    </ObjectDictionary>
     other description information...
</profile>
```

Fig. 6. XML description of a CANopen device profile (extract)

5 Conclusion

The introduction of WWW based management solutions for fieldbusses combines two of the major trends of last years' developments. WWW integration won't yet provide real-time control systems, but offers solutions for local and remote data access and management in process information systems. The users participate from the integrative features of Internet browsers, allowing a single application to act as a homogeneous user interface to compound documents containing heterogeneous data derived from distributed systems. The growing number of reusable ActiveX components and JAVA beans or applets with included download and security methods makes it easy for the customers, to create user specific environments without having to know specific details of the current installation's components. The existing networking environment can be integrated, for example Directory Services. This will reduce installation costs, and efforts for training and preparation. The developers' benefits of WWW-based management concepts are reductions in software developments. This is dedicated to the use of controls and their development, allowing to reuse previously written code or third-party solutions. This is supported by adopting new technologies, like XML. In combination with new methods in software support over the net, versioning and license management will be enhanced.

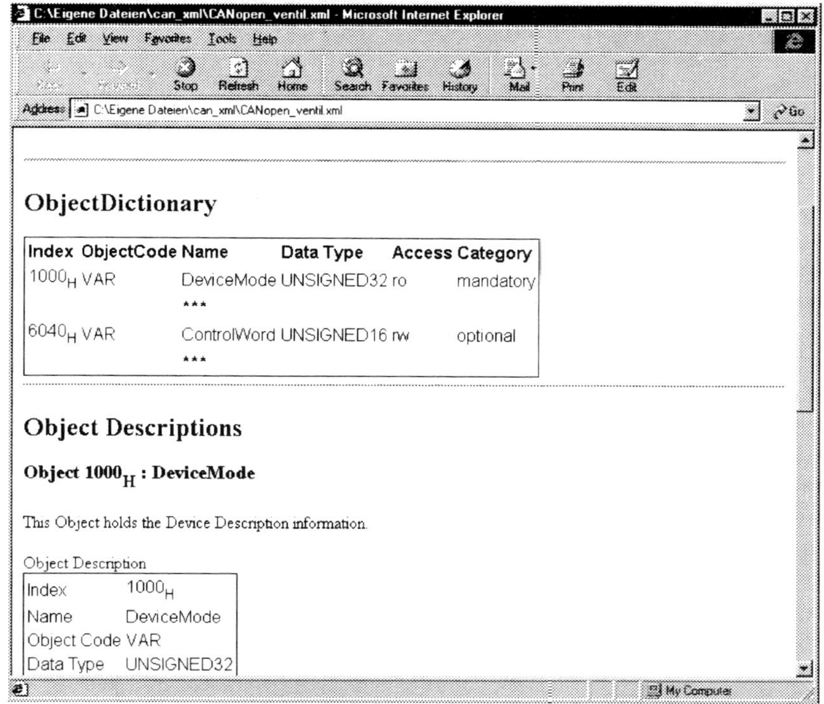

Fig. 7. The XML file from Fig. 6 displayed in a browser using a style sheet

References

1. Grimes, R.: Professional DCOM Programming. Wrox Press, 1997.

2. http://java.sun.com/products/api-overview/index.html

3. n.n.: OPC Data Access Automation Specification, Version 2.0. OPC Foundation, October 14th 1998.

4. Wollschlaeger, M.: Planning, Configuration and Management of Industrial Communication Networks using Internet Technology. IEEE GLOBECOM '98, Sydney, 08.-12.11.1998, pp. 2235-2240.

5. Bray, T.; Paoli, J.; Sperberg-McQueen, C. M.: Extensible Markup Language (XML) 1.0. 1998, http://www.w3.org/TR/REC-xml

6. Cramer, O.: VIGO, a Fieldbus Management System. 4th International conference on the P-NET Fieldbus system, Oporto, 02.-03.05.1996, proc.

7. Wollschlaeger, M.: P-NET Management with Java based Components. 6th International conference on the P-NET Fieldbus System, Vienna, 25.-26.05.1999, proc.

8. Wollschlaeger, M.; Wehrmann, S.: Intranet-Based Management of CAN Devices. 5th International CAN Conference (iCC'98), San Jose, 03-05.11.1998, proc. "Application II" pp. 12-02 to 12-09.

Integration of Fieldbus Objects into Computer-Aided Network Facility Management Systems

Thomas Bangemann, Ralf Dübner, Arne Neumann

ifak Institut für Automation und Kommunikation e.V. Magdeburg
Steinfeldstr. 3, 39179 Barleben, Germany

Abstract. Although currently there are several activities that are focused on a unified way to project and administrate fieldbus systems the planning and management of these systems in the sense of a comprehensive infrastructure is not possible yet. In this paper an approach for the integration of Building Automation Systems using different fieldbus systems into the technical Facility Management shall be demonstrated. This integration is realized using standardized mechanisms like OPC technology and common object definitions for different fieldbus systems based on international standards. The paper gives an introduction into the basic models and introduces a way to solve this task.

1 Introduction

In the field of computer aided planning and management of buildings, spaces and technical equipment the so called Computer-Aided Facility Management Systems (CAFM) have been established at the end of the 80th. Computer- Aided Network Facilities Management (CANFM) is a specialized form of these systems.

CANFM are used to increase efficiency of planning, documenting and managing of data and telecommunication networks [1]. All active and passive components of a network (e.g. computer, telecommunication equipment, switch boards, ...) as well as all connections between them are managed. This enables a complete documentation of networks used in that field of application [2].

In the range of CANFM fieldbus systems are currently not considered as an essential network facility (although they are widely used in building automation systems). It becomes obvious that there is a growing need to consider this trend when preparing new systems for a comprehensive facility management.

2 The Integration Idea

Integration of FB Systems into CANFM systems means that:

- Certain management information of the field devices as device information (vendor, revision number, release,...) and status information (device state, status of a field value,...) has to be provided to the Facility Management for specific purposes.

- CANFM systems influence the behavior of the automation system. The global view of the CANFM system can support the control system in setting field device parameters and reference inputs for control loops (e.g. temperature regulation for a room in dependence of the presence of personal)

- The maintenance of fieldbus systems and it's components can be supported based on the management of the position of field devices, tags assigned to transmitters and actuators, device types and their device management information and the installation information for connections (medium, connectors, bridges, ...).

- Furthermore the CANFM system can support planning and routing of fieldbus systems using internal sets of data and algorithms. Optimization of the routing (network layout) becomes possible. The topology and the installation layouts can be documented in an automated way.

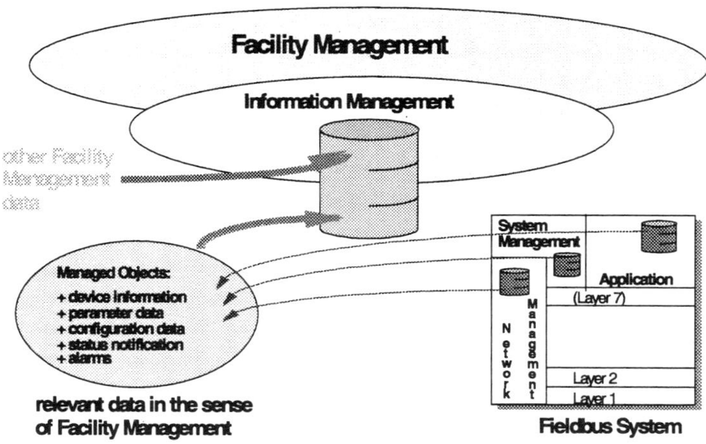

Fig. 1. Information Management for Facility Information

To reach the demonstrated advantages it is obvious to combine the very complex CANFM systems with the fieldbus systems that are currently in the process of standardization. Standardization guaranties a high degree of reusable algorithms and basic data that is required for efficient facility management. The use of dynamic information requires the implementation of a well defined interface to exchange data between both systems [3].

3 Definition of Managed Objects

The Facility Management does not substitute the task of the network management. However it means the migration to manage the network components. In this context the network components are managed objects. Managed objects can be monitored, controlled and are capable of reporting events. Since the Facility Management owns management functionality it monitors the managed objects and passes new interactions to the underlying management or control systems.

The managed objects are described based on the object-oriented modeling. They are characterized by [4]:

- Attributes: each managed object owns attributes that contain management information. The attributes can be protected by password.

- Operations: in order to get the information about the object operations are needed. Operations can be read and write of attributes.

- Notifications: a managed object is able to send notifications to a remote process if a value of an attribute crosses a defined limit or the state of an object has been changed.

- Behavior: the behavior of a managed object describes the connection to the real component. Changes of the behavior of the component cause changes of attributes as well as vice versa changes of attributes influence the behavior of the component.

4 How to access Fieldbus Data

4.1 Use of OPC to access Fieldbus Data

A primary result of the use of OLE (Object Linking and Embedding) for Process Control™ [7] is to eliminate the dependence between manufacturers of application programs and manufacturers of device drivers for peripheral automation devices. This enables a common access method to objects although there exist no standardized fieldbus management objects common to different fieldbus systems.[8]

Semantic aspects are considered by OPC only to a certain extend. Not defined are :

- start-up of the fieldbus system that is accessed by the OPC server,

- the meaning of OPCItem object and OPCGroup object in their concrete context,

- the addressing of OPCItems (definition of naming hierarchy)

The main purpose is to depict these semantic aspects with the functions available with OPC and not to define additional interfaces. So the openness of the solution can be guarantied.

Some of the semantic aspects listed above are specific for every single fieldbus system that is considered. It is the task of the server to encapsulate these specific functions from the interface to the OPC client. Especially the start-up and the maintenance of logical connections to remote field devices must be hidden. Also there is no common definition of application objects, protocol stack objects and resource objects. A mapping of specific objects to generic one's (items) must be done by the OPC server [9].

4.2 Use of HTTP and HTML to access Fieldbus Data

The capability for accessing device information through browser technology is in line with a general trend in automation. The fieldbus data are stored within a database which is located on a remote HTTP server. The running HTTP server software can be a conventional application like Microsoft's Personal Webserver shipped with the Internet Explorer package. A HTTP client like the CANFM system can access the fieldbus data via function calls embedded in a special URL to an ISAPI (internet server application programming interface) extension DLL. This DLL implements a database management process which generates HTML pages with the results requested by the HTTP client. After sending the request to the HTTP server the HTTP client can poll a special directory at the HTTP server for the relevant HTML page. A time control within the HTTP client process is necessary for detecting failures, e.g. the server refused the connection or there is a breakdown of the server connection. After receiving the HTML page from the server the HTTP client can start the decoding of the information transmitted within the page.

The database management process generates HTML pages for every object stored in the database during the systems start-up. If an attribute changed by a write access the corresponding HTML page of the database object will be updated.

5 Steps to integrate Fieldbus Systems into CANFM

In the sense of Facility Management management information basically is such information that is used by the CANFM system to manage the fieldbus equipment as a facility. Consequently the managed objects (MIB) are selected based on requirements from Facility Managers. The object instances used are distributed in the complete fieldbus system. To have an unique access to objects dynamically gained from different fieldbus systems the OPC server has to realize the task of decoupling fieldbus system and CANFM system. Beside universal methods the server interfaces contain methods with functionality specific for every single fieldbus system.

The approach for the integration is divided into the following sub-tasks a) to e):

a) *requirements study from the view point of the CANFM*

Requirements stated by the different types of users (building administrator, network administrator, fieldbus engineer, economist) have to be collected. This is the basis for the definition of data that has to be exchanged between the fieldbus system and the CANFM.

b) *definition of data to be exchanged between field devices and CANFM*

Currently effort is taken to build standardized object in so called Profiles to enable interoperability. A Profile basically is developed for a single fieldbus. Profiles are defined with the purpose of selecting parameters from the optional one's of a superset. There exist profiles for protocols, applications, industrial sectors, devices.

An European standardization activity (CENELEC TC 247 WG4) focussed on communication within building automation systems is just on the way to define a common profile for field level objects on top of field level protocols.

The profile definitions are the starting point for the definition of objects to be handled by the information management of the CANFM. They are completed by the resulting attributes as requested as a result of task a). These additional attributes are associated to the different types of objects (Device objects, Datapoint objects).

c) *definition of a common object description for different fieldbus systems*

The result of this combination are object definitions as illustrated below. Every Object is composed of static and dynamic data. Static data are set in an off-line procedure. Dynamic date is acquired using the OPC mechanism as described in the next sub-chapter e).

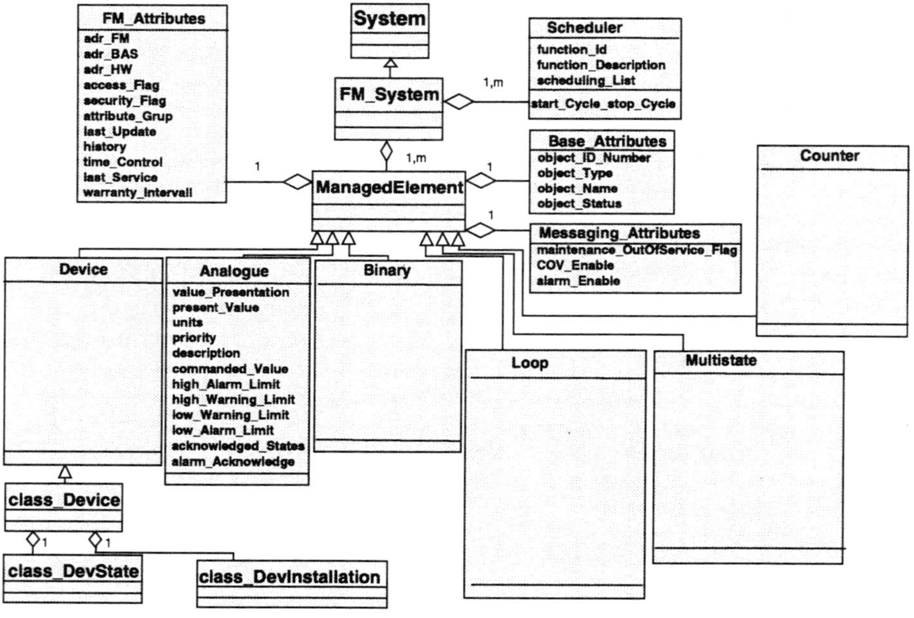

Fig. 2. Object Definition

d) *mapping of object definition to OPC specific conventions*

Due to the hierarchy of OPC classes a structuring of information acquired from the fieldbus system is possible based on facility management restrictions. The structuring of field information in various dimensions (e.g. entity related, location related and function related grouping) is possible in parallel.

e) *realization of the OPC client and integration into CANFM*

With it's three levels OPCServer, OPCGroup and OPCItem OPC provides a very plain hierarchy to present data. The hierarchy of the managed objects of the CANFM system will be more complex. This is an other reason why mapping fieldbus data, acquired via OPC, has to be done dynamically. The following figure describes the principle of acquisition of dynamic data.

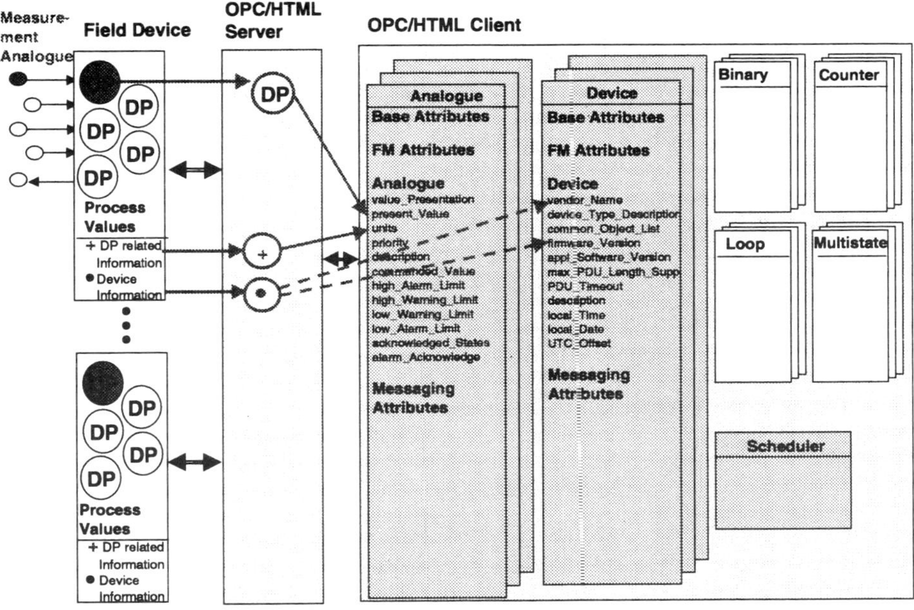

Fig. 3. Mapping of Field Level Data to Managed Objects

To realize the independent interface generic objects have to be build. Generic means that any time a FM-client updates an attribute of a managed object it gets the same type of object whether it is defined with different type for different fieldbus systems. The client does not care about the procedure the object is accessed by the OPC server from the fieldbus system.

f) Implementation of MIB tree (alternative to use of OPC)

The MIB structure is realized by specific subdirectories at the HTTP server. For every object type a subdirectory with one entry page and a number of object instance pages exist, e.g. for Analogue Objects (AO), Binary Objects (BO), Counter Objects (CO), Loop Objects (LO), Multistate Objects (MS). The browser access to the subdirectories is granted by a central entry page in the root directory of the HTTP server. A special case is a subdirectory containing so-called transaction pages. These pages are generated by requesting information from multiple object instances.

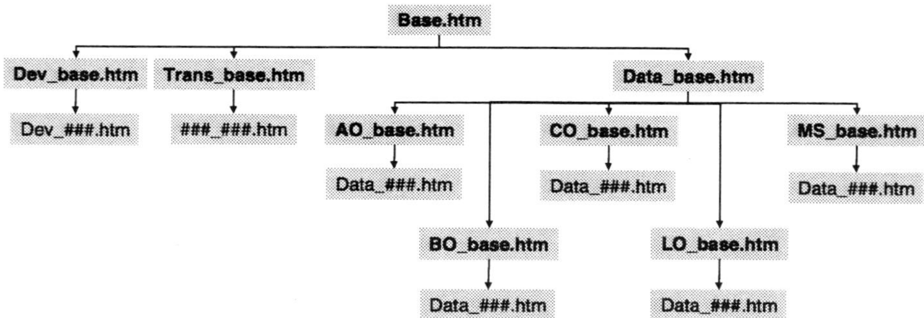

Fig. 4. Directory Structure of the HTTP server

f) Realization of HTTP client and server (alternative to use of OPC)

The HTTP client is a Dynamic Link Library that can be used by the CANFM system for requesting data from the database server. This DLL also provides a dialogue interface for viewing the database contents and changing the parameterization of the stored objects. Especially the storage class is an important criteria which can be changed using dialogues. The background is that some attributes can not be evaluated from the fieldbus, other attributes are polled from the field device on request and the process values are cyclic updated by the external OPC client process.

The HTTP server is a commercial application that will be extended with a specific DLL which provides the functionality needed for accessing the database.

6 Summary

The paper presented an approach of how to integrate building automation systems into a Computer-Aided Network Facility Management system. Because of the general characteristics especially of the OPC technology or use of HTTP and HTML the approach is not restricted to Facility Management systems. It is also possible to implement clients for different applications to access the same objects with another purpose. For these applications other OPC groups or HTML pages may be defined containing corresponding objects.

The paper presents actual results of a project sponsored by German Ministry of Economics with the reference 11461 B. Final results of the work will be available at the end of this year.

References

1. D.H. Traeger: Die Nutzung von Planungssystemen im Ingenieurbüro als Basis für Facility-Management Anwendungen. *Tagungsband des 2. GI-Workschop „Entwurf und Dokumentation im rechnerunterstützten Facility-Management"* (Hrgb: A. Iwainsky). 09.-11.Oktober 1996, auf der Wartburg

2. S. Döring, S. Nitz: Facility-Management im Bereich komplexer Netze. *Tagungsband zum GI-Workshop „Entwurf und Dokumentation im rechnerunterstützten Facility-Management"* (Hrgb: A. Iwainsky). 11.-13. Oktober 1995 auf der Wartburg

3. Th. Bangemann, J. Hähniche, F. Iwanitz, B. Lehnert: Einbeziehung von Feldbussen in Facility-Management-Systeme. *Tagungsband des 2.GI-Workshop „Entwurf und Dokumentation im rechnerunterstützten Facility-Management"* (Hrgb: A. Iwainsky). 09.-11.Oktober 1996, auf der Wartburg

4. H. Kerner: *Reducarnetze nach OSI*; Addison-Wesley Verlag (Deutschland) Bonn, 1993, 2. Auflage, ISBN 3-89319-632-3

5. IEC/TC65/SC65C: *Fieldbus: Part 7 System Management* (Clause 2), Committee Draft, IEC 61158-7-1/f2/Ed.1, 13.3.1998

6. IEC/TC65: Industrial-Process Measurement and Control WG6: *Function Blocks, Working Draft Part 5 Management Services. 30.11.1995*

7. *OLE for Process Control Standard*. Interface Specification. OPC Fondation

8. P. Neumann, F. Iwanitz: Integration of Fieldbus Systems into Distributed Object-Oriented Systems. *IEEE International Workshop on Factory Communication Systems, Barcelona (Spain), 01.-03.10.1997. Proceedings*, pp. 247-254.

9. P. Neumann, F. Iwanitz: Feldbussysteme und OLE for Process Control. *INet '97, Wiesbaden. Kongreßband*, S. 184-190.

Enabling e-Services through Resource Management API's on Multi-Vendor EIB Building Networks

Marc Goossens, Heinrich Reiter

EIB Association s.c., Tinklaan 5, B-1160 Brussels, Belgium

Abstract. Appropriate standardization of network resource management is a prerequisite for accessing this same functionality remotely, e.g. via Internet. In turn, this enables all kinds of e-services such as remote diagnostics and maintenance of building control networks – these can be offered on a commercial basis and constitute new markets for professional contractors and system integrators. On an open multi-vendor network (with multiple system implementations from several system providers and based on different microprocessor families) such as EIB, some degree of polymorphism has to be taken into account to cover variations in implementation. Defining the abstract methods for resource management requires some clever modeling. This contribution investigates the procedural modeling techniques which may be successfully applied. Once established, these methods can be mapped to Network Management API's for configuration tools, management stations and e-services.

1 E-services for the networked home and building

Combined with internet connectivity, the emergence of networked homes and buildings present service providers with the possibility to offer remote services to a building's occupants, owners or "facility managers". These electronic or e-services range from communication and entertainment through security, energy management, home care etc. [1] Many of those may take advantage of the networked functionality furnished by the control network applications. Security and energy management are but two obvious examples.

Going one step beyond these applications, the remote operation and maintenance of the building control network as such provides another level of potential e-services: tele-diagnostics and repair, remote application update, etc. It is important to distinguish between the applications of building automation, and the network infrastructure they use. In this contribution, we investigate the modeling and handling of the control network's infrastructure resources. The aim is to define the right level of abstraction which allows the definition of Network Resource Management API's. These may then be implemented not just on local configuration tools, but also on e-service platforms.

2 The need for implementation-independent modeling

On open network systems for building automation, run-time interworking across several applications in a multi-vendor environment has received broad attention. [2,3]

In contrast, the configuration-time *management* of resources on distributed peer-to-peer networks has mostly been implicitly assumed.

Open systems such as EIB [4] compound the challenge of defining all aspects of network management in an implementation independent fashion - particularly at those points where the nitty-gritty of specific microprocessor families enters into the equation. The point is that different system providers often base their implementations on different such families, or use different architectures depending on the functionality (e.g. complex controller vs. simple sensor). Experience moreover shows that implementations evolve as open systems are maturing (as is the case in EIB). This indicates the need for an appropriate form of polymorphism to allow a (limited) number of different implementation 'flavors' corresponding to distinct device models. Ultimately, it is this polymorphism which introduces the required factor of openness into the system. So rather than specifying such things as say, a standard processor architecture, the modeling approach outlined below encourages an open system market with a variety of customized implementations without sacrificing unified management.

3 Modeling network resource management

3.1 Network resources

For the purposes of this contribution, a network such as EIB may be seen as a pool of (data, memory, ...) resources, which can be controlled or *managed* remotely across the network. Some of these resources are shared, i.e. owned by the network as a whole. This is typically the case for unique addresses:

- domain identifiers (to separate neighboring installations on open media such as Powerline or Radio Frequency);
- physical addresses (including subnetwork addresses);
- shared variable identifiers (group addresses).

Other resources are local to the individual device or node:

- OS "system" parameters (e.g. internal address and variable identifiers);
- application parameters and their values.

In both cases however, the resource ultimately resides or "is stored" in one or more devices connected to the network.

3.2 Procedures for client-server management

Remember that we want to streamline the management and handling of resources, including tasks such as:

- scanning the network for available identifiers (e.g. a free address);
- assigning an identifier to a resource (e.g. a group address to a shared variable);
- loading (writing) an identifier into the device(s) requiring it;
- storage allocation;
- loading system or application parameters into a device;
- reading any of the above back from the device where they were stored.

On the network, this is realized by procedures, consisting of a sequence of frames exchanged between two partners: the Management Client and the Management Server[1]. The Management Client is typically a powerful device with 'controller' functionality, often PC-based. Except for network-oriented management, the Server is always one particular 'target device'. In the former case, it is in fact the network as a whole which acts as partner or server. Ultimately of course, the response to a client request is always generated by the individual devices connected to the network, whether one or many. In addition to its run-time behavior (based on group communication), each EIB device moreover supports a rich management server profile - precisely for this purpose.

In this contribution, we will further consider only the management by a *management client* as a (single) dedicated management instance. From the above, it is clear that this could be a subset of a multi-client or even a fully distributed approach[2]. Whereas the (unique) management client initiates and steps through any particular management procedure, the *management server* instances in each device respond to the client's requests on the network and locally, within "their" device, carry out any corresponding management operations.

On the device implementation side (= *management server*) itself, object-based mechanisms certainly hold greater promise than memory-mapped DMA, which is still in widespread use [6, 7]. Yet the generic model to be found has to include both. This is but one illustration for the evolutionary nature referred to above. We may expect this process to accelerate with the introduction of UART-like building blocks in EIB for link-layer medium access; this is expected to lead to a proliferation of microprocessor families being used for EIB implementations.

A further requirement to the models is that they also accommodate the management description of a number of specialized "system" devices, e.g. routers. If this flexibility can be attained, one may even expect them to cover the management of application-specific resources.

---

[1] Note that in general, one single device may well implement both client as well as server functionality. For and during the execution of one particular transaction procedure however, one device takes on one single role.

[2] Indeed, the approach to network resource management as presented here, neither assumes nor excludes the presence of either:
- a Building Management Information Base, which (possibly among other things) mirrors the network configuration and resource allocation in a database, whether on-line or off-line [5];
- a central "proxy server" which interprets client requests and forwards them to the devices; in this case, the relationship takes on the following shape: [proxy client] ↔ [proxy server / management client] ↔ [management server = typical device]

4 Network entities, procedures and API methods: the overall picture

The server entity reacts rather directly to the requests made by the client entity, with some combination of local actions and a network response. The client actively decides subsequent steps according to the (lack of) response received. So it is this *networking procedure* which is takes on a central role, and which determines a corresponding client-side plus server-side "profile". In both, it implies a behavior, possibly comprising timing aspects, (distributed) state machines and exception handling. Examples given will illustrate how the algorithms for client-side handling are obtained directly from the procedures (assuming a particular server-side implementation). As Figure 1 illustrates, the client-side view of a given management procedure (seen as a whole) is a prime candidate for mapping to a method of a Service API.

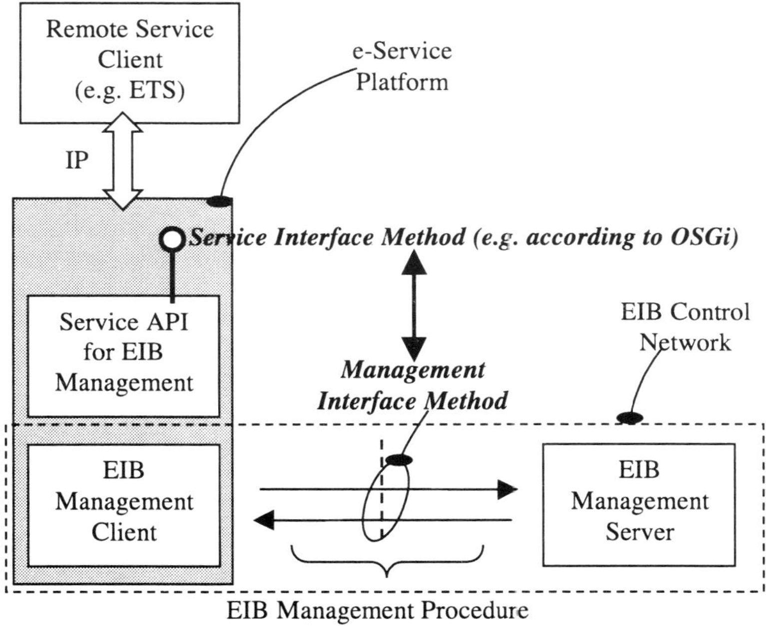

Fig. 1. Mapping a management procedure to a Management API method.

Though the figure sketches the situation for a service API implemented on an e-service platform, the same holds for building management stations or dedicated configuration tools[3]. Such tools, like the standard ETS (EIB Tool Software) utility [5], may also be adapted as remote clients – again as the figure suggests.

---

[3] Corresponding to this API, EIB defines a simple Active Device Management (ADM) scripting syntax to control management procedures on a management client. All standard PC-based EIB tools such as ETS can interpret this control syntax and perform management client operations accordingly.

192

When dealing with remote services, there is clearly an interest in an open and independent standardization. For Java enabled platforms, such an effort is currently underway in the OSGi (Open Service Gateway initiative) framework. A first public Request for Comment is expected by summer 1999; at the time this contribution is submitted, insufficient detailed and concrete information about OSGi is available to allow any in-depth treatment of this aspect.

5 Management Procedures on EIB

Returning to the Network Management Procedures themselves, we will analyze a couple of concrete examples in more detail using UML-style sequence diagrams (see for example [8]). Each diagram models the remote asynchronous interaction between the application layers of the management client and the management server threads, via the EIB Application Layer Protocol Data Units (APDU) [4]. The vertical lines represent the threads, the horizontal arrows show the sequence and direction of the PDU's exchanged between them.

We will use the method DM_RunStateMachine_Read illustrate the concepts and to sketch implementation polymorphism. This method is used to read the state of a run state machine of a management server; we give its abstract syntax without further explanation (the prefix DM stands for Device Management): *DM_RunStateMachineRead (dataBlockStartAddress, flags, stateMachineType, stateMachineNo, state)*. Instead, we focus on two standardized implementations of the actual procedure.

5.1 Procedure: DMP_RunStateMachineRead_RCo_Mem

This procedure uses connection oriented remote communication to implement DM_RunStateMachine_Read. The control and state of the run state machine are located in the memory of the management server and can be accessed via direct memory access (DMA). It is assumed that a transport connection to the target device has already been established, using the DM_Connect procedure.

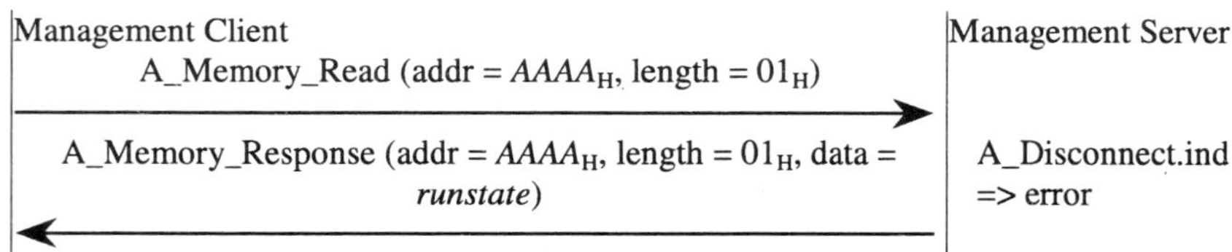

Fig. 2. Reading a run state machine according to the DMA model.

5.2 Procedure: DMP_RunStateMachineRead_R_IO

This method uses either the connection oriented or connectionless remote communication. This time, the control and state of the run state machine are located in Interface Objects (IO) [7]of the management server and can be accessed via property services. The management client searches the correpsonding interface object in the management server.

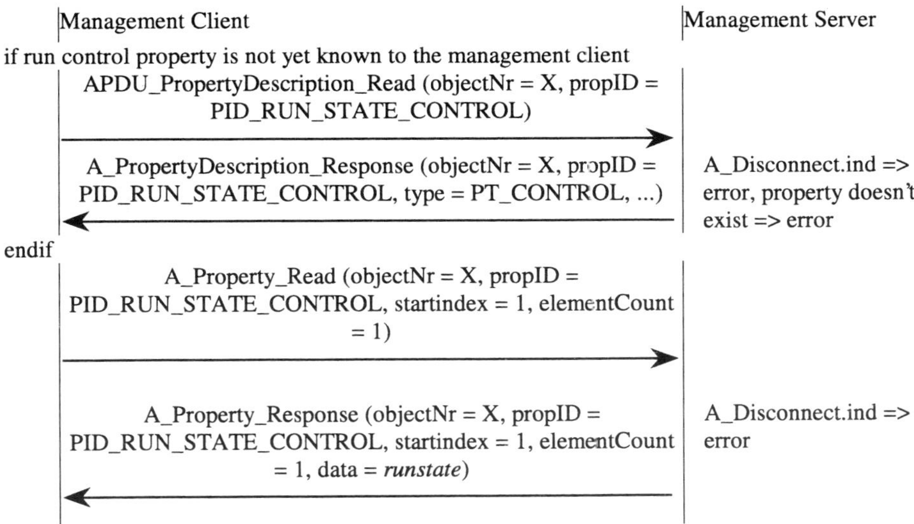

Fig. 3. Object-based access to run state information.

Managing a device's local resources always consists of some form data to be to be read from or written to memory. As the example shows, different implementations may exhibit different addressing modes for the same resource (e.g. an address table). Moreover, the equivalent resource itself may well become more concise, richer or more general as compared to earlier models (e.g. by adding additional structure elements, defining an enhanced load state machine etc.).

Here, we see a detailed illustration of the management implementation polymorphism mentioned above. Comparing the examples given with Figure 1, it is apparent that the task of coping with it may be shifted to the management client (with normally one instance per installation). This keeps the devices (of which an installation contains dozens or even thousands) lean. Moreover, based on a powerful platform (often a PC or equivalent, such as a e-service platform), the typical management client can easily deal with the implementation overhead this entails. The only derived requirement for the device (i.e. server), is a System Implementation Identifier, which allows the client to dynamically select the proper procedure implementation.

6 Outlook

The scope of this contribution does not permit a complete survey of EIB network management, nor an accurate description of the API mapping. Some of this work is currently ongoing as a community standardization process in the EIB Association. Still, our summary above does indicate the logic and validity of the overall approach, specifying precise management procedures and allowing a polymorph mapping to service API's. The realization and commercial availability of corresponding interfaces – also e-service enabled – is scheduled for 1999 and 2000.

Acknowledgement

This contribution is based in part on the modeling work on network management done by the EIB Association System Group in 1997-1999, and the resulting detailed procedural analysis by Klaus Adler from Siemens in Regensburg (Germany).

References

1. Idermark T., Lilliestråle M., Vasell J.: "Ericsson's e-box system – An electronic services enabler", Ericsson Review No. 1, 1999.

2. Dietrich, Loy, Schweinzer Ed.: "LON-Technologie", Hüthig Buch Verlag, Heidelberg, Germany, 1997; Ch. 10, 11.

3. Dietrich, Kastner, Sauter Ed.: "EIB Gebäudebussystem", Hüthig Buch Verlag, Heidelberg, Germany, 1999 (in preparation).

4. EIB Association "The EIB Handbook for Developers" Release 3.0, 1999.

5. Goossens M., "Component-based Project Engineering" in "Feldbustechnik in Forschung, Entwicklung und Anwendung", Dietrich, Schweinzer (Ed.), Springer, 1997.

6. ANSI/ASHRAE 135-1995 "BACnet, A Data Communication Protocol for Buidling Automation and Control Networks", ASHRAE, Inc., Atlanta, USA, 1995. 1. Section 12, "Modeling Control Devices as a Collection of Objects" 2. Section 16, "Remote Device Management Services"

7. Goossens M., "Object-based Distributed Application Design" in "Feldbustechnik in Forschung, Entwicklung und Anwendung", Dietrich, Schweinzer (Ed.), Springer, 1997

8. Fowler, M., Scott, K., "UML Distilled, Applying the Standard Modeling Language", Addison-Wesley, 1998.

Modulares Agent-Design für Feldbusmanagement

Ronald Haidvogl, Martin Knizak, Thilo Sauter

Institut für Computertechnik, Technische Universität Wien,
Gußhausstraße 27-29, A-1040 Wien
T. ++43-1-58801-38430 F. ++43-1-58801-38499
E-mail: {knizak, sauter}@ict.tuwien.ac.at

Abstract. Die Anbindung von Feldbussen an globale Netzwerke (und damit auch an das Internet) eröffnet neue Möglichkeiten für einen Fernzugriff auf Daten in Automationssysteme. Als Basis hierfür sind Methoden und Protokolle für das Management von IP-basierten Netzwerken gut geeignet. Der Zugang zum Feldbus erfolgt über ein entsprechendes Gateway. Wir stellen ein modulares Konzept vor, bei dem das Gateway in einen Internet- und einen feldbusspezifischen Teil zerlegt wird. Vorteilhaft ist dabei die weitgehende Plattformunabhängigkeit und daß die Module in einem LAN verteilt werden können. Zudem kann ein Gateway auch mehrere Feldbusse bedienen.

Abstract. An interconnection between fieldbus systems an global networks (including also the Internet) opens new possibilities for remote access to data in automation systems. A good basis for such a combination are the methods and protocols known from the management of IP based networks. The actual access is provided by an appropriate gateway. We suggest a modular approach to divide the gateway into an Internet- and a fieldbus-specific part. This concept is appealing in that it is largely platform independent and allows for the distribution of the modules in a network. Furthermore, one gateway can also serve more than just one single fieldbus system.

1 Einleitung

Das Internet ist derzeit eines der sich am raschesten entwickelnden Gebiete. Auch in der Automatisierung gewinnt Vernetzung über den Feldbus hinaus an Bedeutung. Es ist daher nicht verwunderlich, daß es zahlreiche Projekte gibt, die die Verbindung von Feldbussen und dem Internet zum Ziel haben. Dabei muß es nicht notwendigerweise um das Internet selbst gehen, auch die Anbindung von Feldbussen an nicht globale, IP-basierte Intranets ist lohnend und eröffnet neue Möglichkeiten des Fernzugriffs auf Automatisierungsnetzwerke.

Der hier gewählte Ansatz setzt auf ein in der Netzwerkadministration bewährtes Protokoll auf. Obwohl es etliche Konzepte gibt, hat sich in diesem Bereich SNMP (Simple Network Management Protocol) als Standard etabliert [1]. Wir verwenden es, um ein Abbild eines Feldbusses für das Internet zu erstellen. Die Umsetzung zwischen dem Feldbusprotokoll und SNMP geschieht in einem entsprechenden Gateway [2]. In der Sprache des Netzwerkmanagements wird jene Applikation, die aus dem Internet auf den Feldbus zugreift, als *Manager* bezeichnet. Das Gateway nennt man *Agent* oder, weil es auch als eine Art zentraler Zwischenspeicher für die Objekte nicht SNMP-fähiger Feldbusknoten dient, *Proxy Agent* (Abb. 1). Der Vorteil von SNMP liegt unter anderem in der klar vorgegebenen Konvention für die Benennung der zur Verfügung gestellten Objekte, die in der MIB (Management Information Base) zusammengefaßt werden. Bemerkenswert ist dabei, daß die MIB in einer hierarchischen Baumstruktur organisiert ist und nicht als Datenbank im herkömmlichen Sinne verstanden werden darf. Vielmehr stellt sie die Organisation der Daten nach außen sicher, nicht aber wie diese tatsächlich intern gespeichert werden.

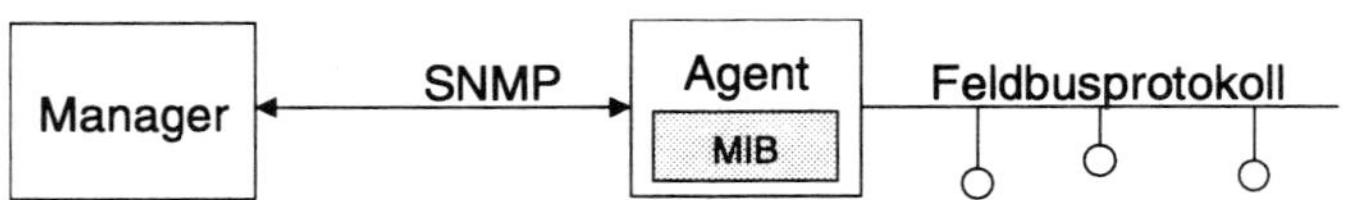

Abb. 1. Agent als Feldbus-SNMP-Gateway

2 Konzept

Ein Konzept für ein Gateway zwischen Internet und Feldbus sollte möglichst universell verwendbar und nicht auf einen speziellen Feldbus zugeschnitten sein. Jener Teil des Gateways, der die SNMP-Verbindung mit dem Manager bearbeitet, ist von vornherein feldbusunabhängig. Der andere Teil allerdings, der die tatsächliche Anbindung zum Feldbus bereitstellt, muß natürlich für jedes Bussystem eigens implementiert werden. Der Schluß liegt also nahe, eine modulare, zweiteilige Struktur zu verwenden. Den dem Inter- bzw. Intranet zugeordneten, unabhängigen Teil bezeichnen wir als *Master-Agent* (MA), den feldbusspezifischen als *Sub-Agent* (SA). Ein Master-Agent kann mehrere Sub-Agents verwalten, und die Module können auch in einem Netzwerk verteilt sein (Abb. 2).

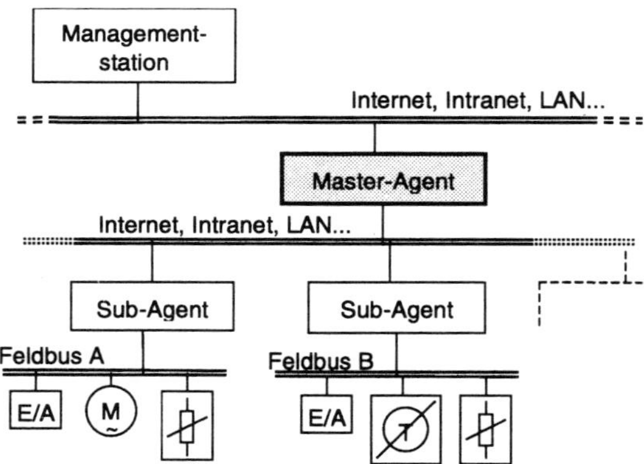

Abb. 2. modulare Struktur des Agents

Die Vorteile dieser Struktur sind offensichtlich: Wenn ein neuer Sub-Agent eingebunden werden soll, müssen keine Änderungen im Master-Agent durchgeführt werden, weil dieser mit den Sub-Agents über eine einheitliche Schnittstelle kommuniziert. Lediglich der neue Sub-Agent muß programmiert werden. Würde man eine monolithische Struktur ohne Trennung zwischen den beiden Teilen verwenden, müßte der SA-

Teil direkt in das MA-Programm eingefügt werden. Bei Verwendung einer geeigneten Schnittstelle wäre dies an sich noch kein Nachteil. Probleme könnten sich jedoch durch bestimmte System-Anforderungen von Werkzeugen ergeben, die zur Entwicklung des neuen Sub-Agents nötig sind. Schwerer noch wiegt die Tatsache, daß auch die verfügbaren Anbindungen zwischen Feldbus und Rechner wechselseitig inkompatibel sind oder zumindest nicht mit allen Betriebssystemen funktionieren. Die Wunschvorstellung *eines* Rechners, der als Gateway zwischen mehreren Feldbussen und dem Internet dient, ist daher nicht immer realistisch. Durch die Trennung der Plattformen bleibt jedoch zumindest ein zentraler Knoten als Verbindung nach außen erhalten. Weitere Vorzüge der verteilten Sub-Agents bestehen darin, daß man auf diese Weise auch Information über Feldbusse sammeln kann, die räumlich weit voneinander getrennt sind, und daß der Ausfall eines Rechners nicht mehrere Feldbusse beeinflußt.

Um ein Höchstmaß an Portabilität zu erhalten, sollte der Master-Agent in einer möglichst plattformunabhängigen Sprache implementiert werden (etwa ANSI C++ oder JAVA). Für die hier behandelte Implementierung wurden MS-Windows NT 4.0 und MS Visual C++ 6.0 verwendet.

Die Verbindung zwischen Manager, Master- und Sub-Agent ist nicht zur Echtzeitsteuerung von Prozessen auf den Feldbussen konzipiert. Dies wäre auch nicht möglich, weil bei einer Übertragung über das Internet die Datenrate ständigen Schwankungen unterworfen ist und mit Wartezeiten gerechnet werden muß. Auch in Intranets, die das Internetprotokoll (IP) verwenden, gibt es dieses Problem.

Die Aufgabenteilung zwischen Master- und Sub-Agent sieht vor, daß der Sub-Agent lediglich die Verbindung zum Feldbus herstellt und dem Master-Agent auf Anfrage Daten zur Verfügung stellt. Der Master-Agent hingegen erfüllt alle Funktionen, die für einen SNMP-Agent notwendig sind: Er verwaltet die MIB, beantwortet Anfragen des Managers und liest bzw. setzt Werte von MIB-Variablen. Außerdem überwacht er Alarmbedingungen und schickt gegebenenfalls eine entsprechende Meldung (einen SNMP-Trap) an den Manager.

Die Informationen, welche Daten der Master-Agent dem Manager zur Verfügung stellen kann, erhält dieser durch eine Konfigurationsdatei im ASN.1-Format (Abstract Syntax Notation) und eine nachfolgende „Abtastung" der MIB.

3 Kommunikation zwischen Manager, Master- und Sub-Agent

Das für die Kommunikation zwischen Manager und Master-Agent verwendete SNMP bedarf keiner gesonderten Erläuterung. Für die hier beschriebene Implementierung wurde die am weitesten verbreitete Version eins als ausreichend erachtet. Diese Version hat allerdings den Nachteil, daß sie kaum einen wirksamen Zugriffsschutz erlaubt. Derartige Funktionen wurden erst in die zweite und besonders in die dritte Version aufgenommen. Für Anwendungen in einem abgeschlossenen Intranet ist SNMPv1 aus sicherheitstechnischer Sicht vertretbar, für Anwendungen hingegen, die das in jeder Hinsicht offene Internet einbeziehen, muß eine aktuellere Version, vorzugsweise SNMPv3, verwendet werden [3].

198

Die MIB ist in vier Teilbäume gegliedert. Die ersten drei Zweige enthalten Typdeklarationen, die tatsächlichen Daten der verfügbaren Objekte und statistische Daten über das Gateway und die MIB selbst [4]. Zusätzlich wurde ein Zweig vorgesehen, der den Umgang mit Traps regelt. Hier kann beispielsweise die IP-Zieladresse für Trap-Nachrichten angegeben werden. Ob diese Adresse auch verwendet werden kann, hängt leider von der Implementierung des SNMP-Stacks ab. Unter Windows-NT beispielsweise sind die Trapziele in der Registry vorgegeben und müßten somit hier geändert werden (was über einen Zugriff auf die MIB nicht möglich ist). Aus diesem Grund wird dieser Teilzweig gegenwärtig nicht ausgewertet. Er wurde aber vorgesehen, um das Konzept für Betriebssysteme offen zu halten, die eine komfortablere Festlegung des Trapziels erlauben.

Die Kommunikation zwischen Master- und Sub-Agent geschieht über Sockets, einheitlich definierte Schnittstellen, die auf allen Systemen gleichermaßen vorhanden sind. Prinzipiell könnte auch eines der bestehenden Protokolle zu diesem Zweck verwendet werden (wie SMUX, DPI oder AgentX [5]), diese Protokolle sind jedoch komplizierter zu implementieren und verlagern mehr Kommunikationsaufwand in die Sub-Agents, als dem hier vorgeschlagenen Konzept entspricht. Daher wurde ein eigenes, einfacheres Protokoll entworfen, das den Anforderungen besser entspricht.

Dazu wurden Primitive definiert, die den Auf- und Abbau einer Verbindung zwischen Master- und Sub-Agent sowie den Datenaustausch ermöglichen. Im Gegensatz zu SNMP gibt es auf dieser Ebene nur mehr zwei Arten des Datenzugriffs, nämlich *Get* und *Set*. Der Get-Next-Befehl wird vom Master-Agent intern in ein entsprechendes *Get* umgewandelt, und Traps sind vom Sub- zum Master-Agent nicht vorgesehen, weil der Master-Agent die Alarmbedingungen verwaltet.

Im folgenden werden die einzelnen SNMP-PDUs (Protocol Data Units) und ihre Umsetzung in der Kommunikation zwischen Manager, Master- und Sub-Agent behandelt. Mit Ausnahme des Trap-Befehls besteht jeder SNMP-Befehl aus einem Paar von Request/Response-PDUs, wie Abb. 3 anhand des Get-Befehls zeigt.

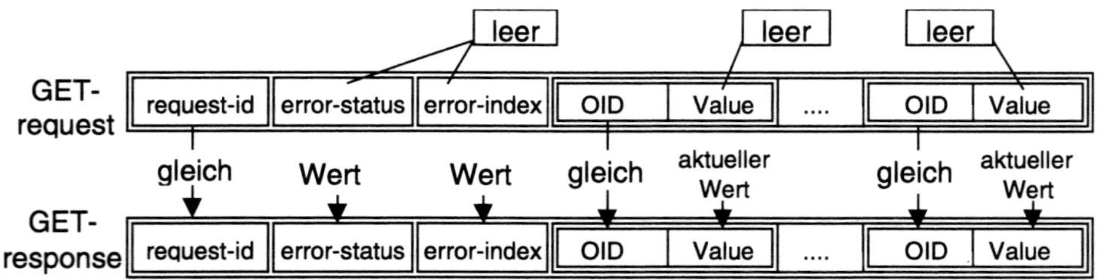

Abb. 3. Struktur einer SNMP-PDU

3.1 Get-Request

Beim Lesezugriff auf Feldbus-Variablen ist die Maxime, die Antwortzeit für den Manager kurz zu halten. Zur Erfüllung der ersten Forderung werden die Werte der Feldbus-Variablen vom Master-Agent autonom aktualisiert und mit einem Zeitstempel versehen in der MIB zwischengespeichert [4]. Dazu ist für jedes Objekt individuell festgelegt, in welchen Zeitabständen es aktualisiert werden soll. Der Master-Agent geht die MIB zyklisch durch und entscheidet aufgrund des Alters, welches Objekt erneuert werden muß.

Die Behandlung von Anfragen des Managers funktioniert ähnlich. Zu jeder Variablen existiert in der MIB eine relative Zeitspanne, die das maximale Alter des gewünschten Wertes definiert. Als Referenz gilt dabei die lokale Zeit des Master-Agents. Bei einer Anfrage des Managers rechnet sich der Master-Agent zunächst das Alter des derzeit gespeicherten Wertes aus. Ist das „nicht älter als"-Kriterium erfüllt, erhält der Manager den gespeicherten Wert. Andernfalls fragt der Master-Agent beim Sub-Agent nach dem aktuellen Wert, der gegebenenfalls erst vom Feldbus geholt werden muß.

3.2 Get-Next-Request

Mit dem Get-Next-Befehl kann die MIB Schritt für Schritt ausgelesen werden, ohne daß der Manager ihre Struktur a priori kennt. Da dieser Befehl nur für das Scannen der MIB bedeutsam ist, hat er keine Entsprechung in der Kommunikation zwischen Master- und Sub-Agent.

3.3 Set-Request

Beim Schreibzugriff auf Feldbus-Objekte muß auf die Besonderheit von SNMP Rücksicht genommen werden, daß mehrere Set-Befehle in eine PDU verpackt sein können, der Manager aber für alle diese Aktionen nur eine summarische Bestätigung erhält. Schlägt nur ein *Set* fehl, gibt es eine negative Rückmeldung ohne die Möglichkeit, die Fehlerquelle zu eruieren. Somit kann der Manager auch nicht unmittelbar feststellen, welche Variablen verändert wurden und welche nicht. Um nicht die Verwendung solcher Mehrfach-Zugriffe verbieten zu müssen und damit eine Inkompatibilität zu SNMP zu provozieren, wird in einem solchen Fall ein Zwei-Phasen-Commitment Protokoll verwendet (Abb. 4).

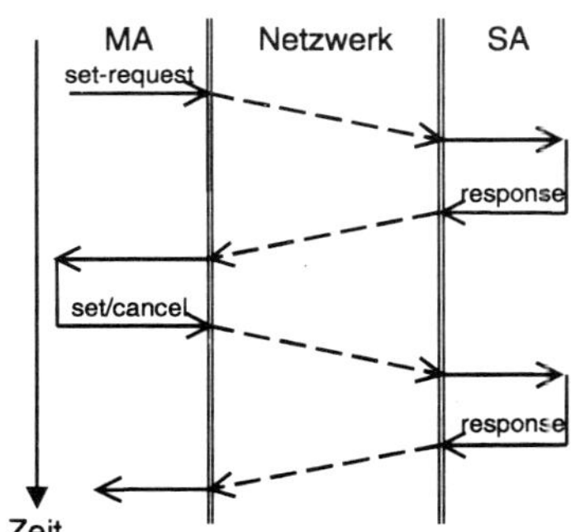

Abb. 4. stufenweise Ausführung eines mehrfachen Set-Befehls

Um bei einem Set-Request vom Manager sicherzustellen, daß das Setzen des Wertes auch tatsächlich durchgeführt werden kann, übermittelt der Master-Agent dem Sub-Agent zunächst alle relevanten Daten mit der gleichzeitigen Anfrage, ob dieser die Werte auch wirklich setzen kann oder ob ein dem Sub-Agent bekanntes Problem ein *Set* verhindern wird. Erst wenn hier eine positive Antwort zurückkommt, wird der Sub-

Agent angewiesen, die Werte tatsächlich zu setzen. Andernfalls wird die Aktion abgebrochen. Für „normale" Set-Anfragen des Masters ist der Protokollablauf natürlich einfacher.

Bei diesem mehrstufigen Protokoll wird davon ausgegangen, daß ein Sub-Agent, der die Anfrage, ob er den Wert setzen kann, positiv beantwortet hat, die Aktion dann auch wirklich durchführen kann. Eine absolute Sicherheit dafür gibt es nicht, die Annahme scheint für unkritische Anwendungen jedoch gerechtfertigt. Wenn Wert auf die exakte Ausführung eines Set-Befehls gelegt wird, darf ein Mehrfach-Zugriff von seiten des Managers nicht angewendet werden.

3.4 Trap

Ein Trap wird nur geschickt, wenn ein unvorhergesehenes Ereignis eingetreten ist, z.B. wenn ein Sub-Agent nicht mehr antwortet, oder der Wert bzw. die Wertänderung einer Variablen einen Grenzwert übersteigt. Die Grenzwerte können für jede Variable in der MIB festgelegt werden und werden vom Master-Agent überwacht, der dann bei Bedarf den Trap auslöst.

4 Verwendete Konzepte auf Programmebene

Beim Master-Agent wurde der SNMP-Teil durch Verwendung des in Windows-NT enthaltenen „extensible Agent" und eine entsprechende Anpassung an die Erfordernisse implementiert. Den Grundstock an Information (welche Sub-Agents gibt es, welche Objekte stellen diese zur Verfügung,...) bekommt der Master-Agent aus den Dateien, die mit Hilfe eines Konfigurationsprogrammes, dem MA-Konfigurator, erstellt wurden. Hier wird sozusagen von Hand zunächst die Verbindung zu jedem Sub-Agent hergestellt, dieser übermittelt seinen Datenbestand, und der Benutzer kann nun entscheiden, welche Variablen von Interesse sind und in die MIB aufgenommen werden sollen (Abb. 5). Beim Konfigurator wurde keine Rücksicht auf plattformunabhängige Programmierung genommen, weil ohnehin davon ausgegangen werden kann, daß für jedes Betriebssystem, abhängig von den zur Verfügung stehenden Entwicklungstools, unterschiedliche Programmier- und Visualisierungsmethoden zur Anwendung kommen.

Trotz des Ziels einer plattformunabhängigen Programmierung kommt man auch beim Master-Agent-Programm nicht ganz ohne betriebssystem-spezifische Funktionen (in diesem Fall die Besonderheiten von Threads unter Windows NT) aus. Bei den Ausgaben beschränkt sich der Master-Agent auf das Standardausgabegerät (stdout), direkte Eingaben vom Benutzer erhält das Programm nicht.

Die Verwaltung der MIB erfolgt durch den Master-Agent. Jeder Variable wird bei der Konfiguration vom Anwender ein „Sample Time"-Wert zugewiesen. Mit Hilfe dieser Abtastzeit wird vom Master-Agent in regelmäßigen Abständen der Wert vom entsprechenden Sub-Agent „geholt". Dazu muß die Tabelle zyklisch durchlaufen und bei jeder Variable ein Vergleich der aktuellen Zeit mit der ebenfalls in der MIB enthaltenen Auslesezeit des letzten Wertes durchgeführt werden. Ergibt diese Auswertung eine Notwendigkeit einer Wertaktualisierung, so wird vom Master-Agent ein Thread erzeugt, der diese Aufgabe erledigen soll. Zusätzlich wird ein neuer Eintrag in die Liste der aktuell laufenden Sub-Agent-Anfragen gemacht (Abb. 6). Nach erfolgter Abfrage wird der entsprechende Eintrag aus dieser Liste wieder gelöscht.

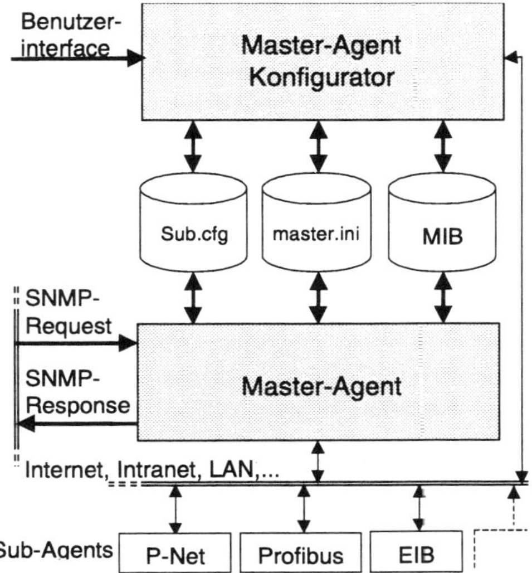

Abb. 5. Konfiguration des Master-Agents

Außerdem wird bei jeder Variable überprüft, ob eine Trapbedingung erfüllt ist. Ist dies der Fall, so wird ein Trap an den Manager, und eventuell auch an das SMS-Ziel (kann ebenfalls bei der Konfiguration angegeben werden) geschickt.

Bei einer Anfrage vom Manager wird zunächst in der internen Anfragenliste kontrolliert, ob bereits eine entsprechende Operation läuft. Ist dies nicht der Fall, so wird unter Zuhilfenahme des bei der Konfiguration einzugebenden Wertes „nicht älter als" überprüft, ob der aktuelle Wert „aktuell genug" ist. Um Konflikte mit der Vorgabe für die zyklische Aktualisierung zu vermeiden, wird dieser Wert in Prozenten des Aktualisierungsintervalls angegeben. Ist der aktuell in der MIB enthaltene Wert beispielsweise älter als die Auslesezeit des Wertes plus 15% des Aktualisierungsintervalls, so muß der Master-Agent den Wert vom Sub-Agent neu erfragen. Zusätzlich gibt es noch die Möglichkeit, in jedem Fall den aktuellen Wert vom Sub-Agent auszulesen, indem auf ein in der MIB-Tabelle enthaltenes, spezielles Feld, das „Update-value"-Feld, zugegriffen wird.

Die bis jetzt dargestellten Formulierungen beziehen sich auf eine Get-Abfrage. Beim *Set* wird natürlich unabhängig vom Alter des in der MIB gespeicherten Wertes gearbeitet. Hier muß lediglich auf etwaige laufende Abfragen geachtet werden. Eine interne Abfrage wird abgebrochen, eine externe zuvor beendet.

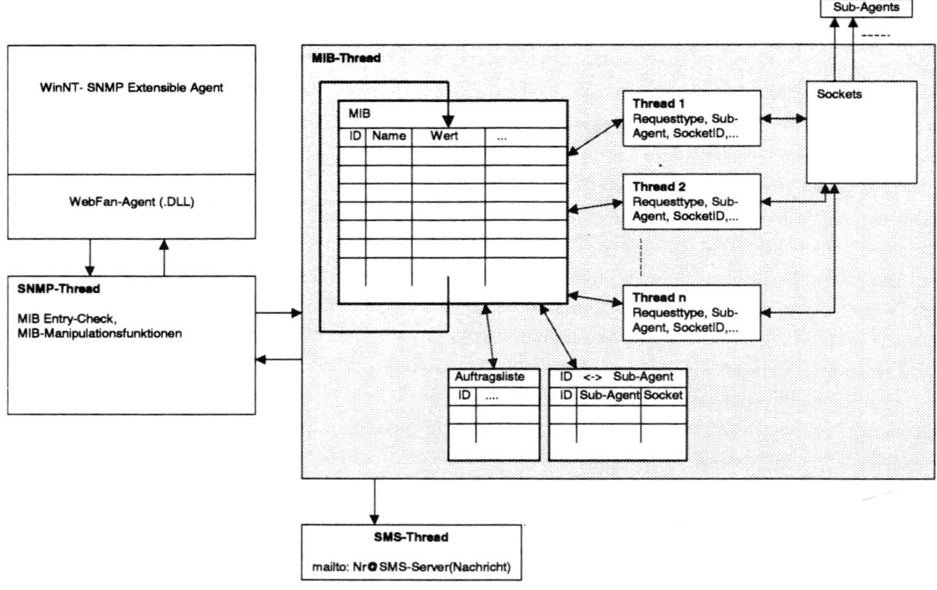

Abb. 6. Programmstruktur des Master-Agents

5 Zusammenfassung

Der modulare Aufbau des Agents erleichtert die Adaption des Agents für weitere Feldbussysteme, da nur der Sub-Agent-Teil neu geschrieben werden muß. Auch die Wahl von SNMP als Zugang vom Internet her erweist sich als vorteilhaft, da eine Vielzahl von Programmen für die Managementstation schon existiert. Diese reichen von einfachen Konsolen-Programmen bis hin zu HTML-basierten Tools. Derzeit sind der Master-Agent für Windows NT und Sub-Agents für P-Net bzw. Profibus implementiert. Weitere Sub-Agents für EIB, LON und Interbus sind geplant.

Literatur

1. Miller, M.: Managing Internetworks with SNMP, M&T Books, 1993.

2. Knizak, M., Kunes, M., Manninger, M., Sauter, T.: Applying internet management standards to fieldbus systems. Proceedings of the WFCS'97, Barcelona 1997, S.309-315.

3. Stallings, W.: SNMPv1-3 and RMON 1+2, Addison Wesley, 1999.

4. Sauter, T., Knizak, K, Manninger, M.: Einbettung von Feldbussen in das Internet-Management, in: Dietrich, D., Schweinzer, H. (Hrsg.), Feldbustechnik in Forschung, Entwicklung und Anwendung, Springer, Wien NewYork 1997, S.223-230.

5. RFC 1227, RFC 1228, RFC 1592, RFC 2257, ftp://ftp.univie.ac.at/netinfo/rfc.

Linux-Gateway zur Fernwartung von Profibus-DP-Geräten

Wolfgang Kastner

Technische Universität Wien
Institut für Rechnergestützte Automation
Treitlstraße 1, A-1040 Wien

Abstract. Feldbusse und geeignete Soft- und Hardware zur Vernetzung solcher Systeme mit übergeordneten Netzwerken sind die aktuellen Schlagworte der modernen Automatisierungstechnik. Dieser Artikel soll ein erster Schritt dahingehend sein zu beweisen, daß das frei verfügbare Betriebssystem Linux als Plattform zur Erstellung derartiger Applikationen bestens geeignet ist. Ebenso können Rechner unter Linux zur Steuerung technischer Prozesse verwendet werden, erfüllen sie doch ihre Aufgabe mindestens genauso gut wie jene immer häufiger eingesetzten Rechner, deren Betriebssysteme a priori für diesen Bereich kaum konzipiert wurden. Um diese These zu untermauern, wurde ein Feldbus-Gerätetreiber für Linux entwickelt, der in weiterer Folge zur Realisierung einer Internet-basierenden Fernwartung herangezogen wurde. Dieser Artikel stellt Design und Implementierung des Treibers und der darauf aufbauenden Fernwartung vor.

Abstract. Since its humble beginnings Linux has grown from strength to strength. On one hand it has evolved into an operating system for business, personal productivity and education, on the other hand it is an ideal platform for software developers. Since in our opinion Linux's breakthrough in factory automation is imminent we decided to implement the worldwide first Linux device driver dedicated to fieldbus systems: a Profibus-DP driver. This paper presents the device driver's implementation and its application programming interface and shows one possible application rest upon it: a network interconnection from Profibus-DP to the Internet using a Java-based gateway.

1 Einleitung und Motivation

Die klassische Steuerungstechnik im Maschinen- und Anlagenbau befindet sich seit mehreren Jahren im Umbruch. Etablierte Systeme sind längst an ihrer Kosten- und Leistungsgrenze angelangt. Anwender und Entwickler in weiteren Bereichen der Automatisierungstechnik verlangen zudem nach größtmöglicher Flexibilität und Standardwerkzeugen. Auch will jedermann an dem rasanten Innovationszyklus der Datenverarbeitungswelt, die heute mehr denn je eine PC-Welt ist, teilhaben. Ein Rechner unter dem Betriebssystem Windows 95/98/NT, sowie Zubehör wie Feldbuskarte, I/O-Koppler und der Option Echtzeitdatenverarbeitung unter diesen Gegebenheiten betreiben zu können scheint daher schlichtweg ein "Muß" zu sein.

Im Schattendasein kommerzieller Produkte hatte das frei verfügbare Unix-Derivat Linux [4] in den letzten Jahren den größten Zuwachs an Installationen – selbst wenn sich die exakten Zahlen bei Freeware nie genau bestimmen lassen. Linux war primär als Programmierübung geplant. Da der Source Code für jedermann von Anfang an frei und kostenlos zugänglich gemacht wurde, entwickelte sich rasch eine weltweite (Fan-) Gemeinde, welche die Weiterentwicklung vorantrieb und aufgetretene Fehler effizient beseitigen konnte. Was einst als akademisches Projekt gestartet wurde, ist Dank des effektiven Kommunikationsmediums Internet und der freiwilligen Arbeit einer riesigen Programmiergemeinschaft zu einem stabilen Betriebssystem gereift. Linux erfreut sich zunehmender Beliebtheit, ist nach wie vor eines der erfolgreichsten Software Projekte und macht heute kommerziellen Systemen nicht nur am PC-Sektor – ursprünglich nur für den PC entwickelt läuft Linux nun auch auf Digital-Alphas, Sparc-Workstations und IBM Großrechnern – ernsthafte Konkurrenz. Auch bieten namhafte Firmen neuerdings Linux-Support (vgl. Ausstellerliste der diesjährigen CeBit). Günstige Standardhardware und nicht vorhandene Lizenzkosten machen Linux somit zu einer idealen Entwicklungsplattform – nebenbei erwähnt werden sollte auch die hohe Betriebssicherheit und Stabilität dieses Betriebssystems.

Um diese These praxisnahe zu untermauern, wird in weiterer Folge über den aktuellen Stand eines Projekts berichtet, welches das ehrgeizige Ziel verfolgt hat, einen Feldbus-Gerätetreiber für Linux auf einer Intel x86-Plattform zu erstellen und darauf aufbauend die Fernwartung eines technischen Prozesses zu betreiben. Konkret wird die Software-Architektur des Treibers, sowie seine Integration in den Linux-Kernel vorgestellt. Mit Hilfe einer C/C++-Bibliothek soll es dem zukünftigen Linux-Benutzer fortan möglich sein, den vollen Leistungsumfang des gewählten Feldbussystems zu nutzen und Automatisierungsapplikationen unter Linux zu erstellen. Als erste Testapplikationen wurden im Rahmen des Projektes komfortable Werkzeuge für die notwendige Initialisierung und Parametrierung des Feldbusses entwickelt, die über graphische Benutzeroberflächen einfach bedient werden können. Aufbauend auf dem Gerätetreiber wurde in weiterer Folge die Realisierung einer Client/Server basierenden Fernwartung in Angriff genommen. Einzige Voraussetzung für ihren Einsatz ist auf Server- und Client-Seite die Internet-Anbindung der Rechner. Die Fernwartungsanwendung läuft auf Client-Seite innerhalb eines WWW-Browsers als Java-Applet. Man beachte: somit ist es nicht mehr notwendig, die Anwendung explizit auf dem jeweilig zur Verfügung stehenden Rechner zu adaptieren/compilieren/linken, geschweige denn zu installieren. Vielmehr wird diese Aufgabe – für den Benutzer transparent – automatisch von der Browsersoftware durchgeführt. Der Benutzer muß nur noch die entsprechende Seite, die vom Browser geladen werden soll, angeben. Schlußfolgerungen und ein kleiner Ausblick auf zukünftige Erweiterungen des Systems (Stichwort: Remote Fieldbus-Monitoring via WWW) beenden den Artikel.

2 Linux in der Automatisierungstechnik

Ziel des Projektes war die Konzeption und Realisierung einer Fernwartung für einen technischen Prozeß, der über Feldgeräte gesteuert wurde. Die Fernwartung sollte potentiellen Benutzern die Möglichkeit bieten, die Zustände einzelner Feldgeräte zu erfassen, sowie steuernd in das Feldbussystem eingreifen zu können. Dazu war es unweigerlich notwendig Vorgänge im Feldbereich ansprechend zu visualisieren. Als primärer Feldbus wurde Profibus-DP gewählt. In das Profibus-DP System wurde zusätzlich der in der Industrie gerne verwendete ASi-Installationsbus eingebunden. Als Profibus-Schnittstelle zur übergeordneten Steuerung (also jenem PC unter Linux) wurde eine handelsübliche Profibus-Schnittstellenkarte verwendet. Der komplette (kleine) Versuchsaufbau wurde als Mono-Master System konzipiert.

Als Entwicklungsrechner (und letztlich steuernde Instanz des Feldbussystems) diente ein Intel Pentium PC unter dem Betriebssystem Linux. Die Software wurde grob gesagt in zwei Schritten entworfen. Zuerst mußte ein Gerätetreiber für das Feldbussystem unter Linux realisiert werden [5]. Als Programmiersprache wurde zu diesem Zweck C++ eingesetzt und der GNU Compiler `g++` eingesetzt. Im zweiten Schritt wurde die Fernwartung in Client/Server Architektur implementiert [2]. Im Serverbereich kam wiederum die Programmiersprache C++ zur Anwendung, die Client Software wurde unter Java entwickelt, wobei die Programmierumgebung hauptsächlich aus dem Java Development Kit (JDK) bestand [3]. Als WWW-Server, der die Clientsoftware (Java-Applet) über das Internet zur Verfügung stellt, wurde das Softwarepaket Apache verwendet [1].

2.1 Gerätetreiber

Erster Schritt bei der Realisierung dieser Vorgaben war die Entwicklung eines Gerätetreibers, mit dessen Hilfe der Versuchsaufbau vorerst lokal gesteuert werden konnte. Gerätetreiber (und damit Geräte) werden unter Linux über sogenannte Special Files angesprochen. Für den Softwareentwickler bedeutet das, daß der Zugriff auf ein Gerät genauso wie der Zugriff auf eine Datei erfolgt. Möchte man ein Gerät benutzen, so muß man dieses vorweg öffnen. Dies geschieht mit dem Systemaufruf `open()`. Danach kann man mit den Befehlen `read()`, `write()` und `ioctl()` vom entsprechenden Gerät Daten lesen, Daten ausgeben oder (sofern vorgesehen) das Gerät konfigurieren. Mit dem Systemaufruf `close()` wird das in Beschlag genommene Gerät sodann wieder freigegeben. Aus Sicht des Programmierers sieht es nach dem Öffnen eines Gerätetreibers also so aus, als hätte er auf eine herkömmliche Datei zugegriffen.

Eine knifflige Angelegenheit, die es vorweg im Rahmen des Projektes zu lösen galt, war die Implementierung und Einbindung eben dieser Datei/Geräte-Funktionen. Dazu war es notwendig, die Hardwareansteuerung der Profibus-Schnittstellenkarte genauer zu untersuchen. Die konkret zum Einsatz gekommene Profibus-Schnittstellenkarte enthielt einen integrierten Mikroprozessor, auf dem das Profibus-DP Protokoll (in Firmware) implementiert war. Die Kommunikation zwischen Einschubkarte und Hostsystem wird über ein Dual-Port-Ram (und genau genommen vier Mailboxen) abgewickelt. In diesem Dual-Port-Ram finden sich grob gesagt ein Eingabe-/Ausgabe-Datenbereich, in dem die aktuellen Abbilder der vorhandenen Sensoren und Aktoren festgehalten sind, und die Slave-Status-List, welche die Diagnosezustände der einzelnen Feldgeräte beinhaltet. Hauptaufgabe des Treibermoduls war diesen Speicherbereich in den Hauptspeicher des PCs zu projizieren, um ihn so für Applikationsprogramme zugänglich zu machen.

Gerätetreiber sind Teil des Kernels und können nur von privilegierten Benutzern (z.B. `root`) in das System eingebunden werden. Dies geschieht entweder zur Laufzeit über sogenannte Kernel-Module (einer Neuerung des 2.ẋ-Kernels; man beachte: ganz im Gegensatz zu üblichen Betriebssystemen muß dazu das Gesamtsystem nicht erneut gestartet bzw. rebootet werden) oder durch direkte Einbindung in den Kernel, was allerdings eine (Neu-)Compilierung des Kernels erfordert.

Einmal in den Kernel eingebunden kann man mit den zuvor vorgestellten Systemaufrufen auf die Profibus-Schnittstellenkarte zugreifen. Um den Umgang mit dem Profibus-DP System weiter zu erleichtern, wurde im Rahmen des Projektes zusätzlich eine Bibliothek erarbeitet, die es dem zukünftigen Anwendungsprogrammierer ermöglicht, auf die zuvor erwähnten recht unhandlichen Systemaufrufe zu verzichten und statt dessen komfortable Funktionen verwenden zu können. Die Bibliothek wurde als C/C++-Shared Library ausgeführt und stellt im wesentlichen die Klasse `pboard` zur Verfügung (im Falle der C-Library sind die Funktionen eigenständig ausgeführt). Die wichtigsten Methoden der Klasse sind: `pboard()`, womit einer Applikation exklusiv der Zugriff auf eine Profibus-Schnittstellenkarte gewährt wird; `reset()` zur (Re-)Initialisierung einer zuvor angeforderten Profibus-Schnittstellenkarte; `send_dp()` bzw. `get_dp()` zur Übertragung bzw. Entgegennahme spezifischer Profibus-DP Kommandos und `get_pdi()`, wodurch eine Applikation einfachen Zugang zur Slave-Status-List und zum Eingabe-/Ausgabe-Datenbereich erhält.

2.2 Fernwartung

Damit die Fernwartung möglichst effizient und natürlich auch benutzerfreundlich gestaltet werden konnte, waren beim Entwurf einige wichtige Punkte zu beachten: (1) die Bedienung sollte mit nahezu jedem Rechner, der Zugang zum Internet hat, möglich sein, unabhängig von dessen Rechnerarchitektur und Betriebssystem; (2) bei Bedarf sollte der Zugang auf bestimmte Personen(gruppen) eingeschränkt werden; (3) falls im Leitsystem mehrere Feldbussysteme verankert sind, sollte es möglich sein, pro Feldbus eine eigene Fernwartung betreiben zu können; (4) die Antwortzeiten sollten trotz großer Entfernungen erträglich sein. Deswegen war darauf Rücksicht zu nehmen, daß die anfallenden Datenmengen, die über Netzwerkverbindungen geschickt werden, auf das Notwendigste zu reduzieren sind; (5) die Fernwartung war möglichst universell zu implementieren. Für unterschiedliche Anwendungen sollte der Kern der Fernwartung nicht wiederholt angepaßt werden!

Aufbauend auf diesen Richtlinien wurde die Fernwartung als Client/Server-Architektur entworfen und die Kommunikation zwischen Client und Server durch ein eigenes Profibus Remote Protocol (PRP) festgelegt. Mithilfe von PRP-Nachrichten kann sich der Client beim Server unter anderem anmelden, authentifizieren und jene Geräte angeben, deren Zustand (zyklisch) überwacht werden soll. Dem Profibus-DP Server kam somit hauptsächlich die Aufgabe zu, ankommende Benutzer-Anfragen entgegenzunehmen und diese gegebenenfalls an das unterliegende Profibus-DP System weiterzugeben. Weiters retourniert er eigenständig PRP-Nachrichten, sofern sich der Eingabe-/Ausgabe-Datenbereich oder die Slave-Status-List geändert hat. Der Server wurde objekt-orientiert entworfen und in der Programmiersprache C++ implementiert. Die Aufgabe des Profibus-DP Clients bestand darin, dem Benutzer die Möglichkeit zu geben, den technischen Prozeß zu beobachten und bei Bedarf steuernd einzugreifen.

Kernstück des Clients ist die Klasse `packet`. Aufbauend auf dieser Klasse gibt es zwei wesentliche Teile, in welche die Software des Clients strukturiert wurde. Ein Teil übernimmt den Empfang von PRP-Paketen und wertet diese anschließend aus. Der andere Teil beinhaltet die Schnittstelle zum Benutzer, wobei im speziellen die Interaktion mit der Maus interpretiert und daraus etwaige PRP-Pakete abgeleitet werden, die sodann an den Server versendet werden. Für den Client war zusätzlich noch ein übersichtliches Benutzerinterface zu entwerfen. Der Client wurde sodann als Applet konzipiert und unter der Programmiersprache Java implementiert.

Der Benutzer der Fernwartung muß nun lediglich eine bestimmte WWW-Seite anwählen (`http://hall.auto.tuwien.ac.at/dp/`), in der das Applet verankert ist. Das Applet wird sodann durch den WWW-Server zum Client transportiert und innerhalb seines WWW-Browsers ausgeführt (Abb. 1). Er kann Veränderungen an den Eingängen der Feldbusgeräte beobachten und Ausgänge durch simple Mausklicks manipulieren. Zusätzlich wird er über den Zustand der angeschlossenen Feldgeräte informiert.

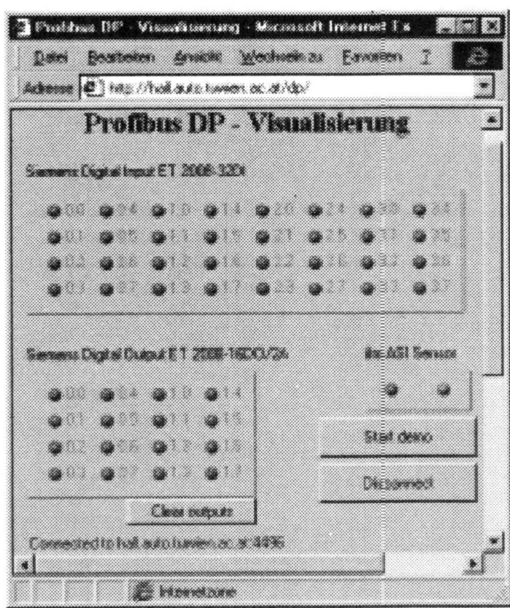

Abb. 1. Java-Applet der Fernwartung

3 Zusammenfassung

Aufgrund steigender Ausbreitung, Verfügbarkeit und Geschwindigkeit der globalen Vernetzung wird die Visualisierung und Fernwartung von eher unkritischen Prozeßabläufen in der Automatisierungstechnik zunehmend wichtiger. Konform dazu beginnt sich Linux – ein frei verfügbares Unix-Derivat – unter den High-Quality Betriebssystemen zu etablieren. Ziel dieser Arbeit war die Entwicklung einer

Fernwartung für das Feldbussystem Profibus-DP, basierend auf einer Client/Server-Architektur. Komponenten am Profibus-DP sollten von einem Rechner, der unter dem Betriebssystem Linux lief, gesteuert werden. Notwendig für die Realisierung dieses Vorhabens war vorweg die Konzeption und Implementierung eines passenden Gerätetreibers. Als Begleiterscheinung entstanden eine C/C++-Bibliothek und zahlreiche unter Zuhilfenahme dieser Bibliothek entwickelte Werkzeuge, die ursprünglich nur zu Testzwecken des Treibers vorgesehen, nun in der Praxis beeindruckende Applikationen darstellen. Damit war ein wesentlicher Bestandteil des Projektes erfüllt: einerseits konnte gezeigt werden, daß Linux eine geeignete Plattform für Softwareentwicklung (in der Automatisierungstechnik) ist, andererseits konnte Linux auch als Betriebssystem für einen heutzutage "klassischen" Prozeßrechner herangezogen werden. Darauf aufbauend wurde die Fernwartung und Visualisierung eines technischen Prozesses in Angriff genommen. Dieser Wunsch folgte dem Trend, die Attraktivität und das Angebot Internet-basierter Anwendungen zu steigern. Als willkommene Begleiterscheinung dieses Teilprojekts konnte gezeigt werden, daß die Programmiersprache Java und herkömmliche WWW-Browser im Automatisierungs-bereich durchaus ihre Berechtigung haben, sofern sie richtig eingesetzt werden.

In weiterer Folge soll der vorhandene Gerätetreiber in einen echtzeitfähigen Linux-Kernel integriert und sein Verhalten in der Praxis getestet werden. Das bestehende Fernwartungssystem soll um einen Proxy-Server erweitert werden, der als Vermittler zwischen Client und Server fungieren wird, so daß Clients nicht mehr direkt mit dem Server kommunizieren müssen. Auch die Implementierung von Fernwartungen für sogenannte Feldbusmonitore werden in Betracht gezogen. Ebenso sind Realisierungen für Gerätetreiber, die den Zugriff auf andere Feldbussysteme ermöglichen, vorgesehen. Dazu wäre eine Kooperation mit Partnern in der Industrie sehr erfreulich und höchst willkommen. Bereits mit diesem Pilotprojekt konnte gezeigt werden, daß Linux als Betriebssystem in der Automatisierungstechnik in Zukunft sicherlich ein Thema ist!

Literatur

.1. Apache HTTP Server Project: `http://www.apache.org/`

2. Csebits C.: *Fernwartung eines Profibus Systems*, Diplomarbeit TU Wien, 1999.

3. JDK Linux Port Project: `http://www.blackdown.org/`

4. Linux Project: `http://www.linux.org/`

5. Mayer M.: *Profibus-DP Treiber für Linux*, Diplomarbeit TU Wien, vor. 1999.

Dieser Artikel soll mit einer Danksagung enden. Mein erster Dank gilt Herrn Dipl.-Ing. Christoph Csebits und Herrn Michael Mayer, die in gemeinsamer Arbeit die vorgestellten Ideen und Konzepte tatkräftig in die Realität umgesetzt haben. Das Projekt konnte nur deshalb erfolgreich durchgeführt werden, da Herr Andreas Vogt (tmg i-tec) Technischen Support geboten, die Firma HTC-Kral Komponenten des ASi-Bussystems bereitgestellt und das Feldbuskompetenzzentrum Wien (und dort im speziellen Herr Dipl.-Ing. Martin Knizak) kurzfristig mit einigen Profibus-DP Geräten ausgeholfen haben. Allen Beteiligten sei auf diesem Wege nochmals herzlich gedankt.

Tele-Diagnosis at Networked Automation Systems

Klaus Kabitzsch, Volodymyr Vasyutynskyy

Dresden University of Technology, Faculty of Computer Science, D-01062 Dresden

Abstract. This paper describes a teleservice - tool, which is used for the remote diagnosis of PLCS at machines. It records the operations in the interior of the PLCS. Thereby all real time - details are measured. To be able to select the measured variables flexibly, a mobile agent is constructed in the far computer. This agent will transfer into the PLCS with the help of the network and begins to work there. The observations at the software are supported by video - and audio - measurements at the machine.

Abstract. Dieses Papier beschreibt ein Teleservice-Werkzeug, das zur Ferndiagnose von SPS an Maschinen benutzt wird. Es zeichnet die Abläufe im Inneren der SPS auf. Dabei werden alle Echtzeit-Details gemessen. Um die gemessenen Variablen flexibel auswählen zu können, wird ein mobiler Agent im fernen Rechner konstruiert. Dieser Agent wird mit Hilfe des Netzwerks in die SPS übertragen und beginnt dort zu arbeiten. Die Beobachtungen an der Software werden von Video- und Audio-Messungen an der Maschine unterstützt.

1 Introduction

Complex machines and equipment are not faultless after their commissioning. Even if all tests were successful, many disturbances and real - time problems remain. Many errors appear only rarely and mostly then, if the maintenance persons has left the construction site already. If errors appear first to a later point in time, it gives large problems particularly at exported machines. While the specialists returned to their office for example in Germany, the machine at the customer is many thousands of kilometers remote. Debugging and repair require high expense and bring following problems:

The journey of the specialists to the customer requires long time. The machine can not work during this time and this causes high costs. Often the specialists find only trivial error causes.

Some error effects exist only occasionally and disappear then again for long time (transient errors). In most cases the customer turns on these machines again and has they worked further. If the troubleshooter arrives at the machine, all tracks are effaced then. Nobody can find then error causes.

The specialists must bring along many resources, gauges and tools. Often they have forgotten an important tool. Then they must wait, until it is supplied later.

210

The journey of the troubleshooters and the transport of the tools cause high costs.

2 Teleservice

It would be a better solution, if the specialists remain first of all in their office. For this reason the computers and PLCS at the remote machine are joined by the telecommunication with the specialists. The troubleshooters can manipulate now at the PLCS and observe their function. All maintenance works are executed over a large distance. But their access to the PLCS still remains not complete:

The troubleshooters can not see into the interior of the PLCS (hardware, software).

They can observe the faulty operations no more with bare eyes, because they are too fast.

If an error appears only rarely, the attention is missing to the right point in time.

If the cause of an error lasts long time, they lose the overview.

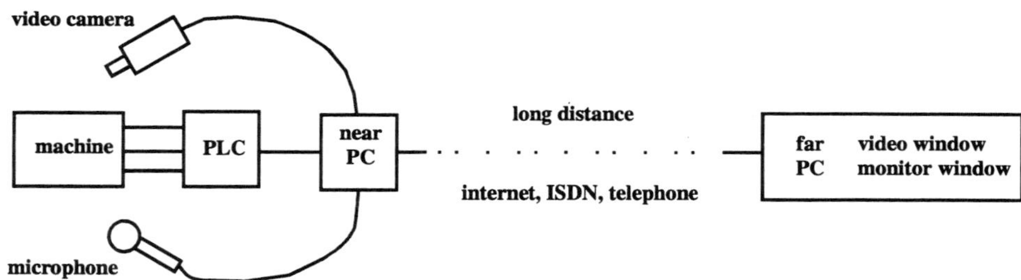

Fig. 1. Principle of the teleservice

3 Monitor - Tools

The teleservice - tool "AutoSPy" solves these problems with the monitor - technology. This tool observes, records and stores all processes in the interior of the PLCS over long time. The stored time course of all program - variables and input - output - connections is marked also as "history".

Figure 2 shows the measurement at the transport system of the high shelf warehouse of a CIM - factory. Sensors measure the position of the sled and send this information to the input - variables E0.1 to E0.6. If the sensor E0.4 refuses in rare cases, this error is recognizable only with the help of a long-time - recording. The duration of the impulses to the input - connection E0.4 is about similarly the cycle time of the PLCS. Therefore the tool must record all variables with high temporal precision. The resolution must be identical with the cycle time.

Error: missed impulse E0.4

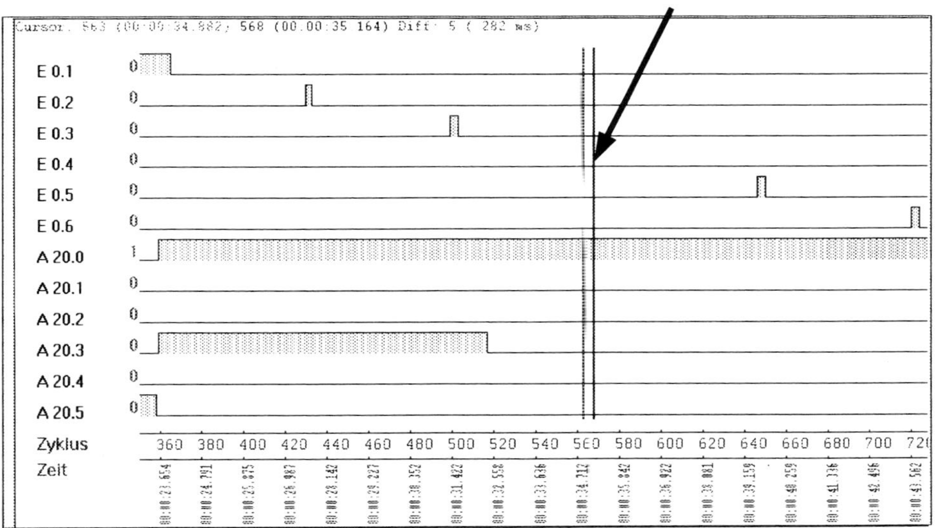

Fig. 2. Measurement with the tool AutoSPy at the transport system of a CIM - factory

It is often necessary, that the contents of variables are recorded with all real time - details. Then all variables may be observed however no more simultaneously, which are founded in a PLC - program. The tool would have to read, store and transfer a mass

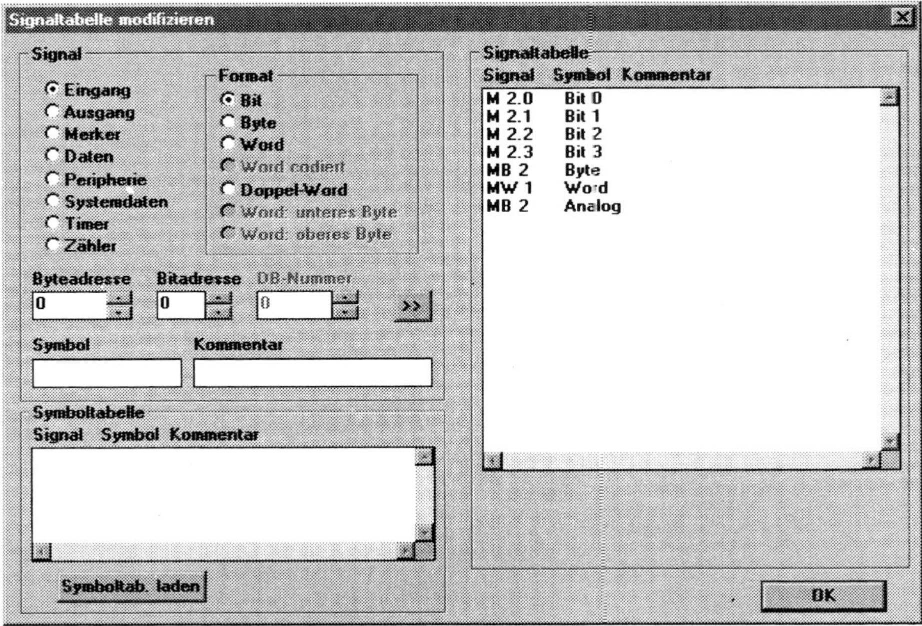

Fig. 3. Selection windows of the tool AutoSPy. In the left window the user chooses names from the symbol - table and data types of the variables. The right window shows the selected variables.

of data in this case. Instead the user must select now some few variables, whose time course he would like to observe precise.

For the selection of the variables the teleservice - tool AutoSPy has important properties:

All variables of the PLCS - software and the input - output - connectors stand at the disposal for this selection.

The uncoded names used in the source code of the PLCS are offered the user actively to the selection. Therefore the far computer shows for example a list with the names of the symbol - table. The user selects the names with the mouse.

The new observation - instruction will transfer from the far computer by the near computer into the PLCS.

To load the new observation - instruction in the PLCS, the user must not switch off the machine or the PLCS. Also a reduction of the work speed, one-step operation etc. is not desired. The new observation - instruction is loaded in the PLCS, while these runs in the real time - operation.

4 Mobile software - agents

To fulfill the demands shown in chapter 3, the technology of the mobile agents is used. If the selection of the variables (figure 3) is terminated in the far computer, the work of a code generator begins there.

This code - generator creates independently a software object (figure 4). The object contains program code, which should read the selected variables in the PLCS. The object contains also an empty memory, which should store the registered history in the PLCS. The object contains also communication - components, which transfer the history to the near computer.

If the agent is created, it will transfer from the far computer to the near computer with the help of the telecommunication (figure 5). The near computer transmits the agent to the PLC. There the agent is installed by the operating system and is integrated in the processing cycle.

As soon as the far computer gives the start - command, the agent begins to work in the PLC. The operating system of the PLC processes the agent at the beginning of each cycle now. The agent observes the variables and puts their values into the scratchpad storage. Subsequently it transfers the history to the near computer. To improve the effectiveness of this transmission, a block-by-block transfer was implemented. In the near computer the history is stored on the hard disk

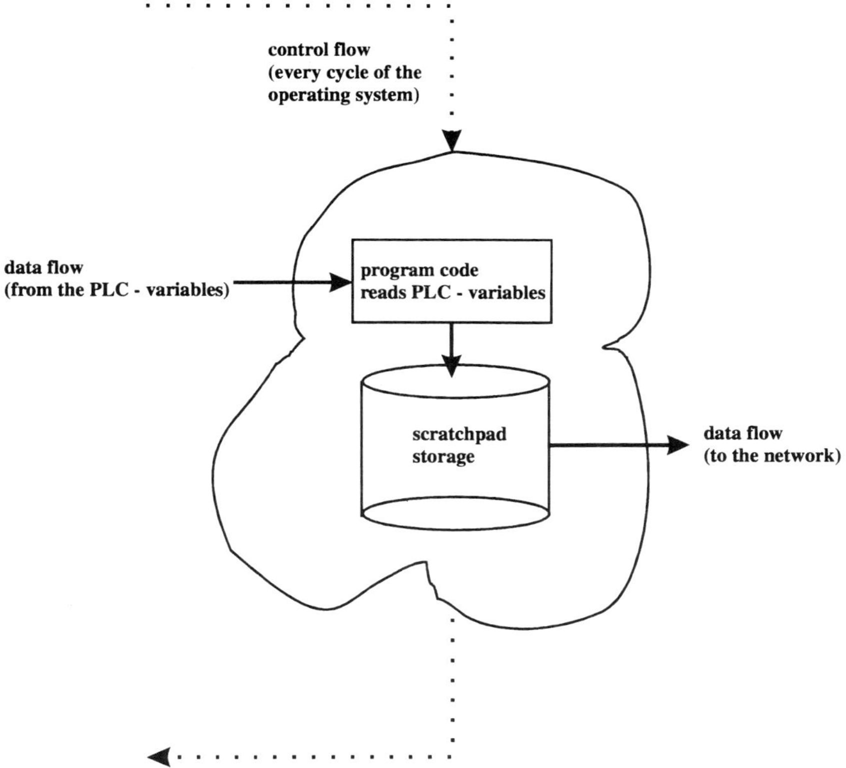

Fig. 4. Elements of a mobile agent

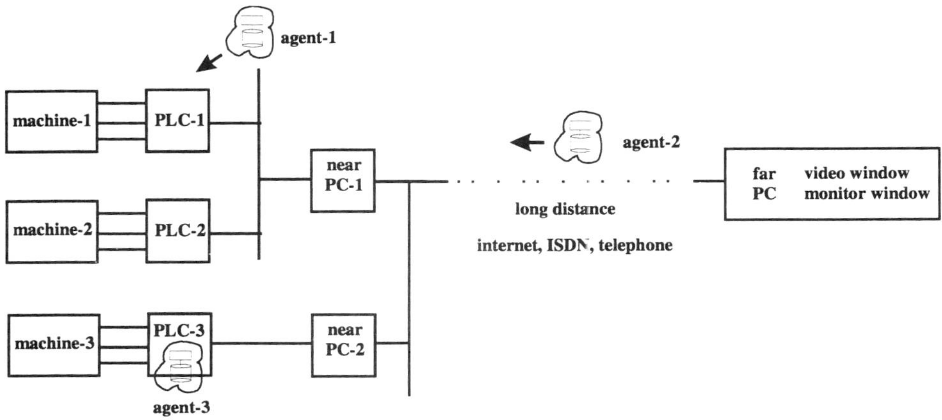

Fig. 5. Mobile agents are transported from the far computer to their target - places: the PLCS

5 Support of video and audio

The preparation of the measurement and the recording of the history require often many hours. Thereby the far computer is remote possibly many thousand kilometers from the machine. The user can not see and hear therefore the PLCS, the machine and their environment. But many users wish an observation of the machine during the measurement. There are different reasons for it:

Security: The user wants to avoid dangers for persons, who stand nearby the machine.

Diagnosis: The user is interested not only in the history, which he has recorded in the interior of the PLCS. He wants to observe also the outer performance of the machine during this measurement. Subsequently he can compare both observations.

Therefore AutoSPy is complemented by a video camera and a microphone. Both are connected to the near computer and transfer images and clays to the far computer. There the images can be observed in a separate window of the tool. Also the noises of the machine are returned by the loudspeaker (figure 1).

References

1. Ribbecke, H.-D.: Fehlerlokalisierung in Automatisierungssystemen. SPS-Magazin (1998) H. 11+12, S. 106 - 107

2. Kabitzsch, K.; Hartenstein, D.: Fehlersuche in Automatisierungssystemen mit Feldbus-Vernetzung. e&i Elektrotechnik und Informationstechnik 114(1997) H.5, S. 274-279, Springer Verlag Wien 1997

3. Dauphin, P.; Hartleb, F.; Kienow, M.; Mertsiotakis, V.; Quick, A.: PEPP: Performance Evaluation of Parallel Programs. Technical Report 5/92, Universität Erlangen-Nürnberg, IMMD VII, 1992

4. Kabitzsch, K.; Hartenstein, D.: Mit Excel gegen Fehler - Fehlern in LON-Netzwerken auf der Spur. Elektronik (1997) H. 19, S. 80 - 85

5. Myers, G.J.: The art of software testing. John Wiley & Sons Inc. 1979

6. Kabitzsch, K.; Hartenstein, D.: Diagnose bei der Inbetriebnahme komplexer Automatisierungssysteme. VDI-Berichte Nr. 1397, S. 429 - 436, VDI-Verlag Düsseldorf 1998

7. Ribbecke, H.-D.; Wittig, E.; Kabitzsch, K.: Fehlern in SPS-Systemen auf der Spur. Elektronik (1998) H. 12, S. 22 - 24

8. TU Dresden, Fakultät Informatik, Institut Informationssysteme, http://www.inf.tu-dresden.de

Diagnose von Feldbussen im Systemverbund

W. Kriesel, T. Heimbold, Th. Minner

Forschungs- und Transferzentrum e.V. an der HTWK Leipzig
Steinbeis Transferzentrum Automatisierungs-, Informations-
und Elektrosysteme Leipzig

Abstract. Embedded Systems are becoming more complex. As both software and hardware become more complex, debugging and testing consume higher proportions of the development budget and project time. The PRACTICAL tools offer one way to reduce this.

The PRACTICAL tools focus on CPU independent hardware assisted debugging tools for embedded systems, both for single CPU systems and for distributed systems. The architecture separates the debugging environment into three groups: GUI (display) and analysis engines, Debugging heads, Targets being debugged.

The first two groups are connected via Ethernet and TCP/IP, whereas debug heads are connected as required by the target hardware. All the tools are independent of the target CPU, providing an alternative to the ever increasing cost of In Circuit Emulation as CPU speeds increase.

Debugging hosts may be any standard PC or Workstation. PRACTICAL tools take advantage of the universal availability of Ethernet connections to allow a wide range of hosts to be connected simultaneously. These hosts may cover simple functions such as target program generation, debug head control, target node simulation and I/O simulation control.

The debugging heads will be programmable by the user, using portable applications, so permitting the wide variety of debugging requirements to be satisfied by a range of small applications held on the debugging host and downloaded as required to the debugging heads. Debugging heads may include device simulators for I/O timing analysis, program loaders, CPU debug access via JTAG or BDM, fieldbus network analysers.

1 Einleitung

Moderne Automatisierungsysteme werden bekanntlich durch den Einsatz einer Vielzahl dezentraler, intelligenter Funktionseinheiten geprägt, die untereinander durch Feldbusse vernetzt sind. Daraus resultiert eine hohe Komplexität sowie die Notwendigkeit zur Entwicklung neuartiger Diagnosewerkzeuge, die den erheblich gestiegenen Anforderungen dieser Systeme gerecht werden.

Die isolierte Betrachtung eines Feldbusses ist somit nicht mehr ausreichend. Vielmehr muß der Feldbus als Verbindungselement innerhalb des gesamten zu diagnostizierenden Systems analysiert werden. Dafür ist es notwendig, auch die Mikrocontroller der intelligenten Funktionseinheiten und ihre Peripherie in den Diagnoseprozeß mit einzubeziehen. Die Vielzahl der sich heute am Markt befindlichen

Controller setzt eine weitgehend von der Hardware unabhängige Realisierung der Analysewerkzeuge voraus. Des weiteren erweist sich die Fixierung auf einen einzigen Feldbus als wesentliche Einschränkung der Flexibilität.

An der Hochschule für Technik Wirtschaft und Kultur (HTWK) Leipzig wurde deshalb innerhalb des ESPRIT Projektes „PRACTICAL" in Kooperation mit Partnern aus Großbritannien und Italien eine neuartige Architektur entworfen, welche die Diagnose von komplexen Systemen als Einheit ermöglicht. Das im folgendem vorgestellte Projekt realisiert ein in sich geschlossenes System an Werkzeugen zur gezielten Diagnose im „embedded"-Bereich. Dies beinhaltet sowohl das „Debuggen" eines einzelnen Controllers wie auch eines verteiltes Systems. Dabei ist neben dem reinen Beobachten auch ein gezieltes Anregen des Systems möglich.

2 Aufbau des Diagnosesystems

Das entwickelte Diagnosesystem ist in 3 Ebenen (im folgenden als Layer bezeichnet) unterteilt:

-        graphisches Benutzerinterface (Host Layer),

-        Diagnosewerkzeuge (Debug Head Layer),

-        zu untersuchendes System (Target Layer).

Die Diagnosewerkzeuge des mittleren Layers sind über ein Netzwerk (Host Network) mit der darüberliegenden Ebene verbunden. Aufgrund der allgemeinen Verfügbarkeit und Akzeptanz wird hierfür das Ethernet mit einem IP-Protokoll verwendet. Dieses Netzwerk findet gleichzeitig zum Datenaustausch der Debug Heads untereinander Verwendung.

Die Verbindung des Debug Head Layers mit dem Target Layer erfolgt je nach Anwendungsfall in Abhängigkeit der verwendeten Hardware. In Abbildung 1 ist die im Projekt entwickelte Struktur dargestellt.

Der **Target Layer** stellt das zu untersuchende System dar. Er beinhaltet alle Mikrocontroller und ihre Peripherie sowie den verbindenden Feldbus. Für Elemente, die nicht zur Verfügung stehen bzw. nicht ohne weiteres in den Diagnoseprozeß mit einbezogen werden können, stehen Werkzeuge zur Simulation zur Verfügung.

Die Anwendung des Target Layers sollte ein abgeschlossenes System darstellen, aber es ist ebenfalls möglich, nur ein Teilobjekt einer komplexen Anwendung zu untersuchen. Dabei ist kein zusätzlicher Aufbau notwendig; es genügt, die Werkzeuge mit dem System zu verbinden. Dies geschieht größtenteils über genormte Steckverbindungen.

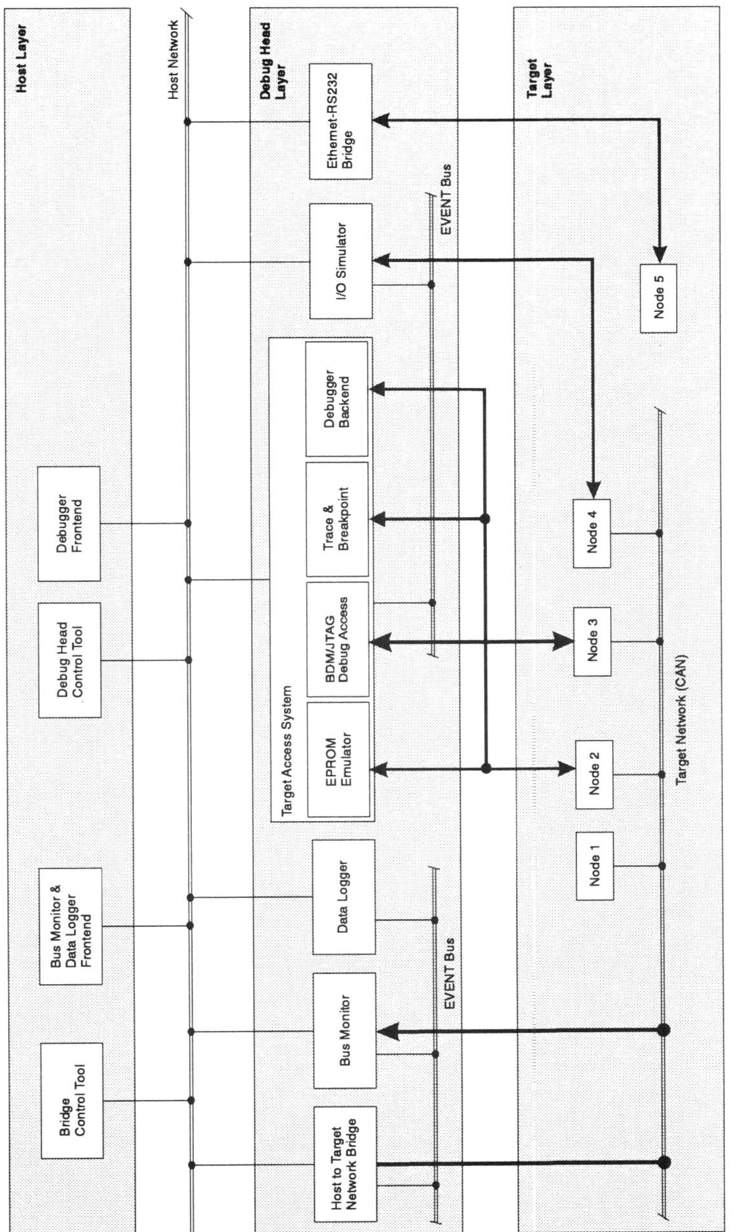

Abb. 1. Struktur des Diagnosesystems mit mehreren Ebenen

Der **Debug Head Layer** stellt die mittlere Ebene dar. Er besteht, wie in Abbildung 2 dargestellt, aus mehreren Einzelwerkzeugen, die über ein ebeneninternes Netzwerk, den „Event Bus" und ein externes, das Host Network" miteinander verbunden sind. Das Host Network stellt gleichzeitig die Verbindung zum darüber liegenden Host Layer dar. Die Werkzeuge (im folgendem als Debug Heads bezeichnet) arbeiten eigenständig und sind jeweils auf eine Aufgabe spezialisiert, können aber über die genannten Netzwerke synchronisiert werden. Es besteht die Möglichkeit, vom Anwender geschriebene Programme im Debug Head auszuführen und so auf spezielle Anforderungen zu reagieren. Dazu stehen ein Compiler sowie ein Simulator zur Verfügung.

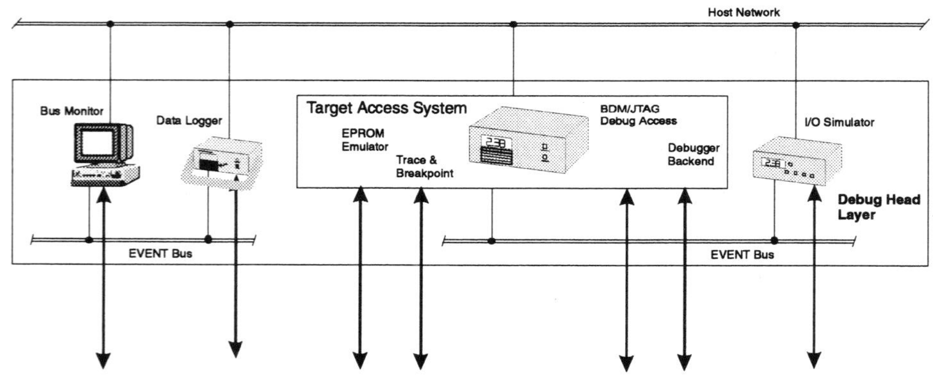

Abb. 2. Debug Head Layer

Die oberste Ebene bildet der **Host Layer**. Er beinhaltet das Benutzerinterface sowie die endgültige Verarbeitung der gewonnenen Daten aus dem Debug Head Layer. Diese Ebene besteht im wesentlichen aus:

- Debugger Front ends,

- Modulen zur Steuerung der Debug Heads (Debug head control tools),

- Modulen zum Fernzugriff (Remote debugg access),

- Test- und Analyse-Modulen.

Die *Debugger Front ends* enthalten alle Anzeigen und Konfigurationsmöglichkeiten zur Planung und Durchführung eines Diagnoseprozesses sowie zur Anzeige und Auswertung der Daten. Sie stellen ebenfalls eine Verbindung zu den notwendigen Debug Heads her.

Die *Debug head control tools* bestehen aus Compilern, Front ends sowie weiteren Modulen. Sie steuern die Aktivitäten der Debug Heads und versorgen sie mit Daten (z.B. Programmcode für einen Mikrocontroller des Target Layers).

Unter Verwendung des *Remote debugg access* besteht die Möglichkeit, den Diagnoseprozeß an einem räumlich getrenntem Ort durchzuführen. Hierfür können Modemverbindungen oder das Internet genutzt werden.

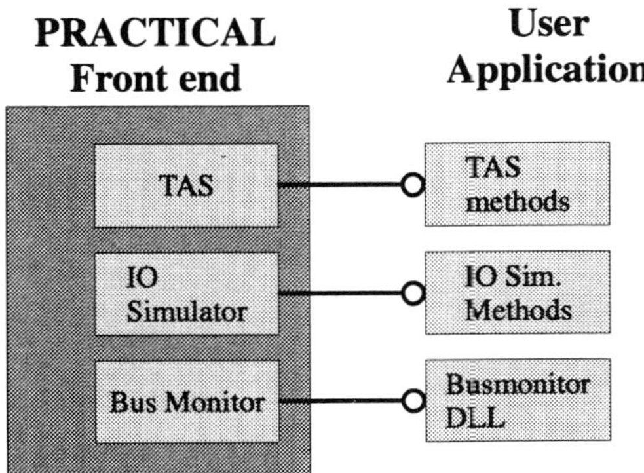

Abb. 3. Front end Interface

Das im Rahmen des Projektes entwickelte Diagnose- und Testsystem stellt ein offenes System dar. Die Schnittstelle der Hard- und Software für die Debug Heads sowie für die Grundfunktionen (API) der Front ends sind verfügbar. Dem Anwender wird durch einen Compiler und Simulator ermöglicht, eigene Programme für Debug Heads zu entwickeln und die Debug Heads somit auf seine individuellen Bedürfnisse anzupassen. Auf Seiten der Front ends kann auf Basis vordefinierter Module (siehe Abb. 3) auf die Basisfunktionen für die Steuerung der Debug Heads zugegriffen werden. Diese Methoden basieren auf der COM/DCOM Technologie und gestatten somit eine einfache Einbindung in eine Vielzahl von Applikationen. Eine anwenderspezifische Aufzeichnung und Auswertung ist somit sichergestellt. Des weiteren ist es ebenfalls möglich, Testsequenzen zu automatisieren und durch eine Anwendung im Host Layer zu steuern.

3 Vorstellung ausgewählter Werkzeuge

Alle Debug Heads basieren auf einem einheitlichem Konzept. Die Hardware besteht aus einem Mikrocontroller-Modul, auf Basis eines Strong-ARM SA 1100 Prozessors, welches im FTZ an der HTWK Leipzig entwickelt wurde (siehe Abb. 4), sowie aus einer aufgabenspezifischen Peripherie und einem Ethernetanschluß. Die im Controller implementierte Software besteht aus zwei Schichten. Die untere Schicht dient zur Ansteuerung und Verwaltung der jeweiligen Hardware sowie zur Kommunikation mit dem Front end. Die darüberliegende Schicht enthält die eigentliche Aufgabe des Debug Heads. Die Software dieser Schicht kann, wie bereits oben erwähnt, durch den Anwender geändert werden, um den Debug Head auf besondere Gegebenheiten anzupassen.

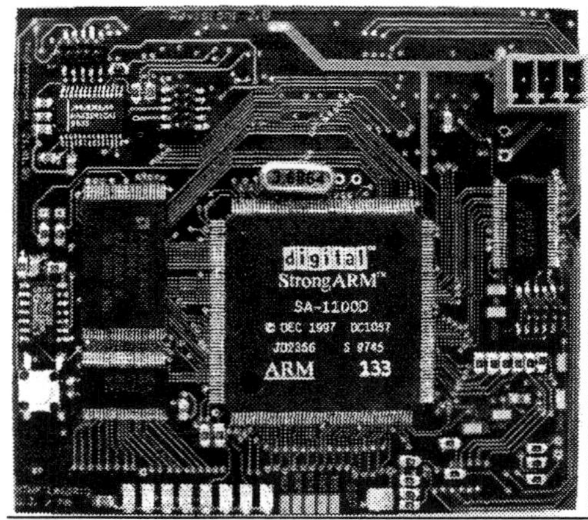

Abb. 4. Mikrocontroller-Modul

3.1 Busmonitor

Der Busmonitor dient zur speziellen Analyse des jeweiligen Feldbusses. Er ist als PCI-Einsteckkarte ausgeführt und besteht aus dem oben genannten Mikrocontroller-Modul, einem Feldbusinterface auf Basis eines FPGA, sowie dem PCI-Interface. Über das Feldbusinterface werden die Daten vom Feldbus gelesen und für den Mikrocontroller aufbereitet. Im Mikrocontroller erfolgt eine umfangreiche Vorbearbeitung dieser Daten, bevor sie mittels des PCI-Interfaces zum PC übertragen werden. Der PC übernimmt die Weiterleitung der Daten zum entsprechendem Front end bzw. fungiert selbst als Front end. Über den Front end erfolgt die Konfiguration des Busmonitors sowie die Auswertung der aufgezeichneten Daten.

Um das Datenaufkommen vom Busmonitor Debug Head zum Front end zu begrenzen, erfolgt im Debug Head eine umfangreiche Vorverarbeitung. Alle Telegramme des Feldbusses (derzeit wird CAN unterstützt) werden hinsichtlich aufgetretener Fehler bewertet. Des weiteren erfolgt eine umfangreiche und vom Anwender frei definierbare Filterung der Telegramme. Zur Aufzeichnung stehen ebenfalls umfangreiche Trigger-Möglichkeiten zur Verfügung. Sie können auch als Algorithmus definiert werden und erlauben so ein gezieltes Aufzeichnen der Busdaten.

Der Busmonitor dient im wesentlichen zum „Abhören" des Feldbusses, ein aktives Zugreifen auf den Bus ist nicht direkt gestattet. Diese Möglichkeit besteht aber über die Zusammenarbeit des Busmonitors mit der Host to Target Network Bridge.

3.2 Data Logger

Der Data Logger Debug Head ist eine Variante des Busmonitors, die für Langzeituntersuchungen konzipiert ist. Sein Funktionsumfang, in Bezug auf die Feldbusdiagnose, entspricht dem des Busmonitors. Zusätzlich kann er im Stand-alone-Betrieb arbeiten, dazu verfügt er über ausreichend Speicherkapazität zur Datenaufzeichnung, eine eigene Stromversorgung mittels Akkumulatoren sowie ein

Bedieninterface zur Vorort-Konfiguration. Die Konfiguration sowie die Auswertung der aufgezeichneten Daten erfolgt über den Front end.

3.3 I/O Simulator

Der I/O Simulator Debug Head dient zum Test des Systems durch Betrachtung der I/O Signale. Seine Aufgabe ist die Aufzeichnung sowie die Aktivierung von Signalen nach Vorgaben des Front ends. Durch die Programmierbarkeit des Debug Heads wird eine höhere Performance gegenüber bestehenden Systemen erreicht, es ist ebenfalls möglich, den Debug Head autonom agieren zu lassen. Innerhalb des Projektes entsteht eine Sammlung von fertigen Debug Head-Programmen für mögliche Anwendungsfälle, die durch den Anwender, falls notwendig, noch modifiziert werden können

3.4 Target Access System

Das Target Access System realisiert den Zugriff auf Mikrocontroller des untersuchten Systems. Es setzt sich aus mehreren Modulen zusammen:

- EPROM Emulator,

- BDM/JTAG Debug Access,

- Trace and Breakpoint,

- Debugger Backend.

Diese Module erlauben:

- die Übertragung von Programmen zum Mikrocontroller,

- Speicherzugriffe (Lesen und Schreiben),

- den Einzelschritt-Betrieb sowie

- das Verwenden von Breakpoints.

Das Target Acces System arbeitet hardwareunabhägig, das bedeutet, daß es nicht an einem speziellen Mikrocontroller gebunden ist. Lediglich die Interpretation des Programmcodes bei der Kompilierung und Anzeige auf dem Front end ist an den verwendeten Mikrocontroller gebunden.

4 Zusammenfassung

Die im Beitrag vorgestellte Architektur sowie die zugehörigen Werkzeuge zur Diagnose von Feldbussen als Bestandteil einer komplexen Anwendung bieten dem Entwickler eine Reihe von Vorteilen Dabei bedeutet die Unterstützung mehrerer Feldbusse ein hohe Flexibilität in der Anwendung der Werkzeuge, sowie eine reduzierte Einarbeitungszeit für den Benutzer. Als die beiden entscheidenden Vorteile bei dem Einsatz von PRACTICAL sind zu nennen:

Kostenreduzierung

Die Integration, Fehlersuche und Testphase eines komplexen Automatisierungssystems beansprucht typischerweise einen großen Teil der gesamten Entwicklungskosten.

Die.Möglichkeit zur schnellen und einfachen Erstellung von komplexen Testsequenzen sowie eine automatische Ausführung tragen zur Reduzierung der notwendigen Zeit und somit der Kosten bei.

Qualitätserhöhung

Die PRACTICAL-Werkzeuge erlauben eine Planung und Durchführung komplexer Testabläufe. Damit kann eine große Anzahl von Testszenarien „durchgespielt" und überprüft werden. Fehler, die unter normalen Umständen nie oder nur sporadisch auftreten, können somit sehr viel einfacher gefunden werden.

Literatur

1. Esprit Project Programme Part II – Description of the RTD Project PRACTICAL, Southampton, 1997

2. PRACTICAL Rationale and Architectural Concepts, Southampton, 1998

3. Villa M., Cassinari F.: Introduction to Concepts and Tools, EP25344 Deliverable, Milano, 1999

4. Heimbold, T.; Pelc, St.: Programmable Remote Analysis Coupling Targets in Control at Low Level EMMSEC99, Stockholm 1999, S. 779-784

5. Kriesel W., Heimbold T., Telschow D.: Bustechnologien für die Automation, Heidelberg, Hüthig 1998

Das Projekt PRACTICAL (Programmable Remote Analysis Coupling Targets in Control al Low Level – EP25344) wurde im Rahmen des ESPRIT-Programmes mit Mittel der EU gefördert.

Kontakt:

FTZ an der HTWK Leipzig
Dr.-Ing. Heimbold
Wächterstr. 13
04107 Leipzig

Tel. 0341-9786178
Fax 0341-9786131
e-mail heimbold@e-technik.htwk-leipzig.de

Engineering of Distributed Automation Systems Based on Novel Information Technologies and Methods

René Simon[1], Jörg Hörger[2]

[1]Institut für Automation und Kommunikation e.V. (ifak)
Steinfeldstraße 3 (IGZ), D-39179 Barleben
Tel.: 039203-81060, Fax: 039203-81100
E-mail: *rsi@ifak.fhg.de*

[2]Siemens AG A&D GT 3
Östliche Rheinbrückenstr.50, D-76187 Karlsruhe
Tel.: 0721-5956675, Fax: 0721-5956728
E-mail: *joerg.hoerger@khe.siemens.de*

Abstract. Hard economical conditions in the Automation and Process Control Industry are making secondary processes as Product Data Management and System Engineering more and more important. The following report describes a method, developed and realized as prototype by the BriteEuRam Project Advanced Control Network (ACORN 1479), which is used for the engineering of automation systems, and based on standardized description methods.

1 Introduction

The following report describes a method, developed and realized as prototype by the BriteEuRam Project Advanced Control Network (ACORN 1479), which is used for the engineering of automation systems, and based on standardized description methods. The prototype consists of four engineering tools (computer aided software engineering tools). The project was started 01/1996 and finished 07/1998. Three important European users (ICI Engineering Technology (GB), ENEL (I), and Electricidade de Portugal SA (P)), three well known research institutes (Fraunhofer-Institut für Informations- und Datenverarbeitung, Karlsruhe, Institut für Automation und Kommunikation ifak, Magdeburg, and the Centro Informationi Studi Esperienze SPA, Mailand), a software house (INTRASOFT, Athen) and the Siemens AG, Karlsruhe as one of the leading vendors of automation components and systems participated in the project.

The project defined the following objectives:

- To develop an engineering concept and to validate it using a prototype, which can be integrated into an overall information integration of the live cycle of an automation system, and

- to support the user with the engineering process for open, fieldbus based control systems (computer aided software engineering tools), to make this process more easy, more cost effective and less error susceptible.

The engineering process, looked at in detail with prototype tool realizations by ACORN consists of the design phase (logical system design) and operation phase (identification of components and solutions).

The following requirements to the ACORN tools were basically defined by the users from the chemical industry and the energy industry. However, they are not restricted to these application domains, but are valid for all areas where automation technologies are used.

a) The concept must not be restricted to the engineering process, but must integrate a information flow with other tools in all phases of the live cycle (design, installation, operation, recycling). The concept must support a „Single Source Principe" over all phases in future.

b) A high degree of re-usability of engineering / design results by criteria controlled, computer supported assignment of system functions (result of the design phase) to control system components (result of the operation phase). Criteria can be: performance, price, preferred vendor, ...

c) High degree of automation. Computer supported extraction of knowledge with intelligent technologies from available information, to create optimized solutions for the user (e.g. proposals for the choice of components).

d) Highest quality standard of the engineering results. Statical consistency checks and verification of the timing behavior, e.g. timing of control loops.

Before the solution developed in ACORN is presented, common problems in the engineering process of automation systems are shown.

2 Engineering and description languages

As already mentioned, besides the functional view on an automation system, the engineering process becomes more and more important. This engineering process becomes more and more complex, because more components with increased functionality are working together in one system. The following aspects influence the complexity of the engineering process:

- Increasing number of *different devices* in an automation system.

- The consideration of *different features* (electrical, mechanical, functional, ...).

- The use of devices from *different vendors*.

- The integrated information handling of the complete *live cycle of an automation system*.

Independent, for each view and objective optimized description languages are, at the first view, in contradiction to the requirement of a (from an information point of view) non-interrupted engineering concept. This engineering concept is characterized by a information exchange between the live cycle phases without losses.

For such an exchange a common transfer syntax must be agreed. Furthermore, the semantic of the information must not changed. Today's interface engineering (smaller and bigger soft- and hardware adapters) including the necessary commissioning and adaptation processes is, even if implementable, complex and expensive. It is desired to replace this process by integrated approaches (modeling, formal / formalized descriptions, automatic implementations)

A product, and an automation system is also product, runs through several phases of the live cycle (e.g. design, realization, operation, and recycling). During each phase, information is needed. This information is provided by the user, if it is not already available from earlier phases. The tools for information input and presentation are often restricted to a single phase and represent therefore the former separation of the work flow (e.g. design, operation).

The tools are often working with tool and vendor specific information presentation. This hinders the continuous information transport as companion of the product over tool and organization borders. This is especially true, when co-operation via company borders is necessary (i.e. car suppliers). An interrupt of the information flow always means a loss of information, which is expensive. Even if it can be replaced by manual re-input, which is not always possible, resulting problems as wrong inputs, inconsistencies, ... are remaining. The best solution for those problems is the use of vendor and live cycle independent product descriptions, which integrate existing and proven methods of all live cycle phases.

The *Standard for the Exchange of Product Data* (ISO 10303, STEP) offers the potential to fulfill the above mentioned requirements. Especially the mapping and therefore the integration of different description technologies is possible.

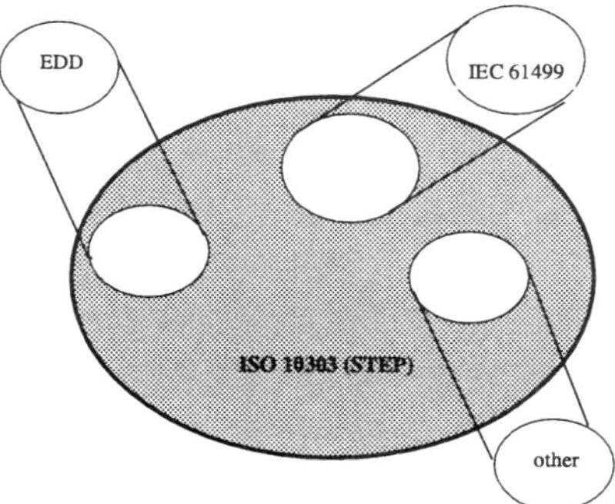

Fig. 1. STEP (ISO 10303) integrates description technologies

So STEP is a basis for the integration of description methods and tools for the whole life cycle of automation systems.

2.1 Standard for the Exchange of Product Model Data (STEP)

The Standard ISO 10303 consists of data description languages EXPRESS and their variants (e.g. EXPRESS-G as graphical presentation), implementation technologies (e.g. file exchange, data base), domain specific data models (Application Protocols) and conformance requirements.

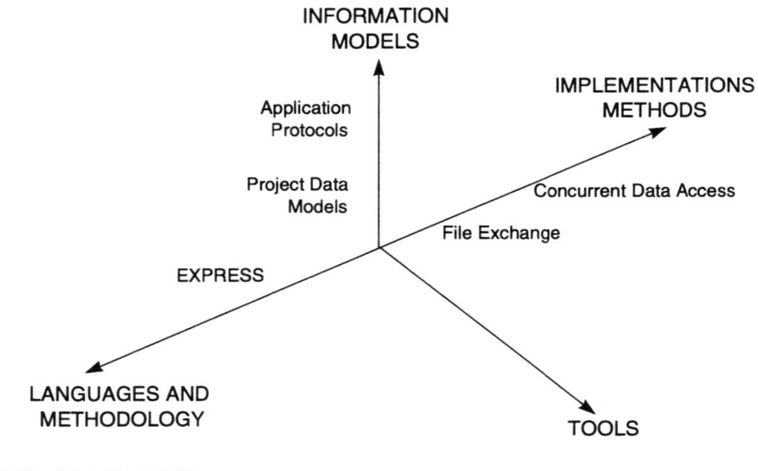

Fig. 2. STEP aspects for the description of domain specific applications

ACORN handles an automation system as product in the definition of ISO 10303. So STEP as method with an already existing infrastructure is available. The problem of describing automation systems is put down to the problem of describing product data.

2.2 IEC61499

Function Blocks are often used in automation, but several variants are existing. The Standard IEC61499 contains a model, which can be used in the process control and manufacturing industry.

Within ACORN, IEC61499 is used to describe the automation functions of a system using the Application Design Tool (ADT). For the algorithms IEC61499 does not contain specifications. The languages of IEC61131 can be used, which is done in ACORN.

2.3 ACORN Device Description (ADD)

To describe components (devices, fieldbus systems) in a machine readable way, ACORN uses ADD. ADD is a enlarged sub-set of EDD (Electronic Device Description). The target of EDD is the integration and operation phase, the later is outside the scope of ACORN. Using ADD/EDD makes it possible to create electronic data sheets and provide them for use in the whole life cycle of the component.

Within ACORN, the components are described using the *Resource Management Tool* (RMT) based on ADD.

3 The ACORN Tool Kit

The ACORN tool kit consists of four tools, which are matching with each other: the *Application Design Tool* (ADT), the *Configuration Management Tool* (CMT), the *Simulation Tool* (ST), and the *Resource Management Tool* (RMT). The information exchange between the tools is realized by a shared data base.

The *Application Design Tool (ADT)* supports the function planning in the design phase of an automation system. The process control oriented, control system independent function planning is using generic Function Block types based on IEC61499 and if needed additional project specific *Instrumentation Requirements*. Using both description methods, the relevant aspects of the system function (Function Blocks), the interaction of the Function Blocks, the requirements to the timing behavior in form of Instrumentation Requirements as well as criteria for the choice of devices are described vendor and device independent.

The separation between the functional design of a system (ADT) and the criteria oriented, device dependent system realization (in a later step using the Configuration Management Tool CMT) makes the design result portable and provides for several realizations. The functional description is done based on IEC 61499, function oriented generic operation timing requirements are added. The Configuration Management Tool (CMT) is used to find suitable components (devices, communication systems, ...) according to the functions and Instrumentation Requirements determined by the ADT, and to choice the fieldbus architecture. If more than one solution is found, the user makes the final choice between the offered possibilities.

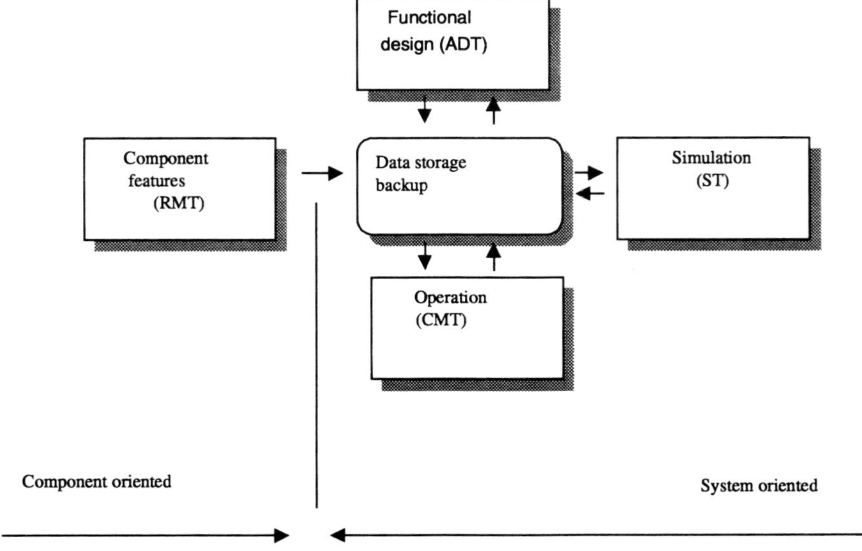

Fig. 3. Tool kit realized by ACORN

228

The component types (with their relevant features), which can be chosen from, are described in a machine readable library. This library is created using the Resource Management Tool.

With the use of the CMT, the transformation from the generic, vendor independent function planning to the concrete system planning and therefore the consideration of device and vendor specific solutions is done. Using the *Resource Management Tool (RMT)*, the relevant features of automation components (devices, fieldbus systems, ...) are described. These descriptions are the basis for the component choice done by the CMT. The description is done in a standardized way as ADD, which can contain vendor or device specific function code. The *Simulation Tool (ST)* provides for the validation of the runtime behavior in an early planning phase. The simulation can be done based on the generic function design (ADT) or on the concrete system design (CMT). Basis of the simulation is a functional description of the Function Blocks according to IEC61131.

The use of the ADT, ST, CMT is repeated for each system. The RMT is different. The description of component types is the basis for the component choice by the CMT, but those descriptions are the same for all systems and are modified only if e.g. the component is changed (new version). In the future, such component descriptions should be provided by the component vendors or other organizations.

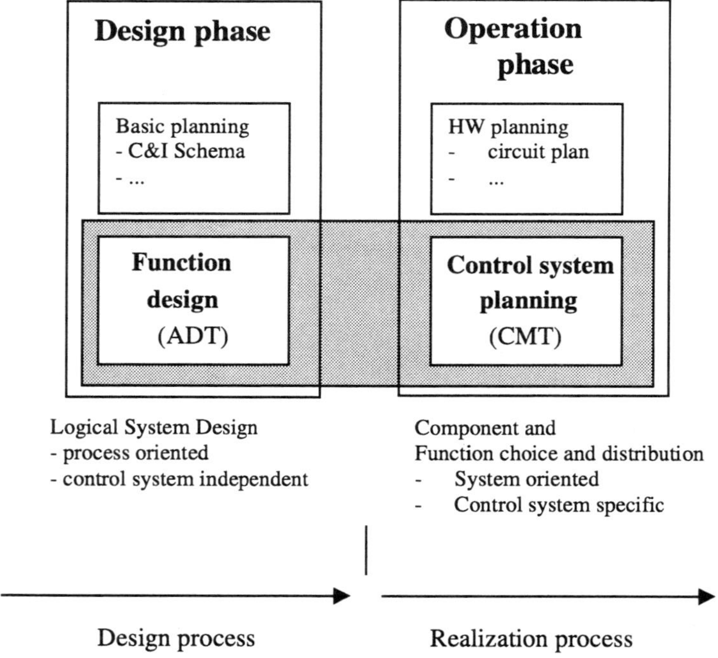

Fig. 4. Separation between functional and system design within the engineering process

4 Summary and Prospects

The method for designing distributed automation systems described within ACORN is using to a large extend techniques and methods already known today. The proposed solution is characterized by a sharp distinction between the design of automation functions and the choice of devices used for processing those functions. This makes it possible to do the functional design without looking at concrete devices. The mapping of the functional design onto concrete devices is done computer supported in a second step according to well defined criteria using Electronic Device Descriptions. By means of simulation of parts of the control system, it is possible to determine during the design process optimal device parameters and to find engineering errors.

The information exchange between the software tools is done according to the STEP Application Protocol (AP212). The tools are working internally with different methods and technologies.

The results created by ACORN are used and continued e.g. in the European CENELEC Project NOAH (Network Oriented Application Harmonization).

References

1. GMA: GMA-Fachbericht, Standortsicherung für die Leit- und Automatisierungstechnik, VDE-Verlag GmbH, Berlin, Offenbach, 1994

2. Olberding, W.: Objektmodellierung zur besseren Wiederverwendbarkeit von Leitsystemsoftware, in 1.

3. Peters, R. W.: Strukturen und Austauschformate für technische Gerätebeschreibungen: Anforderungen an die Hersteller, in 1.

4. Polke, M.: Prozeßleittechnik. Oldenbourg Verlag München. 1994

5. Industrial Automation Systems - Product Data Representation and Exchange, ISO 10303

6. West, M.: STEP: The Future of Engineering Information, SHELL Ltd., 1995

7. Svemar, B.; Koechlin, J.-B.; Simon, R.; Zubimendi, E.; John, K.-H.; Brandl, Th.; Dahlström, L. E.; Gustavsson, L.; Salfati, R.: SIMPLE: Standardization of Interfaces between multiple Programming and Control Environments, Proposal for the Domain of Integration Manufacturing, ESPRIT, 1995

8. Hörger, J.; Simon, R.; Nerke, R.: Field Architecture Reference Model, BriteEuRam Project BE95-479 ACORN, Deliverale D03.1, 1996

PROFIBUS goes Microsoft - Herstellerunabhängige Integration von Feldgeräten in Engineeringsysteme

Harald Bruns[1], Ulrich Hempen[2], Wolfgang Ott[3],
Rolf Vahldieck[4], Karl-Heinz Niemann[5]

[1]ABB Utility Automation GmbH, Kallstadter Str. 1, 68309 Mannheim
[2]Endress + Hauser GmbH & Co., Hauptstr. 1, 79689 Maulberg
[3]Siemens AG, Frauenauracher Str. 85, 91056 Erlangen
[4]ABB Automation Products, Schiller Str. 72, 32425 Minden
[5]ABB Automation Products, Hackethalstr. 7, 30179 Hannover

Abstract. Bei der Integration von Feldbussen in schon bestehende Systeme gibt es noch einige Aufgaben, die gelöst werden müssen. Dies trifft für Feldbusse im Allgemeinen zu, und wird im folgenden Beitrag am Beispiel PROFIBUS aufgezeigt. Speziell für den Einsatz in umfangreichen und heterogenen Anlagen, typischerweise im Bereich der Verfahrenstechnik, ist die eindeutige Definition von Engineering-Schnittstellen, die für alle Beteiligten einfach zu nutzen sind, und die Einbindung verschiedener gerätespezifischer Werkzeuge in übergeordnete anlagenweite Planungssysteme möglich macht von sehr großer Bedeutung. Zur Bewältigung dieser Aufgabe haben Leitsystemhersteller und Gerätelieferanten im ZVEI einen Arbeitskreis gebildet, um im engen Dialog mit der PROFIBUS Nutzer Organisation (PNO) eine herstellerunabhängige Lösung zu erarbeiten. Hauptziel dieser Aktivitäten ist, ein einheitliches Softwareinterface zu spezifizieren, das einerseits für die Feldgerätehersteller einfach zu realisieren und andererseits für die Anbieter von Prozeßleitsystemen bzw. Engineeringumgebungen ebenso einfach zu implementieren ist.

Abstract. With the integration of fieldbusses into existing systems are some tasks, that still wait for completion. This is valid for fieldbusses in general. The following article will show this at the example of PROFIBUS in special. The integration of field instrument configuration-tools with engineering interfaces, that could be used by all vendors in the plant, is very important. This is especially true for large and heterogeneous plants in the process industry. To resolve this task, the control systems and instrument vendors founded a working group in the ZVEI (German manufacturers association) to develop in close contract with the PNO (PROFIBUS Users organization) a vendor neutral solution. Main target of this activities is the specification of a software-interface. This interface should easy to implement for the field instrument manufacturers as well as for the DCS vendors.

1 Problemstellung

In der Prozeßautomatisierung umfaßt eine Leitanlage oft mehr als 10.000 binäre und analoge Ein-/Ausgangssignale. Bei der Anwendung des Feldbusses werden diese Signale über den Bus geführt. Hierzu werden die Feldgeräte direkt an den Bus angeschlossen oder über Remote I/O erfaßt. Häufig sind mehr als 100 unterschiedliche Feldgerätetypen, von verschiedenen Geräteherstellern im Einsatz.

Die Geräte werden für die jeweilige Aufgabenstellung konfiguriert und parametriert. Die gerätespezifischen Eigenschaften und Einstellungen müssen bei der Konfigurierung des Feldbuskopplers und der Buskommunikation berücksichtigt, und die Geräte im Leitsystem bekannt gemacht werden. Es müssen Funktionsbausteine mit Ein- und Ausgangssignalen erstellt, und in die Funktionsplanung des Leitsystem integriert werden.

2 Stand der Technik

Die große Anzahl verschiedener Gerätetypen und Lieferanten innerhalb eines Leitanlagenprojektes macht diese Konfigurationsaufgabe heute schwierig und zeitraubend. Es müssen unterschiedliche Werkzeuge beherrscht sowie Daten zwischen diesen Werkzeugen ausgetauscht werden. Der Datenaustausch ist nicht standardisiert. Daher sind vielfach Datenkonvertierungen erforderlich, die ein detailliertes Fachwissen voraussetzen. Die Konsistenz der Daten, Dokumentationen und Konfigurationen kann letztlich nur durch einen intensiven Anlagentest sichergestellt werden.

Auch der zentrale Arbeitsplatz für Service- und Diagnoseaufgaben im Leitsystem deckt nicht im vollen Umfang die Funktionalität der Feldbusgeräte ab, da die verschiedenen gerätespezifischen Werkzeuge nur direkt an einer Feldbuslinie oder direkt am Feldgerät angeschlossen werden können.

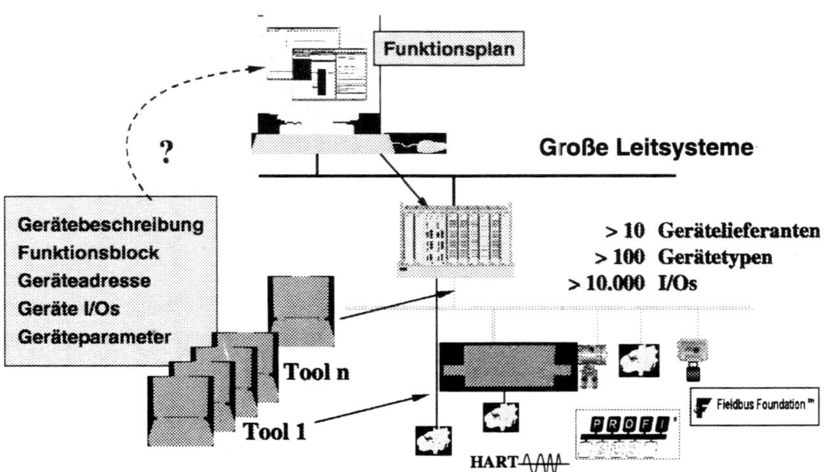

Abb. 1. Unterschiedliche Tools und mehrfache Dateneingabe bestimmten die Feldgeräteintegration bisher

3 Ziele

Um die Durchgängigkeit und Betriebssicherheit der Prozeßleittechnik aufrechtzuerhalten, ist es erforderlich, Feldbusgeräte als eine Teilkomponente der Prozeßautomation vollständig zu integrieren. Die Prozeßleitsysteme müssen den Kommunikationsweg von einem zentralen Engineeringarbeitsplatz über die System- und Feldbusse zu den einzelnen Feldgeräten bereitstellen.

Die wesentlichen Ziele hierbei sind:

- zentraler Arbeitsplatz für Planung, Diagnose und Service mit direktem Zugriff auf alle Feldgeräte

- integrierte, konsistente Gerätekonfigurierung und Dokumentation

- Organisation gemeinsamer Engineeringdaten für das Prozeßleitsystem und die Feldgeräte

- konsistente Konfigurierung des Prozeßleitsystems, der Feldbusse und Geräte

- zentrale Datenhaltung und Datensicherung

- einfache und schnelle Integration von Gerätetypen in das Prozeßleitsystem

Die Integration der Feldgerätetechnik in die Engineeringsysteme der Prozeßleittechnik soll sich nicht nur auf einen kleinen, allgemein gültigen Satz von Konfigurations-, Service- und Diagnosefunktionen begrenzen - dies hieße eine Integration der PROFIL-Definitionen als Basisdefinition für Feldgeräte. Durch die Integration sollen vielmehr die individuellen Geräteeigenschaften, Merkmale und Besonderheiten der verschiedenen Gerätetypen unterstützt werden. Die vom Gerätehersteller bereitgestellten Werkzeuge zur Planung und Service sollen als gerätespezifische Software Komponenten in das Engineeringsystem eingebunden werden. Der Gerätehersteller definiert somit die Konfigurations-, Service- und Diagnosefunktionen für seine Geräte selbst und gestaltet so auch das Erscheinungsbild seiner Geräte in der Engineeringumgebung des Prozeßleitsystems selbst.

Diese Komponenten müssen dann in die Engineeringsysteme aller Leittechniklieferanten integriert werden können. Dadurch reduziert sich der Aufwand für die Gerätehersteller, zu einem intelligenten Feldgerät nur eine standardisierte Softwarekomponente mit allen Konfigurations-, Service- und Diagnosefunktionen anzubieten. Die häufigen projekt- oder leitsystemspezifischen Integrationen, die für einen Gerätetyp mehrfach entwickelt und gepflegt werden müssen, sollen durch eine standardisierte Komponententechnik entfallen.

Der Leitsystemhersteller muß nur einmal die definierten Schnittstellen für die Integration aller Feldbusgeräte implementieren. Hersteller- und/oder gerätespezifische Implementierungen entfallen.

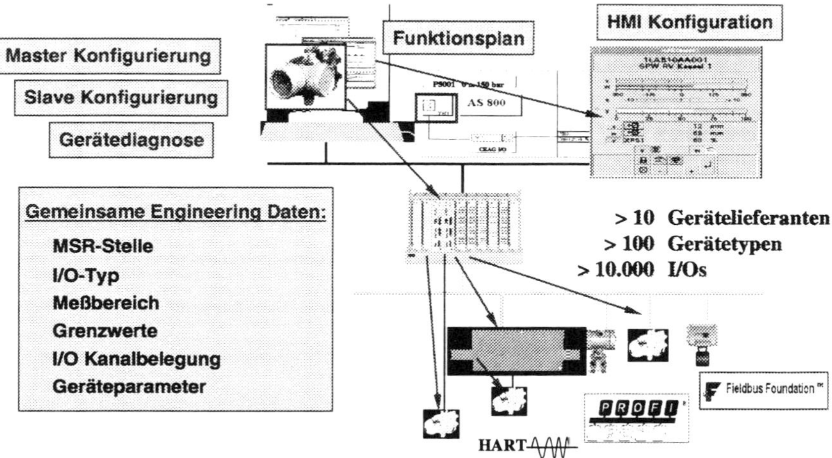

Abb. 2. Erst die homogene Integration des Feldbus in die Engineeringsysteme macht das Potential der Feldbustechnik nutzbar.

4 Technologische Ausrichtung

Durch die Kombination einer Software-Komponente mit einer Hardware-Komponente, ähnlich der Treiber-Software zu einem Drucker, wird das „Plug & Play" Prinzip auch in der Feldbus-Technik eingeführt.

Da Microsoft Windows sich auch als de-facto Industriestandard etabliert hat, ergeben sich bei der Auswahl des Betriebssystems wenig Alternativen. Die von Microsoft eingeführte ActiveX-Technologie erlaubt es Schnittstellen zu definieren, die nicht nur Daten sondern auch Funktionen enthalten. Diese Möglichkeiten wurden bereits erfolgreich bei der OPC Definition (OLE for Process Control) genutzt. Für die Engineering-Komponenten der Feldgeräte werden lediglich Schnittstellen in der ebenfalls von Microsoft definierten COM/DCOM-Technologie (Distributed Component Object Model) festgelegt. Wenn das Engineeringsystem ebenfalls diese Schnittstellen aufweist, übernimmt die ActiveX-Technologie die automatische Integration der Komponenten und sorgt für die Interaktionen zwischen dem Engineeringsystem und den Softwarekomponenten der Geräte, sowie für ein einheitliches Look and Feel im Sinne der Windows-Technik.

5 Stand der Aktivitäten

Um die Lösung und eine einheitliche Vorgehensweise zu konkretisieren, wird demnächst der gemeinsame Entwurf einer Interfacebeschreibung nach dem Muster der OPC Interface Specification vorliegen, die die unabhängige Erstellung von COM basierenden Software-Komponente ermöglicht, die herstellerunabhängig in alle Integrationsumgebungen eingepflanzt werden können, die diese Schnittstellen ebenfalls zu Verfügung stellen. Der Arbeitsname für diese Software ist „Device Type Manager" (DTM).

234

Er beinhaltet gerätetypspezifisch alle Geräteeigenschaften einschließlich der nötigen User-Interfaces zur Konfigurierung, Diagnose und Service.

Natürlich macht die spezifische Erstellung dieser Software-Komponenten vor allem dann Sinn, wenn das Gerät hinreichend komplexe Eigenschaften hat und diese durch anspruchsvolle Konfigurier- und Diagnoseoberflächen unterstützt werden. Die DTMs müssen nicht in jedem Fall neu entwickelt werden. Es ist möglich ein DTM aus einem existierenden Gerätekonfigurationstool zu erstellen oder automatisch aus Gerätebeschreibungen zu generieren. Einfache Geräte sollen nach wie vor durch eine Beschreibungssprache erfaßt werden können, die dann durch einen Interpreter oder Compiler in die Software-Komponente überführt werden kann.

Die Device Type Manager sollen vom Gerätehersteller zusammen mit dem Gerät geliefert werden. Folgende Eigenschaften sind für die DTM kennzeichnend:

- kein Stand-alone-Tool

- eine ActiveX Komponente (COM/DCOM)

- definierte COM Schnittstellen

- alle Regeln des Gerätes bekannt

- alle Anwenderdialoge enthalten

- integrierte Gerätekonfigurierung und Diagnose

- integrierte gerätespezifische Dokumentation

- keine eigene Datenhaltung

- keine direkte Verbindung zu irgendeinem Gerät

- keine Information über die Engineering Umgebung

- Unterstützung von einem oder mehreren Geräte

Ein DTM wird als Komponente eines Engineeringsystems oder eines Stand-alone Tools installiert, das die Geräteinstanzen verwaltet, die Kommunikationsmechanismen bereit stellt und die zugehörige Komponente mit gerätespezifischen Aufgaben beauftragt.

Für das Engineeringsystem oder Stand-alone Tool gelten folgende Anforderungen:

- keine gerätespezifischen Kenntnisse erforderlich

- verwaltet alle Geräteinstanzen und speichert alle Instanzdaten

- stellt die Gerätekommunikation und Verbindung her

- sichert eine anlagenweite konsistente Konfigurierung

- ermöglicht Multi-User und Server/Client Betrieb

- übernimmt die Datenversionierung

6 Lösungskonzept

Je Feldgerät (Aktor, Sensor) wird vom Gerätehersteller ein DTM zur Verfügung gestellt. Folgende, vom Gerätehersteller auf sein Gerät hin optimierte Funktionen sind darin enthalten:

- Benutzeroberfläche (Mehrsprachig einschließlich Hilfesystem)
- Plausibilisierung der Parameter (auch in Abhängigkeit anderer gerätespezifischer Parameter)
- Automatisierte Erzeugung abhängiger Parameter
- Vorgabe der Bearbeitungssequenzen von komplexen Kalibrier-, Abgleich- und Einstellvorgängen bei hochwertigen Feldgeräten
- Lesen und schreiben der Parameter aus/in das Feldgerät
- Auf das Gerät abgestimmte Diagnosefunktionen
- Bereitstellung der Typ-Daten für den Kommunikationsaufbau
- Geräte-/Instanz-spezifische Anlagendokumentation

Die Menge der hier aufgeführten Funktionen stehen in Abhängigkeit vom Funktionsumfang des Gerätes. Ein DTM deckt mindestens ein Feldgerät ab. DTMs können jedoch auch Gerätefamilien (z.B. Druckmeßumformer) z.B. auf Basis der PROFIBUS Profildefinitionen oder die gesamte Palette eines Herstellers abdecken. Kommunikation (über die diversen Bussysteme einer Anlage) und Datenhaltung werden über die Interfaces des Engineeringsystems hantiert. Im Rahmen einer Gesamtanlagenplanung oder Betriebsführung ist ein DTM immer in das entsprechende Engineeringtool einzubinden. Ein paralleler Stand-alone-Betrieb ist aus Konsistenzgründen nur in besonderen Fällen (z.B. Vernetzung in Werkstatt nicht vorhanden) durchzuführen.

7 Von der „GSD" zum DTM

„GSD" steht für „Gerätestammdaten" und beschreibt kommunikationsspezifische Parameter eines PROFIBUS DP Gerätes. Dies sind Angaben zur Geräteidentifikation und Kommunikation. Die GSD ist damit unumgänglich für die Einbindung eines PROFIBUS DP Teilnehmers in ein PROFIBUS Netzwerk [1]. Durch die Einführung konfigurier- und parametrierbarer PROFIBUS DPV1 oder PA Geräte ist nun jedoch nicht nur die Einbindung der zyklischen Kommunikation, sondern ebenfalls die Konfiguration und Parametrierung der Feldgeräte notwendig. Dies erfolgt über azyklische Dienste, die vom Engineering tool oder auch aus der Steuerung gesendet werden.

Die zu beschreibenden einstellbaren PROFIBUS Geräte können dabei in vier Kategorien unterteilt werden:

A: Einfache nur zyklische kommunizierende Geräte, z.B. Lichtschranke

B: Einstellbare Geräte mit fixer Hard- sowie Software, z.B. Druckmeßumformer

236

C: Einstellbare Geräte mit modularer Hardware aber fixen Softwareblöcken, z.B. Remote I/O

D: Einstellbare Geräte mit modularer Hardware und programmierbaren Softwareblökken, z.B. komplexer Regelantrieb

Diese verschiedenen Gerätetypen in einem Bedientool oder einer Engineeringumgebung zu beschreiben und damit bekannt zu machen, übersteigt bei weitem die Leistungsfähigkeit der Beschreibungssprache GSD. Dabei darf die Realität nicht außer Acht gelassen werden:

Die existierende „GSD-Welt" wird auch weiter bestehen bleiben und nicht jeder Feldgerätehersteller ist in der Lage einen DTM zu erzeugen. Eine einfache Beschreibungsmethodik zur Erzeugung eines DTM ist durch die z.Zt. in der Endphase der Spezifikation befindliche PROFIBUS DDL gegeben. . Natürlich hat die PROFIBUS DDL nicht eine unbegrenzte Performance, sie ist jedoch mindestens in der Lage einstellbare Geräte mit fixer Hard- und Software wie z.B. Meßumformer zu beschreiben. Die Migration erfolgt über einen „Default-DTM", der die GSD und die Profibus DDL interpretiert und dadurch die notwendigen Daten der eigentlichen DTM Schnittstelle zur Verfügung stellt. Dieser „Default-DTM" ist nicht gesondert vom Geräte- oder PLS/SPS-Hersteller zu entwickeln. Er steht vielmehr als Standardprodukt zur Verfügung und kann in die unterschiedlichsten Toolumgebungen eingesetzt werden.

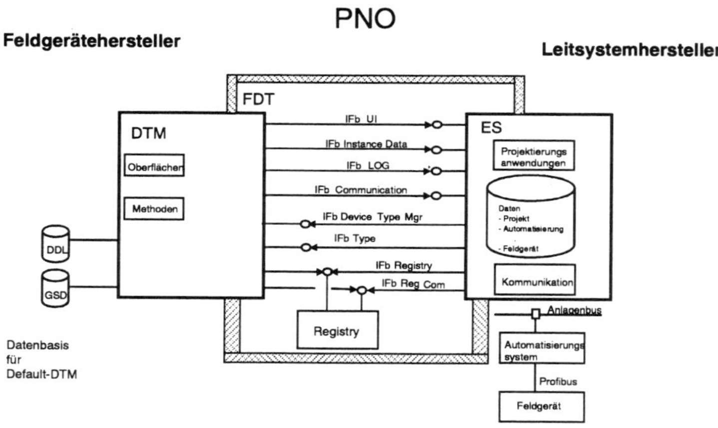

Abb. 3. Das FDT Konzept bildet die Schnittstelle zwischen DTM und Engineering System (ES)

Die Integration eines Gerätes kann somit über die GSD und die PROFIBUS DDL zusammen mit dem „Default-DTM" erfolgen. Die Beschreibung komplexerer Slaves wie unter C.) und D.) angegeben, erfolgt durch ein eigenes DTM, welches in einer freien Sprache entwickelt werden kann.

Die Vorteile dieser beiden Optionen zur Erzeugung eines DTM liegen darin

- Die bestehende „GSD-Welt" zu integrieren und sie auch in der kürzeren Zukunft nutzen zu können. Eine Migration, bzw. parallele Nutzung ist damit möglich.

- Die Akzeptanz für das DTM-Modell durch die Migration der GSD und durch die Möglichkeit einer einfachen Beschreibungsmethode mit der PROFIBUS DDL zu vergrößern.

Durch die Flexibilität des DTM in der Unabhängigkeit zum Kommunikationsprotokoll auch die Offenheit zu anderen Feldbussen besteht.

8 Übertragbarkeit auf andere Feldbusse

Selbstverständlich gilt das bisher beschriebene nicht nur für PROFIBUS, sondern läßt sich auch auf andere Feldbusse übertragen. Dieser Tatbestand wurde von Anfang an berücksichtigt, da die zu lösenden Aufgaben nicht PROFIBUS spezifisch sind, sondern ebenso andere Feldbussysteme betreffen. HART, FF und LON Geräte sind hier vorrangig zu nennen, deren Integration für die Leitsystemhersteller in gleichem Maß notwendig ist.

9 Ausblick

Die homogene Einbindung von Geräten mit unterschiedlichen Bus-Techniken kann künftig nur über gemeinsame Schnittstellen erfolgen. Microsoft mit dem Betriebssystem Windows NT bzw. folgenden Generationen sowie die de-facto Standards wie OPC und COM werden dabei eine Schlüsselrolle und die gemeinsame Integrationsplattform bilden.

Auch wenn Hersteller diesseits und jenseits des Atlantik ihre proprietären Strategien mit verschiedenen Bus-Techniken weiterverfolgen, eröffnet dieses Konzept die Möglichkeit einer einheitlichen Integration dort wo es wirklich zählt: bei Engineering, Diagnose und Service, entkoppelt von busspezifischen Kommunikationstechniken und herstellerspezifischen Engineeringumgebungen.

Referenzen

1. GSD Spezifikation, PROFIBUS Richtlinie, PNO Karlsruhe

Chapter 6: OLE for Process Control OPC

Papers presented in Session 2.3

A Modular OPC-Server Connecting Different Fieldbussystems and Internet Java Applets
Rüping, S.; Klugmann, H.; Gerdes, K.-H.; Mirbach, S.

OPC-Schnittstellen in einer offenen Systemumgebung – Praxis und Erfahrungen
Langmann, R.; Groppe, A.; Ossipov, V.; Stuhrmann, N.

OPC – Making the Fieldbus Interface Transparent
Hadlich, T.; Szczepanski, T.

A Modular OPC-Server Connecting Different Fieldbussystems and Internet Java Applets

S. Rüping[1], H. Klugmann[1], K.-H. Gerdes[2], S. Mirbach[2]

[1]Heinz Nixdorf Institut
University of Paderborn
System and Circuit Technology
Paderborn, Germany

[2]FASTEC GmbH
Decentralized Automation
Paderborn, Germany

Abstract. The standardization of the interface between the office and the field level becomes more and more important in automation systems. In this paper a modular OPC-Server is presented, that combines two features. First, an application software on office level has to be connected to only one server, even if it needs data connections to different fieldbusses. And second, fast and efficient communication between different fieldbusses is possible. Furthermore an OPC-Client is implemented, that is also a Web-Server with Java RMI functions (Remote Method Invocation). Combining the OPC-Server with the OPC-Client results in a system, that enables data access from a standard PC with a Web-Browser into several different bussystems on field level.

1 Introduction

The communication interface between the office and field level in automation systems is an important element, that is needed to guarantee the necessary information flow for efficient working principles. The existence of many different fieldbussystems leads today to application specific solutions, especially concerning the applied software. The introduction of OPC (Open Process Control) [1] was an important development, which standardizes the interface used by a software on office level for data access on field level. There are already several fieldbus-specific OPC-Server on the market.

Additionally communication also between nodes of different fieldbusses becomes more and more important in order to improve the global usage of the infrastructure. A system that enables a simple and fast communication between these nodes, provides more possibilities to find efficient solutions.

In this paper we present our work, that solves both of the described problems. First a modular OPC-Server is implemented, that is able to connect several different fieldbussystems concurrently to a software on office level using a single interface. And secondly our tool provides mechanism for fast and efficient communication between nodes of different fieldbusses.

2 OPC, Open Control and Java

2.1 OPC

The idea to build a modular system based on standards and fixed interfaces has already been used successfully for the design of many technical solutions. Therefore it was obvious, that the software on office level should also follow this principle. The computers on which the software is installed often have Microsoft operating systems. And Microsoft already has interfaces for interprocess communication (DDE, OLE, COM, DCOM [5]) which are used for the implementation.

OPC (OLE for Process Control) [1] is the result of the standardization idea. It specifies, how an application can access the process data. The application does not depend on the system, where the data comes from. This means, that the same application can be used with different fieldbussystems without modifications.

2.2 OpenControl

A further development in this field is OpenControl [2]. The specification has three different parts:

- CALL-Engineering (CALL-E)

- CALL-Runtime (CALL-R)

- CALL-Periphery (CALL-P)

CALL-E is an interface for engineering tools providing a common database. CALL-R implements a communication interface used during operation of the system and CALL-P is the interface for a fast and efficient data exchange on fieldbus level. An extended OPC-Server can be found in the CALL-R specification. CALL-P drivers are used in our system to connect different fieldbusses to the data-kernel, which means we can realize the fast communication between nodes from different fieldbusses.

2.3 Java

Java is an object-oriented programming language, that runs on many different kinds of computers by using a bytecode based on the Java Virtual Machine [3]. A Java applet, which is used in our context, is a piece of software, that is embedded into a html-page. Therefore it can be run by using a standard web-browser and loading the html-page with the Java applet over a network. Today nearly every computer has a web-browser and is connected to a network.

3 The modular OPC-Server

3.1 Basic Concept

An OPC-System can consist of several clients and different server. The communication will take place always between a client and a server. A client can communicate with several server, and several clients can access one server. Therefore it is possible, that an application software can access data coming from different fieldbussystems by using different OPC-Server. The disadvantage of this solution is, that communication between nodes of the fieldbussystems is very slow and must be managed by the application software (OPC-Client). On the other hand, the application software has to connect to different server, which might be more complicated than to use only one server.

Based on the idea to speed up the communication between nodes from different fieldbusses and to simplify the interface for the application software, we developed a modular OPC-Server. In order not to start from scratch, we used a sample implementation of an OPC-Server available in the internet [6], modified the software and added a new data kernel. Fast CALL-P drivers can now be attached to the server to connect the fieldbusses to the kernel. The kernel enables direct data connections without OLE mechanism between the fieldbusses. Figure 1 shows the modules of the server for connecting a single fieldbussystem.

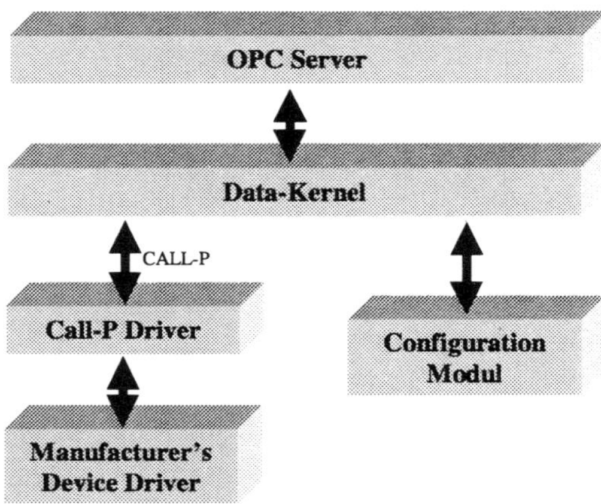

Fig. 1. Modules for connecting a single Fieldbussystem to the OPC-Server

The OPC-Server module on top of the stack implements the functions concerning the OPC-Interface for the application software (OPC-Client). The data-kernel contains the graphical user interface for the configuration of the server and an internal data representation, which provides an unified data interface for the server module.

For each fieldbus connection three modules are needed, the CALL-P driver, the manufacturer's device driver and a configuration module. The manufacturer's device driver provides functions to access data in the fieldbus. The CALL-P driver uses these functions to implement the CALL-P functions, which are called by the data-kernel.

A configuration module has the following tasks:

- Configuration dialogs for all necessary objects

- Input of the object hierarchy

- Reference to the CALL-P driver

- Create, load, modify and save of fieldbus-specific configuration data

- Support of fieldbus-specific data types

3.2 The Data-Kernel

The dafa-kernel is the visual part of the OPC-Server. It is realized as a DLL (Dynamic Link Library), which is called by the non-visual server module on top of the stack (see figure 1). The main tasks of the data-kernel are:

- Add and delete data objects belonging to different fieldbussystems

- User-friendly input and presentation of the available data objects

- Assignment of physical data objects to OPC-Items

- Assignment of data objects to CALL-P channels

- Translation of fieldbus data to OPC-Format

- Data Caching

- Fieldbus-specific configuration (addressing, datatypes, ...)

- Load and save of the complete configuration

Figure 2 shows an example of the data-kernel's main window.

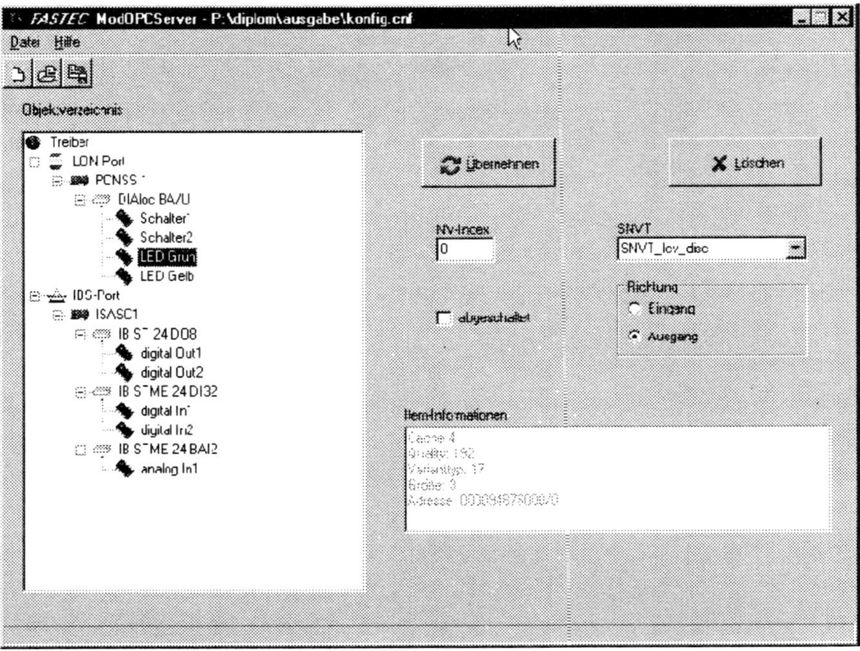

Fig. 2. The main window of the data-kernel

On the left side of the window a hierarchical view of all items is shown. On top level the type of fieldbus has to be selected. The example shows two selections, LON and Interbus. The next step is to specify the hard- and software interface for the busses. In the example it is PCNSS 1 for LON and ISASC1 for Interbus. After that the selected nodes of the systems are presented and on the lowest level the data objects of the nodes are shown. Names can be given to all items which are now the corresponding OPC-Item names.

On the right side of the window the configuration inputs for the currently highlighted item are shown. In the example a LON network variable ("LED Grün") is selected and the Standard-Network-Variable-Type (SNVT), the direction (input or output) and the network variable index can be specified. The right side of the window changes when a Interbus data object is highlighted. Figure 3 shows the configuration parameters in that case.

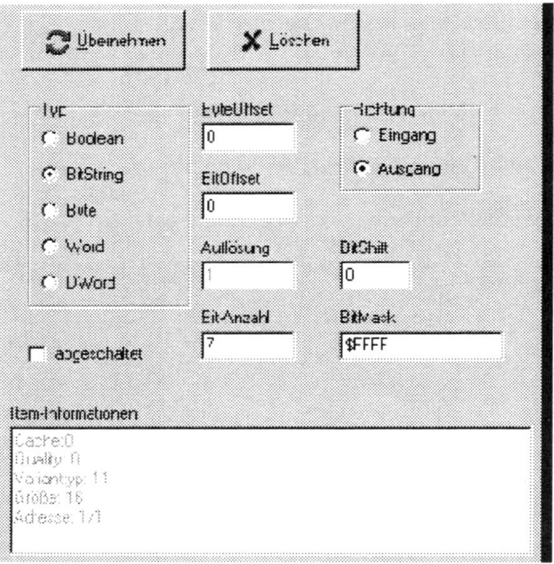

Fig. 3. Interbus data object configuration window

The type, Byte- and Bitoffset, the direction and some other configuration data can be included for Interbus objects. After the input of data objects to the item list on the left side is finished, all items can be read or written by an OPC-Client, which has not to care about the fieldbussystem the data comes from.

Additionally the kernel provides the functions to connect data objects from different fieldbussystems without using the relatively slow OLE mechanism. At the moment this feature has been tested, but a user-friendly front-end is missing. This is one of the current tasks we work on.

4 Connecting Java Applets

With the modular OPC-Server we can provide a standard and simple interface on office level to data objects from different bussystems on field level. The next idea was to implement a OPC-Client, which enables the usage of state of the art Internet technologies. As described in chapter 2, Java applets seem to become very important for applications, that should be accessible over a network and should run on every computer. Java applets are embedded into html-pages that are loaded and run by a standard web-browser. In order to use the Java technology, the following concept has been developed (figure 4).

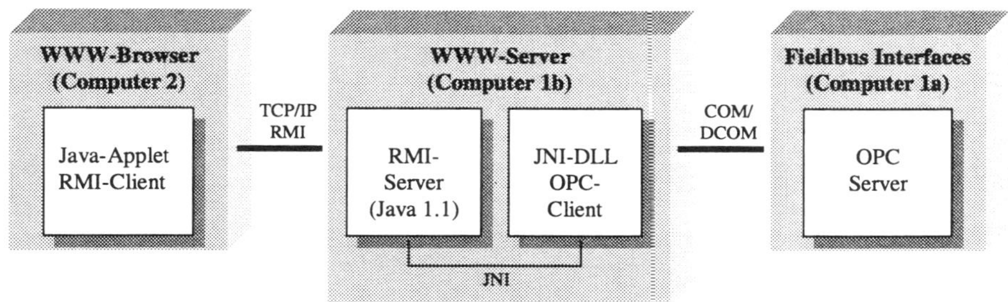

Fig. 4. Accessing data Objects with Java Applets

The OPC-Client uses the COM/DCOM mechanism to connect to the OPC-Server and access all items. The access functions must be written in a programming language (for example C++), which provides an suitable interface to the COM/DCOM mechanism. The Java-Native-Interface (JNI [4]) is used to call these functions from a Java program and so to access OPC-Items. On the other hand the Java program also has to be a RMI server (Remote Method Invocation), which makes it possible, that an applet on a remote computer can communicate with the server. It is required, that both computers have a TCP/IP connection.

A complete system combining both, the modular OPC-Server and the OPC-Client with Java applet capability is shown in figure 5. Several different fieldbussystems can be connected to the OPC-Server, where the data objects are assigned to OPC-Items. Additionally fast communication between different fieldbusses is possible. The OPC-Client connects to the server and provides data access to several Java applets running on remote computers. The remote computers do not need custom application software. They only must have a standard web-browser and a network interface (Internet).

5 Conclusions

In this paper a modular OPC-Server is presented, that makes data from several different fieldbus systems accessible using a single interface. Additionally data objects from the different fieldbusses can be connected to each other. The fast CALL-P drivers and the direct connection of the objects in the data-kernel enables a fast and efficient communication. Furthermore a special OPC-Client is presented, which provides a Java

RMI interface. With this client PCs with standard web-browser can access data of several different fieldbusses over a network (e.g. Internet).

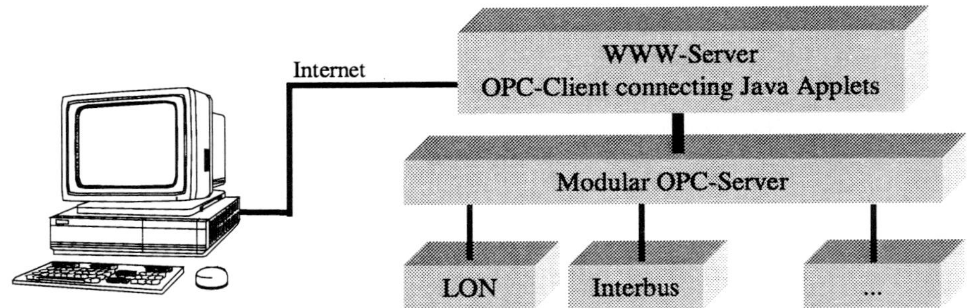

Fig 5. System with different Fieldbussystems and Internet Applets

Future work will be to implement a user front-end, that is used to configure the direct data connections realized in the data-kernel.

<u>Contact:</u> rueping@hni.uni-paderborn.de

References

1. OPC Taskforce: OPC Specification Version 1.0a,
    http://www.opcfoundation.org/opc_spec_document.htm, 1997

2. Interbus-S Club e.V.: Open Control Interface, Version 1.4,
    http://www.open-control.com, 1998

3. The Java Language Specification, Addison–Wesley, 1996

4. Sun: JNI-Specification,
    http://java.sun.com/products/jdk/1.1/download-pdf-ps.html

5. Frank E. Redmond: DCOM, Microsoft Distributed Component Object Model, IDG Books, 1997

6. *http://www.intellution.com/internet_opc_foundation/opcsample.asp*

OPC-Schnittstellen in einer offenen Systemumgebung - Praxis und Erfahrungen

Reinhard Langmann*, Arnd Groppe*, Viktor Ossipov**, Norbert Stuhrmann*

* University of Applied Sciences Düsseldorf, Electrotechnical Department
Josef-Gockeln-Str. 9, D-40474 Düsseldorf
Tel. +49-211-4351308, Fax. +49-211-4351308, R.Langmann@t-online.de
**St. Petersburg State Elektrotechnical University, Department RAPS
Prof. Popova str. 5, 197376 St. Petersburg, Russia
Tel: +812 - 234 9661, Fax. +812 - 346 - 2758, ossipov@raps.etu.spb.ru

Abstract. Der Beitrag präsentiert Ergebnisse einer Untersuchung zum pratischen Einsatz von OPC-Schnittstellen in einer offenen Systemumgebung. Die Untersuchungen basieren auf der Hard- und Software für eine Modellfabrik mit "flacher" Kommunikationsstruktur, die seit diesem Jahr im Labor Prozeßlenkung an der FH Düsseldorf aufgebaut wird. Es zeigt sich, daß OPC-Schnittstellen schnell installiert und sicher angewendet werden können. Mit den gegenwärtig verfügbaren OPC-Schnittstellen lassen sich die Zeitanforderungen im Leitsystem- bzw. Prozeßvisualisierungsbereich gut erfüllen. Die zuverlässige und effiziente Nutzung einer OPC-Schnittstelle erfordert jedoch hinreichend sichere Kenntnisse und Erfahrungen in der PC- und Windows-Technologie.

Abstract. The paper represents any results of a research about usability of OPC interfaces within an open system environment. The hardware and software platform for this works is a plant model using a lean communication structure which is being built up since 1999 at the process control lab of the University of Applied Sciences in Düsseldorf. The results of this works show that OPC interfaces provide quick installation procedures and safe operation. Currently available OPC interfaces are suitable to meet the timing requirements for a process data communication between visualization/supervisory systems on one side and fieldbus/control systems on the other. However, the application of an OPC interface requires a sufficient knowledge and experiences in PC and Windows technology.

1 Modellfabrik mit „flacher" Kommunikationsstruktur

Offene Systeme, Feldbus- und Internet-Technologie sowie die PC-Technik sind heute in der Automatisierungsbranche allgegenwärtige Schlagwörter. Bei der konsequenten praktischen Realisierung gibt es aber noch eine Vielzahl von Problemen zu lösen. Insbesondere für die Konsistenz und Durchgängigkeit der vertikalen aber auch horizontalen Kommunikation und Informationsverarbeitung innerhalb eines offenen automatisierten Systems gibt es bisher zwar eine Reihe von Lösungsansätzen, aber nur wenig Erkenntnisse und Untersuchungen zu integrativen und praktikablen Lösungen.

Das Entwicklungs- und Forschungsprojekt „ Industrielle Kommunikation in einer Zwei-Ebenen-Modellfabrik für die dezentralisierte PC-gestützte Automatisierung von Produktionsprozessen" (kurz: INKODA) an der FH Düsseldorf soll an diesem Punkt ansetzen und Probleme der kommunikations- und informationstechnischen Vernetzung untersuchen. Anhand beispielhafter automatisierungstechnischer Modellstrukturen sollen Lösungsmöglichkeiten für wesentliche Kernelemente mit aktuell verfügbaren und noch zu entwickelnden Produkten aufgezeigt werden. Im Rahmen des Projektes werden u.a. folgende Ziele verfolgt:

- Schaffung einer Modellstruktur mit zwei Automatisierungsebenen für die dezentralisierte Automatisierung von Produktionsprozessen,

- Test und Erprobung fortgeschrittener Kommunikations- und Informationsstrukturen für die dezentralisierte Automatisierung,

- Untersuchung der Fähigkeit komplexer Medien (Video, Audio, Graphik u.ä.) zur Effizienzverbesserung von Automatisierungslösungen,

Die Modellfabrik wird aus verschiedenen PC-basierenden Automatisierungsstationen aufgebaut, die über eine flache Protokollhierarchie mit zwei Ebenen und unterschiedlichen Medien informationstechnisch verknüpft sind. Abb. 1 zeigt die Komponentenstruktur der Modellfabrik.

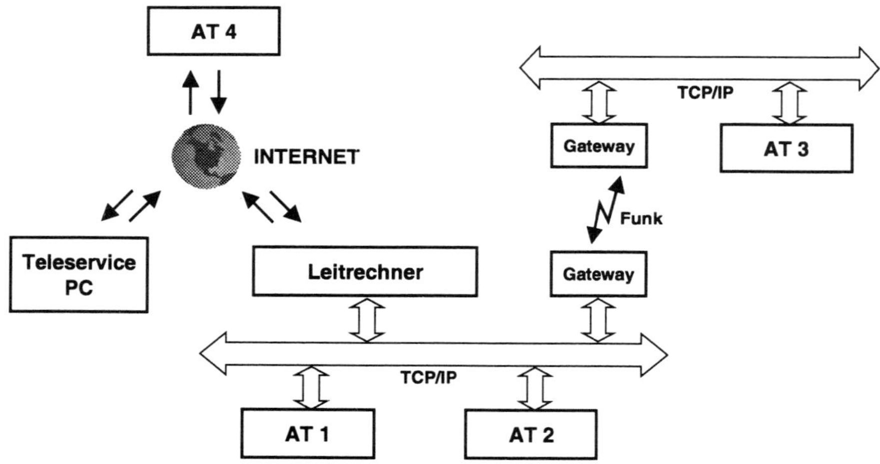

Abb. 1. Komponentenstruktur der Modellfabrik im Projekt INKODA

Als physische Automatisierungsstationen kommen fertigungstechnische Einheiten aus dem Lehrmittelprogramm von FESTO und ADIRO zum Einsatz. Die Steuerung der Stationen erfolgt ausschließlich auf der Basis von PC's bzw. Industrie-PCs. Zur Anbindung der Prozeßperipherie werden Feldbussysteme genutzt.

2 Feldbus und OPC als Basistechniken für die Kommunikation

Aufgrund der im Projekt vorgegebenen Zielstellung der weitgehenden Herstellerunabhängigkeit und Flexibilität der Automatisierungskomponenten sowie der hohen Anforderungen an die Transparenz und Verwaltung der kompletten Modellfabrik wurden zwei Prämissen vorgegeben:

(1) Es gibt nur zwei physikalische Kommunikationsebenen: TCP/IP für die Office/Management/Leitebene und Feldbussysteme für die Plant/Fabrik/Feld-Ebene.

(2) Als Schnittstelle zwischen den beiden Ebenen soll durchgängig OPC eingesetzt werden.

2.1 Feldbustechnik

Vorgabe (1) ermöglicht einen einfachen und konsistenten Zugang zu allen Komponenten der Modellfabrik über Extranets und schafft, unter Berücksichtigung entsprechender Sicherheitsvorkehrungen, Möglichkeiten zu geographisch verteilter Vernetzung von Automatisierungsinseln. Darüber hinaus gewährleistet die Feldbustechnik für PC-basierte, offene und standardisierte Steuerungs- und Automatisierungslösungen eine adäquate Anbindung der Sensoren und Aktoren des technischen Prozesses.

Problematisch war in der Vergangenheit (und ist es noch) die Kopplung der Feldbussysteme an die übergeordneten Steuerungs- und Leitrechner. Verschiedene Projekte und Ansätze zur Schaffung vereinheitlichter Feldbusschnittstellen [1, 2, 3] erhielten bisher nicht die zu einer marktrelevanten Bedeutung erforderliche Anwenderakzeptanz. Erst mit der Umsetzung von OPC ab ca. 1997 scheint sich hier einiges zu ändern.

2.2 OPC-Schnittstelle

Über die *Vorgabe (2)* wird unabhängig von der konkret unterlagerten Steuerungstechnik eine einheitliche automatisierungstechnische Sicht auf das dezentralisierte Gesamtsystem geschaffen.

OPC als OLE for Process Control wurde in den USA von namhaften Steuerungs- und Visualisierungssoftware-Herstellern definiert und umfaßt den Datenaustausch über die Windows COM/DCOM-Schnittstelle. OPC arbeitet nach dem Client-Server-Prinzip und spezifiziert in der ersten korrekten Version 1.0A (veröffentlicht 1997) 113 Funktionen für den Zugriff auf Prozeßdaten. Mit der Version 2.0 wurde Ende 1998 die Behandlung asynchroner Kommunikation vereinfacht und weitere Funktionen zur Ereignis- und Alarmbehandlung integriert.

Abb. 2 zeigt die Grundstruktur einer OPC-Schnittstelle. Die Schnittstelle ist unter Berücksichtigung der Microsoft-Welt speziell für die Nutzung durch C/C++- oder Visual Basic-Anwendungen (OPC-Clients) vorgesehen.

Seit etwa 1998 werden eine steigende Anzahl von Automatisierungsprodukten mit einer OPC-Schnittstelle ausgerüstet (Prozeßleitsysteme, SPS-Steuerungspakete, Feldbussysteme u.ä.). Die praktischen Erfahrungen im Anwendungseinsatz sind aber noch begrenzt, zumal sich die OPC-Spezifikation noch in einer sehr dynamischen Entwicklung befindet. Auch die Informationen zu den Voraussetzungen, zur Konfiguration, Wartung und Diagnose einer OPC-Schnittstelle richten sich eher an einen Windows-Systemspezialisten und sind für automatisierungstechnische Anwender häufig unverständlich.

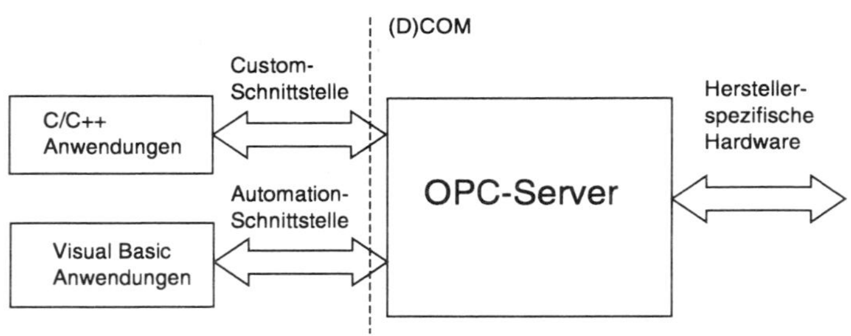

Abb. 2. Grundstruktur einer OPC-Schnittstelle

3 OPC in einer heterogenen Systemumgebung

3.1 Systemumgebung

Eine wesentliche Aufgabe des Projektes INKODA ist aufgrund der o.g. Aspekte die praktische Untersuchung von OPC-Schnittstellen in einer vernetzten und offenen Windows- und Feldbus-Umgebung. Seit 1999 werden deshalb an der in Abb. 3 gezeigten Teilstruktur der Modellfabrik Tests und Untersuchungen durchgeführt.

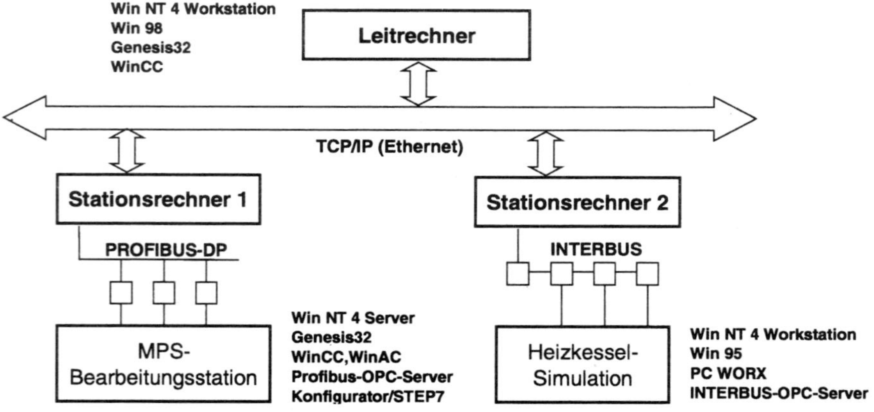

Abb. 3. Teilstruktur der Modellfabrik für die Untersuchung der OPC-Schnittstellen

Als Komponenten sind eingesetzt:

- *Feldbussysteme*: INTERBUS, PROFIBUS-DP.
- *Windows-Betriebssysteme:* Windows 95, Windows 98, Windows NT 4 Arbeitsstation (Service Pack 3), Windows NT 4 Server (Service Pack 3).

- *Prozeßleitsysteme:* Genesis for Windows (ICONICS), WinCC (Siemens).
- Verschiedene OPC-Test- und Diagnoseclients von Phoenix Contact, Siemens, Rockwell Automation, Softing u.a.

Für die Erzeugung der Prozeßein- und –ausgangssignale werden folgende physikalische Automatisierungseinheiten verwendet:

- *INTERBUS-Peripherie*: Angeschlossen ist eine Ausbildungseinheit von Phoenix Contact, mit der in einer Prozeßsimulation der Betrieb eines Heizkessels simuliert wird.
- *Profibus-DP-Peripherie*: Als Automatisierungseinheit ist hier eine MPS-Bearbeitungsstation von Festo Didactic angeschlossen.

Stationsrechner 1 mit Windows NT 4 ist als Server eingerichtet; Stationsrechner 2 und der Leitrechner arbeiten jeweils für beide Windows-Betriebssysteme als Client.

3.2 Vorbereitung einer OPC-Installation

Im Vorfeld einer OPC-Installation sind insbesondere zwei Aspekte zu berücksichtigen:

1. Einarbeitung in das OPC-Grundprinzip.
2. Korrekte Konfigurierung einer Windows-DCOM-Umgebung.

Eine Einarbeitung in die OPC-Problematik ist für eine erfolgreiche Nutzung auch für Automatisierungsanwender zumindest in Grundzügen erforderlich. OPC-Werkzeuge für die Konfigurierung und Diagnose nutzen die bei OPC übliche Begriffswelt und erfordern ein entsprechendes Verständnis. Die meisten Informationen zu OPC liegen z.Z. noch in englischer Sprache vor. Pragmatisch aufbereitete und für einen Einsteiger übersichtliche Erläuterungen zur OPC-Technik finden sich gegenwärtig kaum. Die ausführlichsten Informationen zu OPC liefert die OPC-Foundation (http://www.opcfoundation.org) sowie auch einige der Mitgliedsfirmen. Die OPC-Produkte der Hersteller werden zwar auch mit einer OPC-Dokumentation geliefert, diese ist aber qualitativ sehr unterschiedlich aufbereitet. Ein guter Einstieg findet sich z.B. auf der SIMATIC NET-CD von Siemens.

Neben der üblichen Installation einer Windows-TCP-IP-Netzumgebung erfordert der verteilte Datenzugriff über OPC eine korrekt konfigurierte DCOM-Umgebung. Dazu gibt es unter Windows das Konfigurierungsprogramm `dcomcnfg.exe`. Konfiguriert werden damit insbesondere die Sicherheits- und Zugriffsberechtigungen der jeweiligen DCOM-Anwendungen. Falsche Sicherheitseinstellungen in DCOM können die Übertragung der OPC-Daten über das Netz verhindern.

Der Betrieb einer DCOM-Umgebung mit Windows 95,Windows 98 (Client) gemeinsam mit einem Windows NT-Server konnte in der Testumgebung nicht realisiert werden, da DCOM 95/98 nicht in der Lage ist Anwendungen (z.B. den OPC-Server) auf einem Remote-Rechner zu starten

3.3 Konfiguration und Test einer OPC-Schnittstelle

Aufgrund der vereinheitlichten OPC-Spezifikation werden OPC-Schnittstellen unterschiedlicher Hersteller im Prinzip nach der gleichen Methode konfiguriert und parametriert. Die Bedienung der entsprechenden Werkzeuge ist deshalb sehr ähnlich und bei vorhandenen OPC-Kenntnissen einfach durchzuführen.

Für die Einrichtung einer OPC-Schnittstelle ist nur die Installation des OPC-Servers und eine Konfiguration des OPC-Client notwendig. Anwendungsprogramme, die als OPC-Client arbeiten, können nach der OPC-Server-Installation die entsprechenden Prozeßdaten des OPC-Server sofort nutzen.

Die Konfiguration bzw. Einrichtung des OPC-Client erfolgt in folgenden Schritten:

(1) Festlegung der Prozeßsignale bzw. Prozeßvariablen des unterlagerten Feldbus- bzw. Steuerungssystems, die über OPC sichtbar sein sollen.
(2) Konfigurierung der Refresh-Methoden für die OPC-Prozeßdaten.
(3) Test der Verbindungen zum OPC-Datenzugriff.

Für eine effiziente Konfiguration der OPC-Schnittstelle nach *Schritt 1* sollten die Werkzeuge für die Projektierung der Feldbussysteme und für die Erstellung der Steuerungsprogramme für OPC vorbereitet sein. Für INTERBUS mit der Automatisierungssoftware PC WORX bedeutet dies z.B., daß in den entsprechenden Werkzeugen nur die entsprechenden Optionen aktiviert werden müssen, um Prozeß- oder Programmvariable als OPC-Daten zu deklarieren.

Im *Schritt 2* müssen im OPC-Client die Refresh-Methoden der OPC-Daten festgelegt werden. Dazu erfolgt vorher eine Strukturierung der OPC-Daten in Gruppen mit dann jeweils gruppenweise bezogener Refresh-Rate und –betriebsart (asynchron oder synchron).

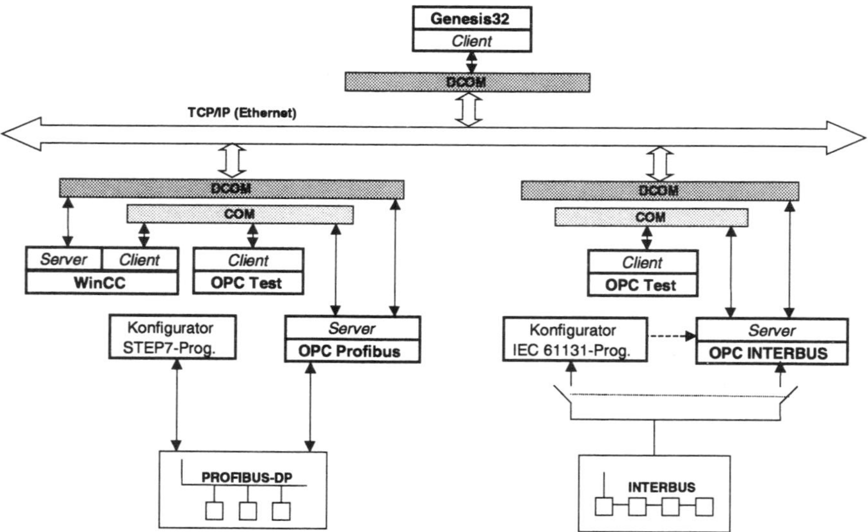

Abb. 4. Kommunikationstechnische Struktur der Testumgebung aus Abb. 2

Nach der Konfigurierung des OPC-Servers kann im *Schritt 3* der Test der OPC-Verbindungen erfolgen. Die Hersteller von OPC-Servern liefern dazu entsprechende

Diagnose- bzw. Testclients mit, die jedoch häufig nur im lokalen Betrieb über COM auf den OPC-Server zugreifen können. Hier ist es m.E. sinnvoller, sofort einen auch für Remote-Tests geeigneten OPC-Client (z.B. WinCC) zum Test zu nutzen.

Nach der Konfiguration der OPC-Server ergibt sich für die Testumgebung aus Abb. 3 die in Abb. 4 dargestellte kommunikationstechnische Struktur.

3.4 OPC-Kommunikation im Betrieb

Nach erfolgreicher Inbetriebnahme der OPC-Server und der DCOM-Umgebung auf jedem im Netzwerk angeschlossenen PC kann nun netzwerkweit durch beliebige OPC-Client-Anwendungen auf die OPC-Prozeßdaten zugegriffen werden. Da als OPC-Client auch jedes Visual Basic-Programm fungieren kann und die meisten Microsoft-Programme die Erweiterung mit Visual Basic-Makros gestatten, schafft OPC eine unkomplizierte Verbindung zwischen der Microsoft Office-Welt und der Steuerungswelt des zu automatisierenden technischen Prozesses.

Zeituntersuchungen zur OPC-Datenübertragung liegen bisher nur wenige vor. Die veröffentlichten Angaben schwanken sehr stark und die Testbedingungen sind nur unvollständig erläutert. Die zwei Testberichte nach [4,5] sind relativ ausführlich und beschäftigen sich mit den *Übertragungsraten zwischen OPC-Server und OPC-Client*. Für die OPC-Schnittstelle von Rockwell Automation [4] ergeben sich z.B. folgende Datentransferraten:

♦ 1000 Items in 25 ms im 100 MB-Ethernet-Netzwerk bei 5 Clients im Netz.
♦ 1000 Items in 20 ms im lokalen Rechner (Pentium 266).

Die Angaben von Intellution in [5] weisen Datentransferraten abhängig von CPU-Auslastung und Art des Zugriffs (Einzelitem oder Gruppe) von ca. 16 ... 300 ms für je 1000 Items aus. Interessant ist hier die Aussage, daß ein In-Process-Server auf 1000 Items in ca. 1 ms zugreifen kann. Ein In-Process-Server befindet sich dabei im gleichen Prozeßraum wie der OPC-Client. Er steht nur diesem Client zur Verfügung und ist als DLL realisiert.

Für den INTERBUS-OPC-Server von Phoenix Contact liegen nach [6] die *Übertragungsraten zwischen OPC-Server und der unterlagerten Hardware* abhängig vom Typ der INTERBUS-Anschaltbaugruppe und der Art des Datenzugriffs sowie der Adressierung bei ca.

♦ 1000 Items in 70 ... 400 ms (Pentium 233/266).

Trotz der stark schwankenden Zeitangaben wird deutlich, daß OPC für den Leit- und Visualisierungsbereich gut geeignet ist, da hier die Forderungen an die Refresh-Rate nur bei etwa < 0,5 s für den Refresh einer Bildschirmseite mit ca. 10 ... 50 Prozeßvariablen liegen. Für den Datenaustausch zwischen Feldbussystem und Steuerungsprogramm müssen jedoch bei geforderten Zykluszeiten von < 5 ms z.Z. noch andere, hardwarespezifische Datenzugriffe oder In-Process-OPC-Server (wenig flexibel) eingesetzt werden.

Berücksichtigt werden muß, daß bei einigen OPC-Clients, die minimal einstellbare Refresh-Rate für synchronen Zugriff auf ca. 250 ms begrenzt ist, und damit bereits die

Datentransferrate über die OPC-Schnitstelle unabhängig vom OPC-Server begrenzt wird

Während des OPC-Betriebes ergaben sich zwei Probleme, die für eine praktische Anwendung der OPC-Schnittstellen berücksichtigt werden müssen:

- Eine Online-Änderung von Prozeßvariablen, die als OPC-Items deklariert sind, sollte vermieden werden, da der OPC-Server u.U. darauf mit Fehlermeldungen reagiert.
- OPC-Server werden implizit durch die Zugriffe von OPC-Clients gestartet und gestoppt. Wird ein OPC-Server über den ersten OPC-Client-Zugriff gestartet und stürzt danach der Client ab, so kann ein Beenden des OPC-Servers nur über einen System-Eingriff erfolgen.

3.5 Entwicklung eigener OPC-Schnittstellen

Aufgrund der veröffentlichten OPC-Spezifikation kann prinzipiell jeder Anwender eigene OPC-Produkte entwickeln. Der dazu erforderliche Entwicklungsaufwand unterscheidet sich bezogen auf OPC-Clients bzw. OPC-Server beträchtlich.

OPC-Client

Zur Entwicklung eines OPC-Client ist ein Visual Basic bzw. C/C++-Programm unter Nutzung der vereinheitlichten OPC-Aufruffunktionen zu schreiben. Grundkenntnisse in OLE2/COM von Microsoft sollten vorhanden sein.

Bei einigen Mitgliedsfirmen der OPC-Foundation finden sich in der jeweiligen Dokumentation zu den OPC-Produkten hilfreiche Beispiel-Quellcodes, mit denen bereits relativ schnell einfache OPC-Clients entwickelt werden können (dazu auch [7]).

OPC-Server

Eine eigene OPC-Server-Entwicklung ist kompliziert und erfordert tiefgründige Kenntnisse der COM/DCOM-Technologie sowie ausreichende C/C++-Programmiererfahrungen. Außerdem muß natürlich eine intensive Auseinandersetzung mit der OPC-Spezifikation und den dazu ergänzenden Unterlagen erfolgen.

Man kann sich die Entwicklungsarbeiten vereinfachen, wenn der Quellcode eines OPC-Servers als Beispielimplementierung gut kommentiert vorliegt. Ein solcher Quellcode ist u.a. Bestandteil von OPC-Entwicklungs-Toolkits verschiedener Anbieter (Softing, Technosoftware, Merz, ...).

4 Schlußfolgerungen

OPC ist eine leistungsfähige und flexible sowie einfach handhabbare Schnittstelle zur Verbindung der höheren Automatisierungsebenen mit den Sensoren/Aktoren sowie Steuerungsprogrammen der Feld- und Prozeßebene. In Verbindung mit PC-basierten Steuerungen schafft OPC die Anbindung des technischen Prozesses an die Microsoft Office-Welt. Durch die nahtlose Einbettung von OPC in die Microsoft-Welt von Windows entstehen offene und konsistente Systemlösungen, die für einen Anwender vernünftig bedien- und wartbar sind. In Verbindung mit der Nutzung von Ethernet und TCP/IP bis in die Feldebene hinein, kann OPC eine Schlüsselrolle als Datenschnittstelle zukünftiger Automatisierungslösungen spielen.

Der Einsatz neuer Automatisierungssysteme mit OPC bzw. die Umstellung älterer Anlagen erfordert zusätzliche Kenntnisse des Inbetriebnahme- und Servicepersonals zur PC/Windows-Technik. Für die Entwicklung eigener OPC-Produkte müssen neben der Nutzung verfügbarer OPC-Toolkits tiefergehende Windows-Systemkenntnisse vorhanden sein.

Literatur

1. *Quade, J.:* Ein einheitliches Feldbus-API. atp, R. Oldenbourg Verlag, München, 39(1997)12, S. 50 – 57

2. *Pritschow, G.; Häberle, U.:* Einheitliche Anwenderschnittstelle für Feldbussysteme. atp, R. Oldenbourg Verlag, München, 38(1996)12, S. 46 – 49

3. *Langmann, R.; Eßer, M.:* Schnittstelle zwischen Anwendungssoftware und Feldbuskommunikation. atp, R. Oldenbourg Verlag, München, 39(1997)7, S. 30 – 38

4. Burke, T.J.: The Performance and Throughtput of OPC. Rockwell Automation, June 11, 1998.

5. Chisholm, A.: DCOM, OPC and Performance Issues. Intellution Inc, 2/3/98

6. Phoenix Contact: Datenblatt zum OPC-Server IBS OPC SERVER, Mai 1999

7. Groppe, A.: OPC in einer Zwei-Ebenen-Modellfabrik. Diplomarbeit, 1999, FH Düsseldorf

OPC – Making the Fieldbus Interface Transparent

Thomas Hadlich, Thorsten Szczepanski

ifak system GmbH, Steinfeldstr. 3, 39179 Barleben

Abstract. The performance of fieldbus systems has been studied more or less thoroughly. With the advent of high speed fieldbus systems as PROFIBUS-DP the bottleneck of systems moves from the communication to the application, especially to the interface between both. Two requirements are placed on such interface in general. First it should be based on an open, accepted standard where secondly it should support real-time operation. As the PC becomes more and more important even in the realm of industrial automation the quality of interfaces based on Microsoft Windows technologies gains importance too. The first solution provided with Windows 3.1 was the Dynamic Data Exchange (DDE). Although an open an finally widely accepted standard its communication performance was not satisfactory. The development of the OPC (OLE for Process Control) architecture aimes at solving the question of performance while providing an open interface standard between applications. Although the openness of this solutions is not debatable, the performance issue has to be investigated. This paper contributes to this by exploring the performance of an implementation of the OPC interface standard.

1 Introduction

In every area of industry there is a move from proprietary solutions to open, vendor-independent standards. As well as reducing costs it allows the choice of components according to their performance and reduces the dependence on suppliers. An impressive example for this move can be found in industrial communications. Today's prevailing systems are all based on open, vendor-independent fieldbus standards. Such open standards were needed in the domain of applications too. The specification of the Dynamic Data Exchange (DDE) protocol provided a first solution for the data exchange between MS-Windows based applications. The main drawback of this solution was its low bandwidth. Therefore DDE was not very well suited for real-time systems, a major requirement as far as automation systems are concerned.

The performance of fieldbus systems has been the subject of countless studies. As revealing as such studies may have been, from the user point of view it is the overall performance of the system consisting of application and communication stack that matters most. This paper presents the results of a performance study of an OPC/COM hierarchy. The investigation of OPCs performance capabilities was based on the especially developed suite of applications. It stands in proxy for a probable software chain from a process control system (OPC test client), OPC server (OPC test server) responsible for the data acquisition from possibly one or more data sources. Based on this, an environment which allows to maintain a given real-time criterion was determined. This environment was defined in terms of the workload of the system as well as a specific netload on the Ethernet connection between the computers involved. The client-/server model as well as the publisher-/subscriber model (both supported by OPC) are taken into the consideration. In the next step the different load scenarios are determined for the case of a concentrated application (one PC running the data server as well as the client application) and a distributed application. The relevant load parameter are changed from best-case to worst case conditions. The presented results are reduced to the most important values system reaction time (client-/server model) and message interval (publisher/subscriber model) respectively. Based on the gained information the definition of a so called working environment for the case of concentrated and distributed application took place. The working environment was set especially with the observation of a given message interval in mind. The workload (resp. netload) which guaranteed a 99% probability of sustaining the interval was chosen. The conclusion discuss the usefulness of OPC in the framework of real-time applications.

2 Overview

From a communication point of view a distributed system consists of service provider and service user, which are logically connected. The kind of logical connection can be described by two different models- the client/server and the publisher/subscriber model. The OPC standard distinguishes only between synchronous and asynchronous services. This section defines the use of these terms within the framework of the paper.

2.1 Communication in Distributed Systems – Client/Server

The data exchange takes places between a client and a server, where the client addresses (polls) the server. A client issues orders or requests to an accessible server. The communication between client and server is determined by protocols, which main tasks are as follows:

- Reception of request or response service primitives from client or server applications and the coding of these primitives in protocol data units (PDU)

- The decoding of PDUs and their report to the client (confirmation service primitive) or the server (indication service primitive) [9].

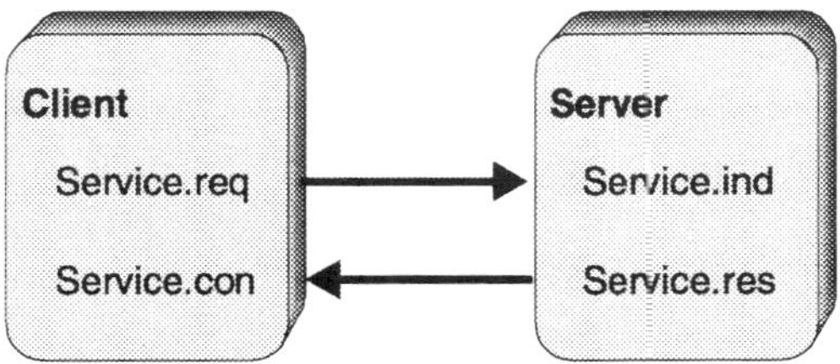

Fig. 1. Client-/Server-Model

The communication between client and servers is acyclic (Fig. 1.). The action rests with the client which sends a service request to the server. This request is answered by the server with a service response. Services are send only according to the need of the client. Typical applications are the transfer of commands between applications or irregular data transfers.

2.2 Communication in Distributed Systems – Publisher Subscriber

The publisher-/subscriber model assumes a cyclic data supply by the publisher. The data transfer from the publisher depends either on an external request or an internal event (e.g. an expiring timer). As shown in the figure, the data send by the publisher can be received by more than one subscriber. Usually there is no explicit addressing of the subscribers.

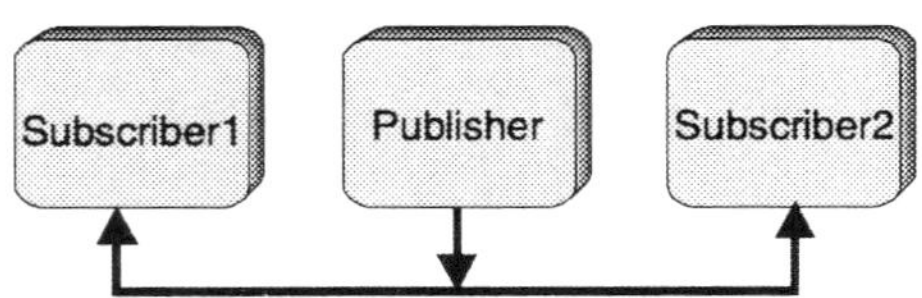

Fig. 2. Publisher-/Subscriber-Model

2.3 The Component Object Model

With the development of the Component Object Model (COM) a new platform dedicated to openness as well as speed was made available by Microsoft. COM defines mechanisms of interaction between objects. Objects in the view of the COM-model are application objects as well as data objects. Internal data of these objects are stored and accessed locally. The communication between objects is limited to so called "interfaces", which are collections of functions. COM has to be considered as replacement of all methods of data exchange like proprietary API calls, DDE or direct calls of Windows API functions.

The basis method of communication as supported by COM is based on synchronous (blocking) calls. The calling object waits until it receives the response of the called method. The employed mechanism corresponds to the client-/server model. But in order to carry through a cyclic data exchange it would be preferable to use a mechanism comparable to the publisher-/subscriber model. It would reduce the system load which stems from the cyclic polling of data which would be necessary in the client server model, to wait for an event controlled message from the data source. COM defines "advisory connections between data object and one or more advisory sinks" [4] which can be used to implement asynchronous communication.

2.4 OLE for Process Control (OPC)

Based on the COM architecture OPC - Ole for Process Control – was created as a vendor-independent standard aimed at the definition of data exchange between MS-Windows based applications as well applications and vendor-specific hardware in the realm of automation. It is maintained by an independent association – the OPC Foundation. OPC defines logical objects and methods tailored to the special needs of the process industry based on COM technology. Several OPC-Clients are able to access one OPC-Server in parallel. In the area of fieldbus communication OPC is the most often installed interface between a fieldbus controller and PC based applications.

OPC uses the mechanisms provided by COM to implement synchronous services (comparable to client/server) and asynchronous services (comparable to publisher/subscriber). For an in-depth study of OPC the reader is referred to [5].

3 Measurement Preliminaries

3.1 Used Applications

OPC test client. The OPC test client serves as a simulation of the communication interface of a SCADA system. The different communication tasks (read/write, cyclic, acyclic) can be controlled by the client.

OPC test server/COM test client The OPC test server/COM testlient provides the middleware. It simulates an application which provides an OPC server interface to the SCADA system, but acquires its data from a underlying COM server.

COM test server. The COM test server represents a low level communication interface to any hardware.

Workload-Generator. The workload generation was done by the Microsoft tool "rprobe.exe", which is part of the Windows NT Resource Kit. The generated load is determined by the variation of the three phases wait, data access and compute of a single thread. A further control over the generated load is given by the variation of the number of parallel threads and processes maintained by the workload generator. For a detailed description see [3].

Netload-Generator. The netload generation used the Linux tool "ping". It is contained within the version 5.1 of the Suse Linux distribution (or any other Linux distribution). The tool uses the ICMP service ECHO_REQUEST. The host called by this request responds with an ECHO_RESPONSE datagram. The frequency and size of the data packets determine the generated netload. The interested reader is referred to [8] for a more detailed description.

3.2 Environment Conditions - Concentrated Application

Best Case. The PC of the concentrated application (Fig. 3.) runs only the OPC/COM-applications. There is no other load on the PC than the load generated by the operating system itself is existent. The additional workload generated by the workload generator is set to 0 (in fact the generator is not running).

Worst Case. The OPC/COM-applications are executed under an additional workload of 100%.

3.3 Workload-Profile. The generated workload is increased in 10% steps.

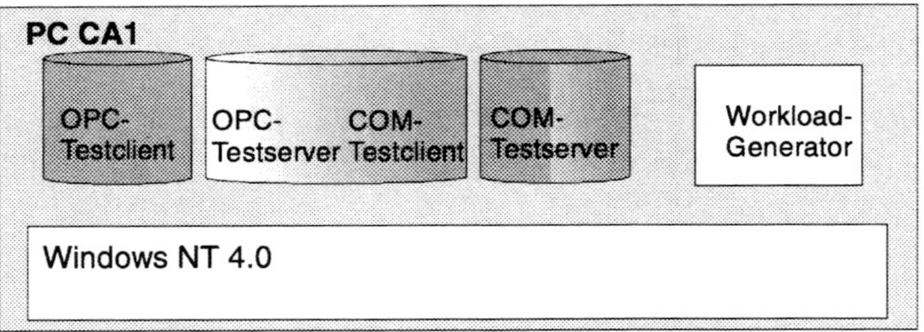

Fig. 3. Concentrated Application – Hardware/Software

3.4 Environment Conditions - Distributed Application

Best Case. The PCs of the distributed application (PC DA1, PC DA2) execute the OPC/COM test applications only. There is no other load on the PC than the load generated by the operating system itself. The additional workload generated by the workload tool is set to 0. The PCs (PC VA1, PC VA2, PC VA3) are connected by a coaxial cable forming a bus topology. The netload consists of the idle communication between these computers as performed by the operating system. There is no additional netload.

Worst Case. The execution of the distributed application takes places under an additional workload of 100%. The additional workload is generated on the computer running the OPC test client (PC DA1) as well as on the computer running the OPC server/COM client/COM server applications. The PCs (PC DA1, PC DA2, PC DA3) are connected by a coaxial cable forming a bus topology. The netload consist of the idle communication between these computers as performed by the operating system plus an additional netload of 60%.

3.5 Workload-Profile. The generated workload is increased in 10% steps.

Netload-Profil. The generated netload is increased in 10% steps.

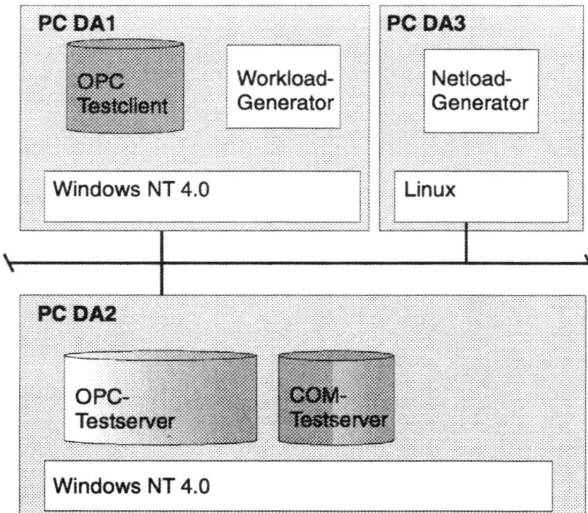

Fig. 4. Distributed Application – Hardware/Software

4 Measurement Results

4.1 General Remarks

The application uses a set of commands which represent a typical working scenario between a client and an OPC server. Table 1 contains an overview of the different communication structures and their short labels as used in this section.

Table 1. Overview of the different communication structures

Short Label	Meaning
KD0	Function call with return state
KD1	Function call with input data (1 Byte) and return state
KDn	Function call without input data but output data (array of four floats) and return state
DT	Cyclic data transfer (array of four floats)

The interpretation of the measurement results in this section focuses on the main parameters needed to assess the system performance from the users point of the view. In the client/server model the time needed to execute a synchronous service in the OPC/COM-Toolchain starting from the request in the test client until the reception of the function result back in the client was chosen. For the publisher-/subscriber model the interval between two message events in the client was considered most important.
Client/Server. The evaluation of the results of the client/server measurements is based on the system reaction time.

System reaction time. The system reaction time allows the assessment of the performance of the complete OPC/COM tool chain, because it exhibits all delays in the chain while executing a synchronous service. The figures contain the mean reaction time to an OPC request (line chart) as well as the standard deviation (bar chart) over a set of 1000 values. The resulting distribution function contains 80%-90% of all measured values within one standard deviation of the mean value.

Publisher/Subscriber. The chosen parameter message interval, the time difference between two consecutive message arrivals at the client, is considered as the most important one, because the observance of this interval matters most to the user.

Message interval. The given real time frame of 100ms in mind a 2-times oversampling rate was considered sufficient for this application. The OPC server sends data messages every 50 ms to the client. The resulting distribution equals a standard distribution, meaning that 95% of all values are within two standard deviations of the mean value. The figures contain the mean message interval (line chart) as well as the standard deviation (bar chart) over a set of 1000 values.

4.2 Concentrated Application

Scenario 1. This scenario investigated the system behaviour while a number of read accesses from the client took place. The processor load was increased in 10% steps. The results as shown in the diagram (Fig. 5.) exhibit the relative robustness of the system below a workload of 70%.

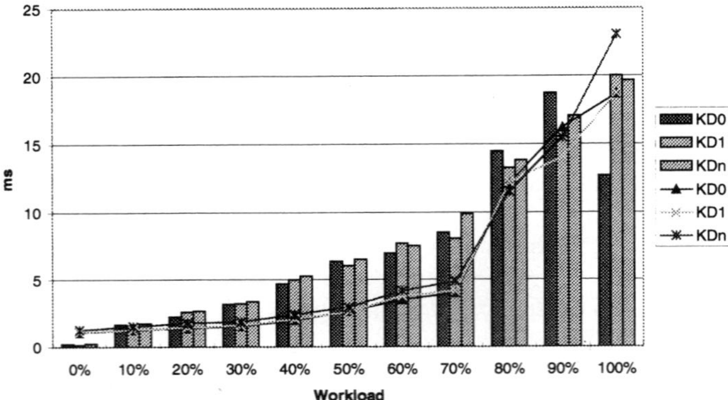

Fig. 5.   System Reaction Time – Concentrated Application

Scenario 2. This scenario investigated the arrival interval of a cyclic telegram transferred from the OPC-server. The relatively high standard deviation (if compared with the previous scenario) results from an error compensation mechanism which was built into the OPC sever. If the server detects a delay in the send process it tries to compensate it by adapting the next interval to the delay. The message interval as shown in Fig. 6. can be observed with a probability of 99% up to a processor load of 40%.

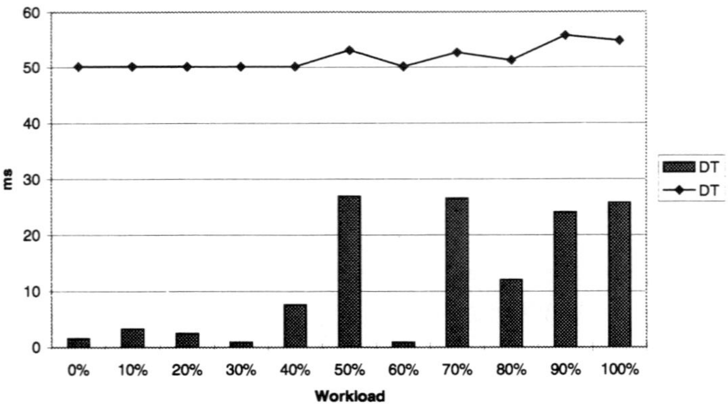

Fig. 6.   Message Interval – Concentrated Application

4.3 Distributed Application

Scenario 3. The scenario corresponds to scenario 1 as far as data access is concerned. The application is distributed as described in section 3.4. The workload on the PC running the OPC client was increased in 10% steps. The results show the indifference of the mean value and standard deviation of the reaction times to the client PC workload. This behaviour could be expected, because the main load in executing the OPC request rests with the server. The singularity at a 60% workload points to the fact that, all efforts notwithstanding, Windows–NT should not be mistaken for a real-time operating system in a very rigid sense. A certain amount of non-determinism remains as far as the task execution is concerned.

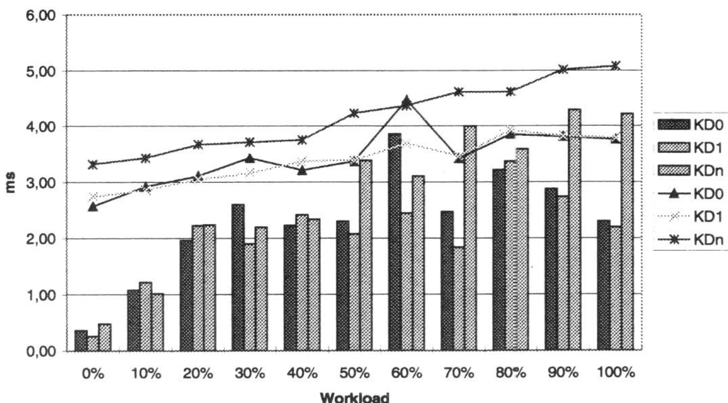

Fig. 7. Distributed Application – System reaction time (Client Dependence)

Scenario 4. Scenario 4 aimed at the test of the dependence of the system performance on workload changes on the server PC. The observed behaviour shows a direct correspondence between the workload and the mean value and standard deviation of the system reaction time. As explained above, it is a behaviour which could be expected because the main load in the execution of the OPC requests lays on the server side.

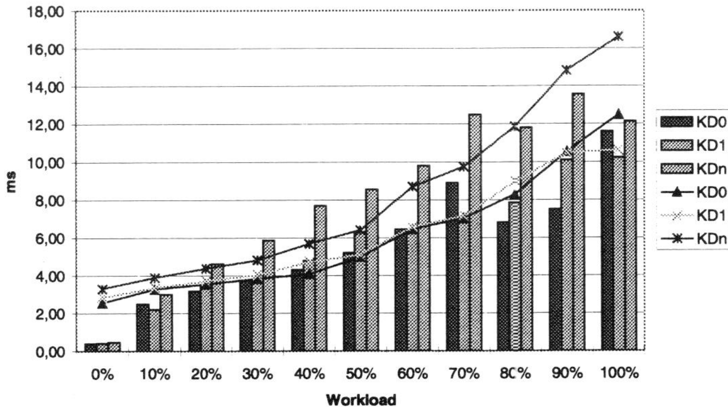

Fig. 8. Distributed Application – System reaction time (Server Dependence)

Scenario 5. Scenario 5 considered the influence of the netload to the overall performance of the system. Up to a netload of 30% the mean value and the standard deviation of the system reaction time are rather indifferent against the load on the network. A further increase in the netload leads to a rapid deterioration of the system behaviour.

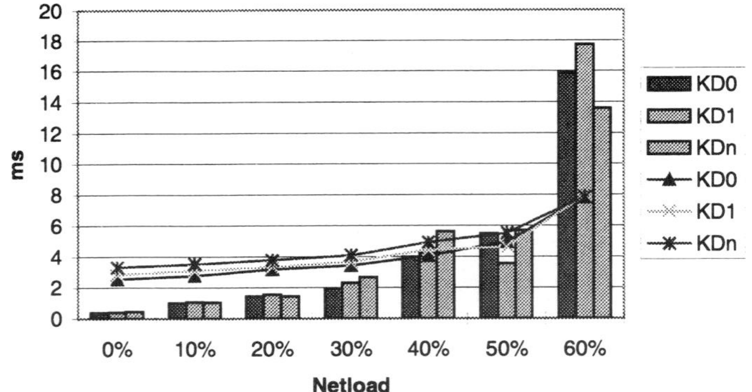

Fig. 9. Distributed Application – System reaction time (Network Dependence)

Scenario 6. The previous scenarios evaluated the change in system performance dependent on one load parameter under the assumption of the remaining of all other parameters at the best case level. Within this scenario an average load (50%) of the PCs executing the OPC/COM application is simulated. The increase of the netload allows the definition of a working set of load conditions.

The chart shows a slow increase in mean value and standard deviation of the system reaction time up to a netload of 50%. The overall slower system reaction follows from the existent processor load.

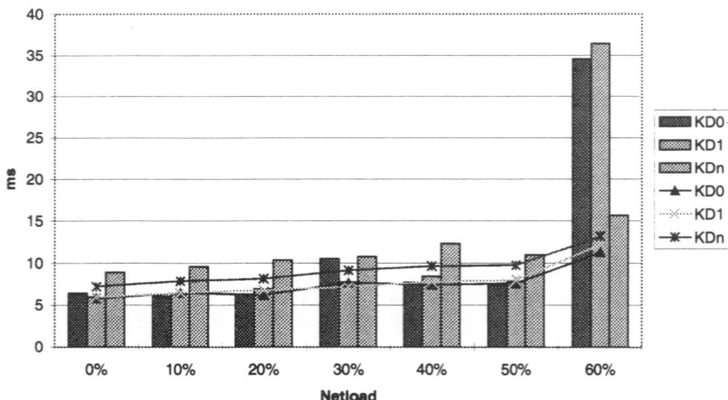

Fig. 10. Distributed Application – System reaction time (Network/Processor Dependence)

Scenario 7. The behaviour of the message interval corresponds to the behaviour of the system reaction time. As seen in scenario 4 there is a direct correspondence between the server load and the mean value and standard deviation of the parameter in consideration. Above a processor load of 80% only 95% of the measured values comply to the given maximum of 100ms.

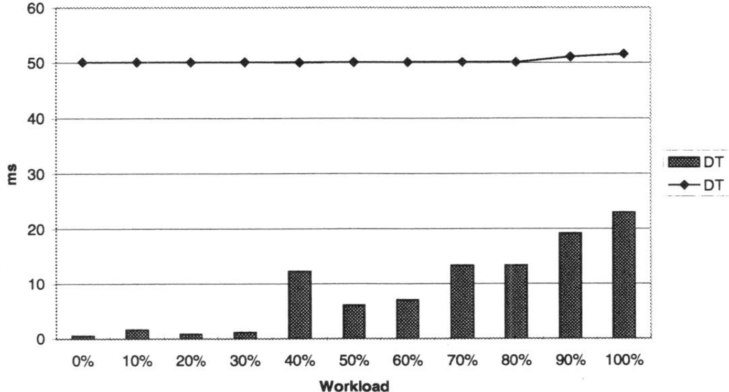

Fig. 11. Distributed Application – Message Interval (Network Dependence)

Scenario 8 . This scenario is concerned with the observance of the message interval under an average processor load of 50% on the client and server PC. The chart exhibits a relative instable behaviour of the standard deviation. This is caused by the error compensation mechanism described in scenario 2, which exerts now a major influence on the values (caused by the existing workload on the server PC). Up to a net load of 20% the given maximum of 100ms is observed with a probability of 99%.

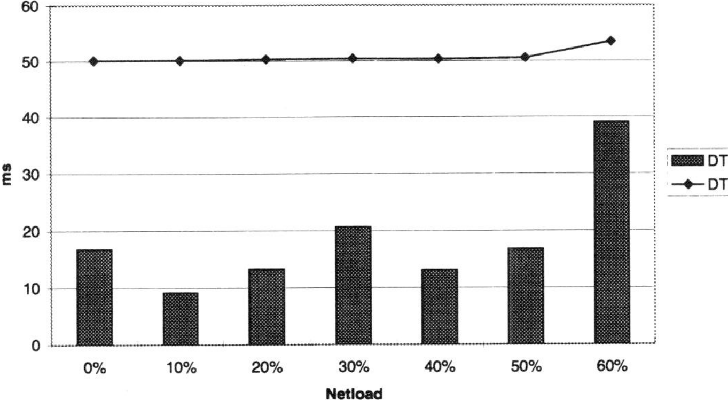

Fig. 12. Distributed Application – Message Interval (Network/Processor Dependence)

5 Conclusions

This paper researched the impact of system design and system load on the performance of distributed applications. All documented scenarios are based on tests searching not only for a "positive" result but especially for a "negative" result, which exhibited the condition under which the application started to work unreliably.

5.1 Concentrated Application

The scenarios of the concentrated application show that the observance of the message interval can be guaranteed (with a certain probability) even while executing all parts of the application on a single PC. In order to guarantee a 99% compliance to the message interval the external workload should not exceed a boundary of 40%.

5.2 Distributed Application

The scenarios which were based on distributed execution of the software investigated the behaviour of the system by changing one load parameter while assuming the others resting in the best case state. The achieved results lead to the following rules of thumb:

- The influence of the workload on a PC executing the client software on the overall performance of the system is relatively small

- The workload on a PC executing the server software determines the overall performance of the system

- The netload on the connection between the computers determines the overall performance of the system

The scenarios 6 and 8 led to the definition of a set of load conditions which would guarantee the message interval of 100 ms. It could be proven that if the upper limit was formed by a workload of 50% on the client and server computer and a netload of no more than 20% the message interval was observed with a probability of more than 99%.

Regarding the general application of OPC within the framework of SCADA or control systems we believe that OPC is well suited for transportation of data for visualisation and data storage. The expected progress in PC soft- and hardware will allow the use of OPC even in control applications in the near future.

Still, the deepest impact on the performance of any distributed system (based on OPC or not) is reached by considerate system design.

References

1. Brockschmidt, K: Inside OLE, Microsoft Press, Redmond, 1995.

2. Burke, Th. J.: The Performance and Throughput of OPC – A Rockwell Software Perspective, White Paper, Rockwell Software Inc., 1998.

3. Microsoft-NT Workstation Resource Kit, Microsoft Press, Redmond, 1997.

4. MSDN Library January 1999 release, Microsoft Corp., 1999.

5. OLE for Process Control – Data Access Standard V1.1, OPC Foundation, 1997.

6. Rogerson, D.: Inside COM, Microsoft Press, Redmond, 1997.

7. Schnittstellen zum Prozeßleitsystem, Draft, GTM, 1998.

8. UNIX System Manager's Manual.

9. Wedekind, H.: Verteilte Systeme, Grundlagen und zukünftige Entwicklung aus der Sicht des Sonderforschungsbereichs 182 "Multiprozessor- und Netzwerkkonfigurationen"; BI Wissenschaftsverlag, 1994.

Chapter 7: System Aspects

Invited Paper presented in Plenary Session 2

IEEE Utility Communications Architecture (UCA) applies mainstream standard
Ethernet
Schwarz, K.

Papers presented in Session 2.6

A Component-Based Architecture for Integrating Fieldbus Systems into
Distributed Control Applications
Bachmann, R.; Hoang, M. S.; Rieger, P.

Decentralized automation concepts based on Ethernet-TCP/IP and CANopen
Ziegler, A.

Considerations on a LonWorks/IP Gateway Implementation
Soucek, S.; Schweinzer, H.-J.

The Importance of Being Competent: the Role of Competence Centres in the
Fieldbus World
Sauter, T.; Felser, M.

IEEE Utility Communications Architecture (UCA) applies mainstream standard Ethernet

Karlheinz Schwarz

Schwarz Consulting Company (SCC)
Im Eichbaeumle 108
D-76139 Karlsruhe, Germany
schwarz@scc-online.de
www.scc-online.de

Abstract. The Utility Communications Architecture (UCA™) is a standards-based approach to utility communications which provides for wide scale integration at reduced costs, and which solves many of the most pressing communications problems for today's utilities. The UCA is designed to apply across all of the functional areas within the electric, gas, and water utilities. These functional areas include customer interface, distribution, transmission power plant, control center, and corporate information systems. The UCA includes detailed object models, which defines the tag, format, representation, and the meaning of utility data. This modeling effort goes far beyond the scope of any other utility communications approach, and provides for an unprecedented level of multi-vendor interoperability applicable in most industries.

1 Introduction

Worldwide, electric utility deregulation is expanding and creating demands to integrate, consolidate and disseminate information quickly and accurately between and within utilities. Utilities spend an ever-increasing amount – estimated $2 billion to $5 billion dollars a year in the USA only – for voice and data communications. There are already strong pressures to find ways of reducing operating costs to improve the utility earnings. In response to this need, IEEE has published a complete set of communication protocols. The international utility industry, comprising Electric, Gas, and Water utilities, is a key participant in the development and use of IEEE standards, along with the manufacturers of equipment conforming to the developed standards. These utilities rely heavily on standards in many areas including: electrical, mechanical, computers hardware, and computer software.

Several years ago, the members of both IEEE and the Electric Power Research Institute (EPRI, Palo Alto, USA) identified the need to understand and properly implement open communication architectures. This need arose from the rapid expansion of interconnection technologies and the consequent demand for better exchange of information between various elements of the power delivery system, the utility's customer service and support staffs, and their customers.

In their role as research and development arm of the international electric utility industry, EPRI funded a research and development effort to create the Utility Communications Architecture (UCA™). ISO, IEEE, and other related communication standards (e.g., 10/100 Mbit/s Ethernet, ISDN, MMS – ISO 9506) were assembled to define UCA which its member utilities can use in meeting their communications needs. EPRI developed an assessment of functional and communications needs for the industry by area, and developed a methodology to map these needs to established and emerging standards developed by organization such as the IEEE. The result is a specification that points to standards that could be used to accomplish work at a utility in an 'open' fashion.

Since the standards comprising UCA come from many areas such as the electric industry, the gas industry, the water industry, the communications industry, and the computer industry, EPRI believed that an IEEE Standards Coordinating Committee (SCC) is the best vehicle to facilitate coordination among all of these groups. Therefore, IEEE set up a new IEEE SCC, the SCC 36 (Utility Communications Architecture, UCA) whose charter is to coordinate the on-going work of refining and expanding the UCA communications protocols, and to ensure that these protocols are developed and/or accepted as international standards through the IEEE, the International Electrotechnical Commission (IEC), and other standards organizations as appropriate. The SCC 36 has unanimously decided to publish the UCA Version 2 specification as IEEE Technical Report (IEEE TR 1550) in July 1999.

2 The Utility Communications Architecture (UCA)

The Utility Communications Architecture (UCA) is a standards-based approach to utility communications which provides for wide scale integration at reduced costs, and which solves many of the most pressing communications problems for today's utilities. The UCA is designed to apply across all of the functional areas within the electric, gas, and water utilities. These functional areas include customer interface, distribution, transmission power plant, control center, and corporate information systems. It is important to note that UCA is an architecture, rather than a simple, protocol. The UCA Version 2.0 incorporates a family of basic communications protocols to meet the requirements of a wide range of utility environments. The selection and organization of these protocols has been designed to provide great flexibility in choosing the appropriate technology to meet a utility's price/performance criteria, while maintaining consistency at the device and data level to reduce integration and vendor product costs. In addition, the UCA includes detailed object models, which defines the tags, format, representation, and the meaning of utility data. This modeling effort goes far beyond the scope of any other utility communications approach, and provides for an unprecedented level of multi-vendor interoperability.

The UCA documents currently specify a set of existing international standards which can be applied to specific communications architectural requirements in the utility industry. Information in the documents can be used to define and implement a wide variety of standards-compliant communications systems such as those required to support Distribution Automation, Demand Side Management, Substations and Control Systems, Power Plant Automation, and Customer Interfaces.

The UCA documents comprise the following documents:

Common parts

- Introduction to UCA
- UCA Profile Specification

Modeling and communication for intelligent devices

- Common Application Service Models (CASM),
- Generic Object Models for Substation and Feeder Equipment (GOMSFE),
- Customer Interface Device Models (under preparation)
- Power Plant Device Models (under preparation)

Real-time data exchange between control centers

- IEC 60870-6-503: TASE.2 Services and Protocol
- IEC 60870-6-802: TASE.2 Object Models
- IEC 60870-6-702: TASE.2 Application Profile

The common parts and the parts for modeling and communication for intelligent devices have been incorporated in the IEC TC 57 committee drafts of the series IEC 61850 (Communication networks and systems in substations) published in 1999. The standards for real-time data exchange between control centers (TASE.2 – Telecontrol application service element 2) have been published as international standards in 1997.

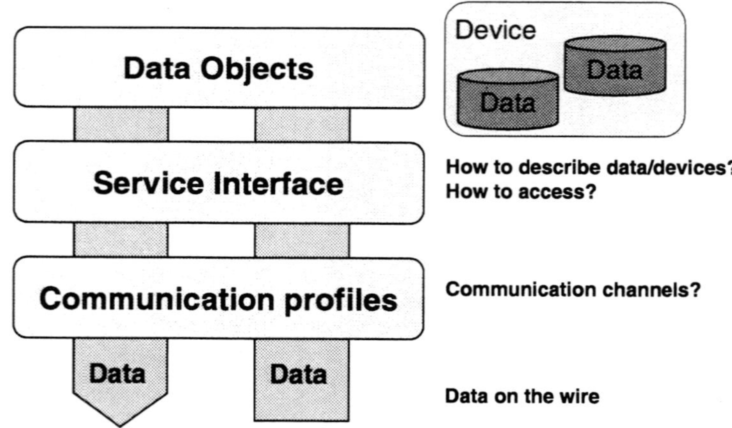

Fig. 1. The three levels of UCA

The UCA comprises the data object models (forming the highest level), the service interfaces to these models (defining, retrieving, reporting, and logging of process data, controlling devices, file transfer etc.), and the communication profiles (see figure 1).

3 The UCA Communications profile

Similar to current Internet solutions, UCA provides a network solution to interconnect data sources within and between utilities.

Ethernet was chosen as the main solution because of its:

- Market dominance;
- Plentiful, low-cost hardware, such as bridges and routers; and a
- Scalability from 10 and 100 Mbit/s, with 1 Gbit/s becoming available soon.

Figure 2 lists the complete communications architecture of UCA. UCA provides TCP/IP as well as full ISO/OSI seven layer solutions. For specific requirements a third block shows the reduced stack variants.

	Full 7 CO	WAN 7 CL	Modified 7 CO	Reduced Stack CO	Reduced Stack CL	LAN-Based FAIS	LAN-Based ** Ethernet	TCP/IP RFC 1006	TCP/IP RFC 1070	TCP/IP RFC 1240
Application	MMS ACSE	MMS CL-ACSE	MMS ACSE	MMS ACSE	MMS CL-ACSE	MMS	MMS ACSE	MMS ACSE	MMS ACSE	MMS CL-ACSE
Presentation	Presentation	CL Pres.	FastByte Pres.					Presentation	Presentation	CL Pres.
Session	Session	CL-Session	FastByte Session					Session	Session	CL-Session
Transport	TP4	CLTP	TP4					TP0 TCP	TP4 CLNP UDP	UDP
Network	CLNP	CLNP	CLNP			Auxiliary		IP	IP	IP
MAC Data Link	LLC1 ADLC FT3 or UCA 1	LLC1 ADLC FT3 or UCA 1	LLC1 ADLC FT3 or UCA 1	LLC1 ADLC FT3	LLC1 ADLC FT3 or Ethernet	LLC3 802.4 Token Ring	LLC3 ADLC FT3*_over_ Ethernet	Ethernet SLIP, PPP (typical)	Ethernet SLIP, PPP (typical)	Ethernet SLIP, PPP (typical)

7 Layer   3 Layer   TCP/IP

Fig. 2. UCA Communications architecture

To allow data access from any device, anywhere, UCA adapted International Organization for Standards (ISO, Geneve, Switzerland) Open Systems Interconnect (OSI) standard and Transport Control Protocol/Internet Protocol (TCP/IP). Protocols for the OSI's seven-layer are implemented. The most important protocol, the application layer, is built on the services of the Manufacturing Messaging Specification (MMS, ISO 9506).

4 Common Application Service Model (CASM)

The UCA Common Application Service Model (CASM) provides a common set of communication functions for data access, reporting, logging, control applications and related support. The use of a common set of services allows for 1) isolation of the models from service and communication details, 2) a high level of application inter-operability, and 3) reduced integration and development costs through the use of common mechanisms for data access and communication establishment. The CASM services are abstract and may be mapped into any number of existing communication application standards. MMS (ISO 9506) is the service specification of choice and mapping of CASM into MMS is included in the UCA document

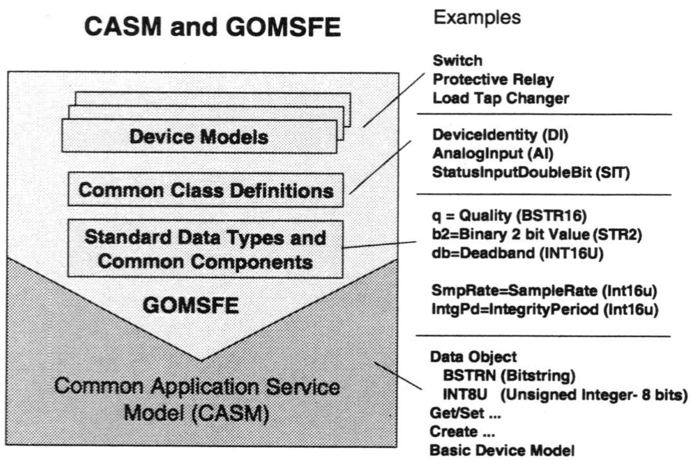

Fig. 3. UCA layering

As depicted in figure 3, the CASM provide services directly to the GOMSFE models.

5 The Generic Object Models for Substation and Feeder Equipment (GOMSFE)

One of the primary tasks has been the development of models for protective relay functionality along with all other anticipated IEDs in the substation. The development of these IED models is known as the Generic Object Models for Substation and Feeder Equipment (GOMSFE). Starting with a base set of models, each of the relay vendors has added draft models for an additional one or two functions, which brought the total to 13 models. These 13 protective relay function models have been reviewed in depth, and two basic building block models were developed (Basic Relay Object and Basic Time Curve Object). The existing models have been reworked to use the basic building block objects, and add extensions as necessary. It was concluded that an additional 23 relays could be modeled using the basic building blocks.

The most important GOMSFE device models are listed below. These models define some 2000 tagged information like vendor name. software revision, switch position status, current phase A measurement, or control a switch.

Exerpt of the UCA device models:

- Generic Input/Output
- Measurement Functions
- Transformer Functions
- Switch Functions
- Reactive Functions
- Protection Functions
- Distance (DIST)
- Synchronizing or Synchronism-Check (SYNC)
- High Impedance Ground Detector (HIZR)
- Directional Overcurrent (DOCR)
- Reclosing Relay (RECR)
- Differential Relay (DIFF)
- Measurement Unit
- Basic RTU Object Models
- Transformer Object Models
- Switch Object Model
- Automated Switch Controller Object Model
- Circuit Breaker Controller Object Model
- Recloser Control Object Model
- Reactive Component Object Models

These object models provide the interoperability of the various devices and systems connected in substations. They define the semantic of operations.

6 UCA substation demonstration initiative

EPRI's UCA Substation Communications Automation project has as its goal to produce industry consensus regarding Substation Integrated Control, Protection and Data Acquisition, and to allow interoperability of substation devices from different manufacturers. To this end, an open process has been followed on this project, to review each major project document and milestone in the open forum of standards-related organizations. The initiative is an excellent opportunity to present the benefits of the (redundant) Fast Ethernet and the device modeling technology.

The UCA 2.0 profiles for field equipment communications are separated into Application Profiles, Transport Profiles, and Data Link Profiles. These profiles are combined to form complete Profiles that can meet different requirements.

By adopting existing standards, the utility can take advantage of the economies of scale of the electric utility and industrial control industry that has made extensive use of these protocols. The substation initiative is now supported by some 30 utilities and 25 Substation device and systems vendors:

<table>
<tr><td>

Utilities:
- American Electric Power
- Arizona Electric Power Cooperative
- Ameren
- Boston Edison
- Baltimore Gas & Electric
- Bonneville Power Administration
- Cinergy
- CFE - Mexico
- ComEd
- ConEd
- Duke
- Duquesne Light and Power
- ESKOM
- Florida Power Corp
- GPU Energy
- Indianapolis Power & Light
- National Grid Co
- Northern States Power
- NUON
- Ontario Hydro
- Potomac Edison Power Co
- Pennsylvania Power and Light
- Southern California Edison
- Tampa Electric
- Texas Utilities
- Tennessee Valley Authority ...

</td><td>

Vendors:
- ABB
- Basler
- Beckwith
- Bitronics
- Cooper Power Systems
- Doble Instruments
- Dranetz/BMI
- Electrotech
- Alstom
- GE/Multilin Protection Sytems
- GE/Harris Energy Systems
- Telegyr
- Schneider Electric
- Sweitzer Engineering Labs
- Siemens
- Tasnet
- RFL
- Omicron
- Avo International
- Bailey Control Networks
- QEI Inc
- Toshiba
- Mitsubishi ...

</td></tr>
</table>

7 Application in the gas industry

UCA was adapted by GRI (Gas research institute, USA) for use by gas utilities. This effort culminated in an evaluation of UCA in a gas utility environment at Pacific Gas and Electric Company, San Francisco. With gas industry operations becoming more complex, as the study shows, the benefits of UCA are significant. With UCA in place, system operators can more easily automate systems, gather operating data, exchange information, and analyze historical statistics. However, despite the potential savings, manufacturers report little demand from gas utilities for UCA-compliant equipment.

The benefits of UCA include:

- The ability to develop integrated business applications across functional areas.
- Simplified implementation of fully integrated communications networks.
- Purchasing alternatives from multiple vendors (compatible hardware and software).
- Reduced operating costs through reductions in installation, maintenance, operation, and training.
- An enhanced ability to respond quickly to the continuing changes of a less regulated, more competitive business environment while still offering value-added customer services.

At Pacific Gas and Electric Company, UCA-compliant equipment was used to collect distribution system data (e.g., pipeline pressures, flow rates, and gas quality) at regulator stations and throughout a distribution piping system, along with information on customer load, weather, cathodic protection, and other conditions. The estimated cost savings demonstrated in the field experiment, extrapolated to the gas industry as a whole, is $133 million, with the potential for an additional $47 million savings ($180 million total) by further integrating and consolidating data collection and monitoring functions into a single "intelligent electronic device" at field sites.

8 Summary

Deregulation will place greater demands for information on utilities than they have experienced before. IEEE's UCA TR 1550 provides a timely, cost-effective, and standardized solution to allow advanced IED functions and distributed systems to form the foundation for 'next Generation' electric utility protection, control, and monitoring systems.

The benefactors of the results of open device data integration span the entire industry and include all of the stake-holders in this industry. The customers are in a position to save large sums of money and time. The vendors who provide solutions that meet or exceed expectations will become very successful. This is an exciting time in the industry with an inexorable move toward practical software components.

The most important issues are the models of the real device data and the rules (service interface) how to access these data. On the other side it is obvious that an appropriate transport mechanism (communication profiles), e.g., the TCP/IP or a point-to-point link, must be used to exchange the messages between devices.

By providing a common communications protocol stack, UCA allows a utility and other industries to "plug and play" equipment from different vendors. The specification of the uniquely tagged semantic of the most important device model data leads to a tremendous cost reduction during engineering, commissioning, operation, and maintenance.

References

1. IEEE Technical Report 1550 (1999): Utility Communications Architecture, UCA

2. IEC 60870-6-TASE.2: Telecontrol application service element 2

3. Committee drafts IEC 61850-7-y: Communication networks and systems in substations – Basic communication systems

A Component-Based Architecture for Integrating Fieldbus Systems into Distributed Control Applications

R. Bachmann, M. S. Hoang, P. Rieger

Dresden University of Technology
Department of Electrical Engineering
Institute for Automation and Control Engineering
Mommsenstr. 13, 01062 Dresden, Germany
fax: +49/(0)351 463 7039
e-mail: bachmann@eatns1.et.tu-dresden.de

Abstract. This paper presents a novel software architecture for the easy and complete integration of fieldbus systems into complex distributed control systems. A combination of well-established technologies with a new component-based approach results in a powerful system architecture suitable to meet today's user requirements. Our architectural concepts allow the rapid, compilerless development of efficient applications for fieldbus-based process control and visualization. Moreover, the assembled application may easily be adapted to changing process or operating requirements during runtime without interrupting active control tasks.

1 Motivation

Today fieldbus systems are increasingly used in virtually all areas of automation. The user of modern fieldbus components expects a great flexibility within the heterogeneous spectrum of existing fieldbus systems like Profibus, Interbus, CANopen and DeviceNet. Moreover, components and systems are required to be as easily manageable as possible. There have been recent efforts to define a flexible, unified software interface (e.g., RACKS[1], NOAH[2] [8]), which is of great importance for the efficient development of software for field devices. There are also specifications for interfaces in the area of PC-based automation (e.g., OPC[3] [7], Open Control [4]). Although these specifications are very important for device and software developers, they do not define overall architectural concepts to fulfill critical user requirements such as flexible and uncomplicated handling of control systems [1].

The concepts presented here aim at the easy and complete integration of fieldbus systems into complex distributed control systems. This can be achieved by combining

---

[1] RACKS - Reusable Application Interface for Communicating Real-Time KernelS
[2] NOAH - Network Oriented Application Harmonisation
[3] OPC - OLE (Object Linking and Embedding) for Process Control

standard fieldbus interfaces (RACKS, OPC), higher level communication architectures (CORBA[4] [6], DCOM[5] [5]) and the component-based software model DIO[6] [3].

2 Fieldbus Integration with DIO

2.1 Background

In recent years, component-based software and distributed object technologies emerged as two mainstreams of the evolution from "classical" object-oriented programming. While component-based technologies have proven successful at enabling rapid application assembly from off-the-shelf, pre-packaged software objects using visual building tools, distributed object technologies can be used to make components interoperable across processes, networks and platforms in an implementation-independent manner.

Meanwhile, the importance of adopting these technologies for developing automation applications, e.g., process control and visualization, has been recognized by many researchers and software-vendors. Unfortunately, existing models and concepts do not provide the specific characteristics required by industrial automation systems. Lack of real-time supports, lack of multi-level engineering supports, poor performance and static composition techniques are only some of the problems to be mentioned here.

DIO – a component-based software model - is capable of solving the problems stated above. This model defines the concepts of "pluggable" components and strongly typed message channels for the dynamic configuring of distributed automation applications as well as for optimizing the runtime performance of this software. Based on two novel design patterns, namely *Abstract Ports* and *Object Imitations*, DIO specifies further mechanisms to fulfill other domain-specific requirements, e.g., real-time support, multi-level engineering, dynamic compositions, etc.

2.2 Architectural Concepts of DIO

The strongly typed ports of components constitute a basic concept of DIO (Fig. 1). These ports help to solve problems of conventional application development. The second step in the scenario

- Definition of object dependencies by the software developer
- Data type checking and implementation of logical connections by the compiler

is omitted because all objects (DIO components) are capable of checking data types. A component may have Outgoing-Ports (O-Ports) and Incoming-Ports (I-Ports). Message channels can be established between an O-Port of a component and compatible I-Ports of other components. An O-Port initiates a communication, an I-Port receives messages

---

[4] CORBA – Common Object Request Broker Architecture
[5] DCOM – Distributed Component Object Model
[6] DIO – Distributed Intelligent Objects

278

and acts correspondingly. Messages always flow from O-Ports to I-Ports, but data may flow in both directions.

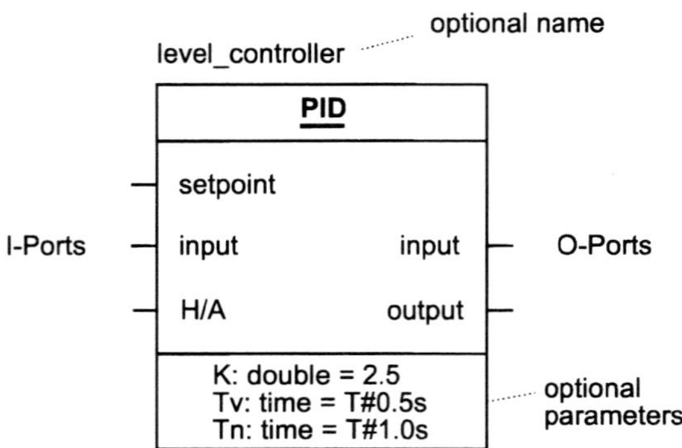

Fig. 1. Symbol of a DIO component (example: PID controller)

A special software tool – the *Configurator* – is used for the development of DIO applications (compilerless). Connections between objects are implemented visually, resulting in DIO charts. Another tool – the *Operator* – is used to start or stop an application. Starting an application may mean starting a simulation or starting a real-time application.

2.3 Synergy of Fieldbus Systems and DIO

Fig.2 illustrates the integration of DIO components and DIO tools into a control system. It can be seen from the figure that DIO charts look similar to conventional function block charts. That helps to prevent difficulties in acceptance by the users. Since DIO charts are not portable across platforms, a scripting language can be used for the application development.

The synergy of fieldbus systems and DIO has the following main advantages:

- FBD like view of the real system
- Easy and intuitive object-oriented application development across fieldbus systems by simply connecting DIO components as proxy objects of real devices (e.g., sensors, actuators, ...)
- Pluggable components allow application modifications in runtime
- Optimal runtime performance

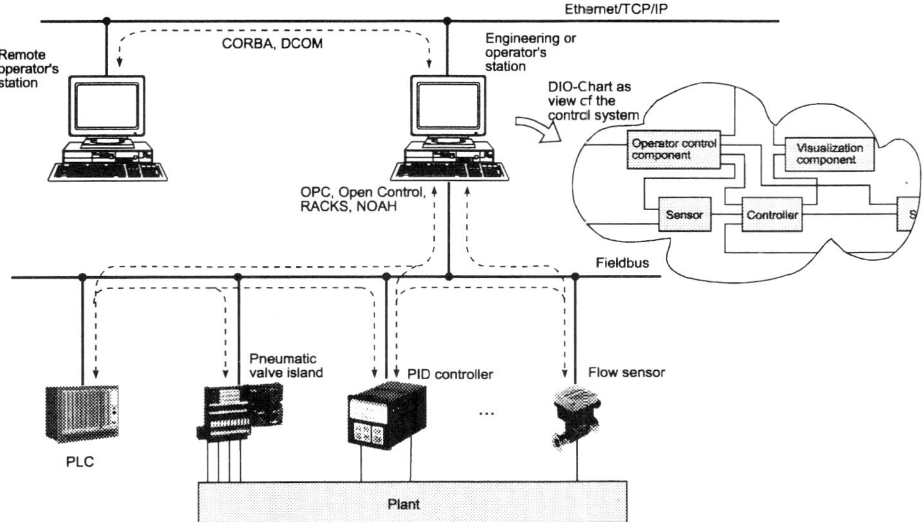

Fig. 2. Use of DIO in Fieldbus Systems

The basic configuring of the fieldbus system (topology, addresses, etc.) is similar to other systems using device data files (e.g., EDS[7] file for DeviceNet, GSD[8] file for Profibus). But the application development may be carried out with an extreme level of abstraction compared to conventional systems (e.g., IEC-1131 programming system). When developing the application, the user does not need to worry about communication, whether via fieldbus or higher communication systems like Ethernet. Moreover, simulation objects are treated like proxy objects of real devices (e.g., PID controller, sensors, actuators).

An important question is how the user of a DIO engineering system obtains the proxy objects (DIO components) of real devices. He can develop the necessary components himself (e.g., in Visual-C++) or the manufacturer of a device provides the user with the respective DIO component. As a consequence, in addition to e.g. a GSD file manufacturers would have to supply one or more DIO components as representatives of their device. Besides these DIO components which can act as proxy objects of devices, several device- and user-specific components for process control and visualization may be implemented.

The integration of DIO into the hierarchy of (standardized) communication technologies and interfaces is illustrated in Fig. 3. It can be seen from this figure that various interfaces may be used to connect an application to a fieldbus system. DIO can theoretically be implemented on the basis of any existing distributed object technologies (e.g., CORBA, DCOM).

---

[7] EDS – Electronic Data Sheet
[8] GSD – Geräte-Stamm-Daten (Device Basic Data)

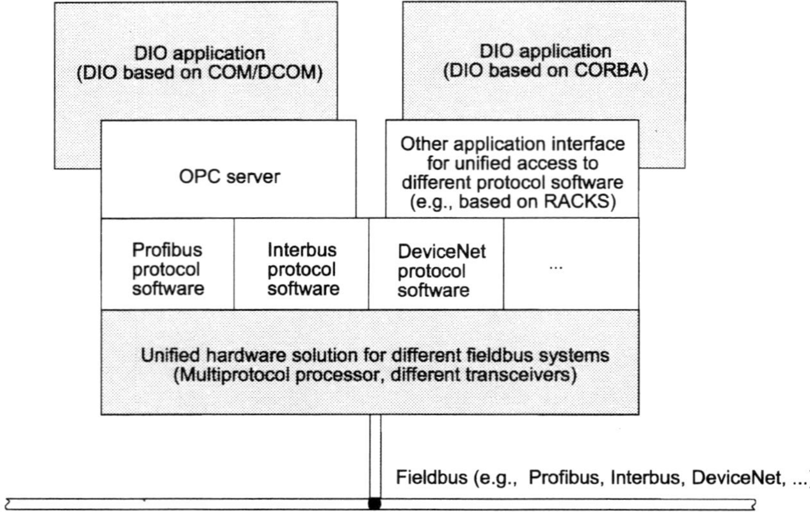

Fig. 3. Relation of DIO to (Standardized) Communication Technologies or Interfaces

3 Sample Application

The example included in this paper is intended to illustrate the use of DIO components based on DCOM/OPC.

Fig. 4 gives an overview of the used hardware components and shows the functional background of this example. A simple control loop consisting of a sensor, an actuator and a PID controller was to be formed. The sensor signal is taken from the output of a plant model and fed into an analog input module of the peripheral I/O device ET200. The actuator signal comes from an analog output module. These two signals are transmitted across a Profibus-DP segment. A Windows NT 4.0 PC is connected to this Profibus-DP segment using a fieldbus interface card which can be accessed by application software via an OPC server. The PID controller algorithm – encapsulated in a DIO component - is executed on the PC. Two other DIO components act as proxy objects of the sensor and actuator. Very simple operating and visualization facilities are included in the example, and it can be shown that DIO components (e.g., trend chart) may be added to the running system without any interference in the control loop.

For the purpose of demonstrating the efficiency (implementation extent, performance) of our approach, this simple application was implemented in three different ways:

- MFC application developed in Visual-C++ using the OPC custom interface
- Visual Basic application using the OPC automation interface
- DIO application

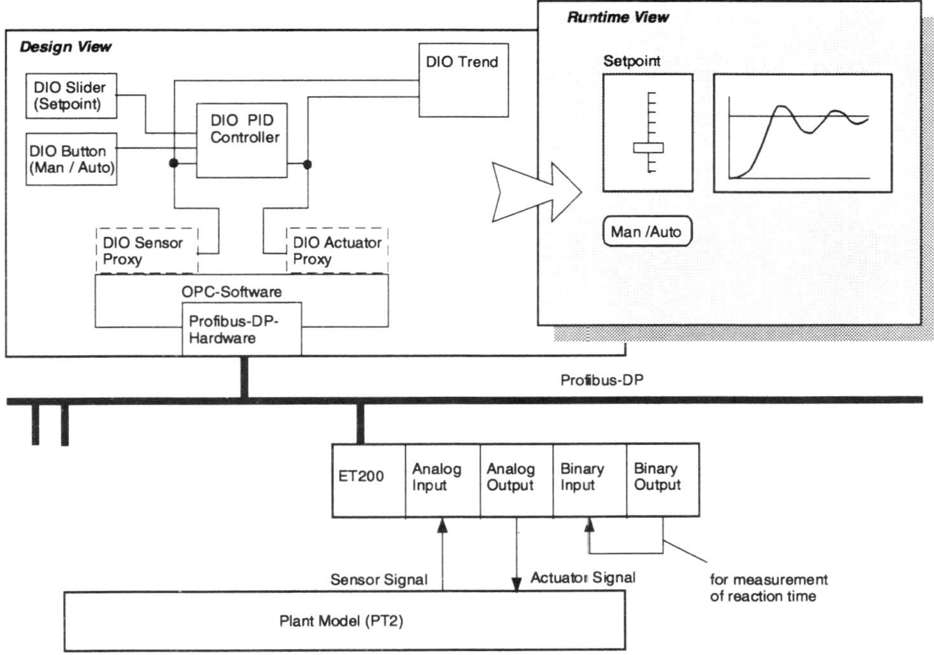

Fig. 4. Overview of the DIO Sample Application

The implementation extent can be seen from two main aspects – the ease of use and the code complexity to be implemented and tested. For the purpose of efficiently evaluating the runtime performance of our approach, further simplifications have been done. A binary signal has been sent as frequently as possible across the fieldbus to a binary output module. The output of this module was fed back into a binary input module on the same device (Fig. 4). The duration of 10,000 of these cycles was measured.

Table 1 shows the results of the comparison including some further evaluations. While the DIO approach yields nearly the same runtime performance as the MFC application, the other results reflect its superiority.

Table 1. Result of the comparison

	MFC/custom interface	VB/automation interface	DIO application
Ease of use / Code complexity	-	0	+
Pluggable components	-	-	+
Reaction time	100 %	362 %	101,7 %

4 Implementation of DIO on OPC

Since many industrial users prefer easy implementation techniques, they tend to use rapid application development systems like Visual Basic and scripting languages including VBScript and JavaScript. Ultimately, an OPC-based solution developer prefers using the OPC Automation Interface.

As shown in the previous section, the use of OPC Automation Interfaces has the following major drawbacks:

- Although these interfaces are fully supported by any automation client (e.g. Visual Basic, Visual C++, and VBA enabled applications), the functionality provided by an OPC server is normally not fully accessible due to the limitations of the respective client application.

- The interfaces described by the specification do not support VBScript or Java Script. As a consequence, the development of web-based automation applications is not directly supported.

- As all data access and event-handling processes are based on the automation dispatch mechanism, the performance could be very poor. Moreover, instead of direct communication between the OPC automation wrapper and a user object (e.g. controller, visualization objects,...), every data access and event-handling operation involves the automation client.

Our DIO implementation on OPC aims exactly at overcoming these problems. Figure 5 gives an overview of the implementation background of this approach.

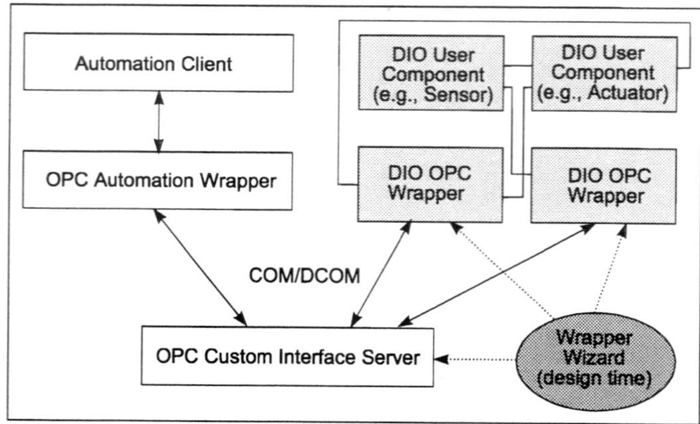

Fig. 5. Preferred Structure of a DIO implementation

5 Conclusions and Future Work

The approach presented in this paper makes full use of standard interfaces and the novel component-based software model DIO. This combination results in an architecture directly addressing major needs of today's users. Applications across fieldbus systems may easily be developed on a high level of abstraction, thus making the fieldbus system transparent to the user. Running applications may be further customized without interference in the system. The runtime performance achievable using the presented architecture is very good compared to popular alternative approaches.

We intend to develop more DIO components which can act as proxy objects of widely spread field devices like sensors, actuators or valve positioners. In addition, more DIO components for process control and visualization will be developed. The advantages of the presented architecture will be demonstrated in more sophisticated applications using a demonstration plant, which consists of real industrial devices and a plant model.

References

1. Arbeitskreis „Systemaspekte" im ZVEI: Die Prozeßleittechnik im Spannungsfeld neuer Standards und Technologien. atp, 1999, H. 1, S. 18ff.

2. Bachmann, R.; Rieger, P.: An Object-Oriented Knowledge-Based System for Intelligent Start-Up and Diagnosis of Fieldbus Devices. *Preprint ECC'99*.

3. Hoang, M. S.: DIO – Ein komponentenbasiertes Softwaremodell für verteilte Automatisierungssysteme. Dissertation, TU Dresden, 1998.

4. Interbus-S Club e. V.: Open Control Interface, Version 1.4, 1998.

5. Microsoft Corporation: DCOM Architecture. White Paper, http://www.microsoft.com/com/dcom.asp, 1998.

6. Object Management Group: The Common Object Request Broker: Architecture and Specification. Rev. 2.0, 1996.

7. OPC Taskforce, OLE for Process Control – Data Access Standard, Version 1.0A, 1997.

8. Quade, Jürgen: Ein einheitliches Feldbus-API. atp, 1997, H. 12, S. 50ff.

9. Rieger, P., Hoang, M. S.: Komponentenbasierte Automatisierungssoftware – Objektorientiert und anwendungsnah. Hanser Verlag, published 1999.

10. Töpfer, H., Bachmann, R.: Automatisierungstechnik aus Herstellersicht. Ergänzungsband 1: Feldbustechnik. Bürkert Steuer- und Regeltechnik, 1998.

Decentralized automation concepts based on Ethernet-TCP/IP and CANopen

Andreas Ziegler

Weidmüller ConneXt, Paderborner Straße 175, 32760 Detmold

Abstract: The advantages, which decentralized automation will make available to the machine and plant designer, are still to be seen. In discussion is a solution that can be realized now because the necessary components and technologies are all readily available, such as: Ethernet-TCP/IP, CANopen, industrial-compatible rail mounted PCs, OPC and IEC 1131-3.

1 Introduction

The driving force behind the decentralization in office automation was and is the continuous doubling of microprocessor power almost every two years. A drastic reduction in costs accompanied this development. This remarkable evolution has made it possible to operate small intelligent units and network them by means of powerful protocols.

This trend can equally be recognized in diverse applications within machinery and plant automation. For example, the input/output functions of a PLC assembly group which consists of a controller and a number of input/output cards in a backplane bus system, will usually be deployed in the field, with the aid of a field bus system, to where the sensor/actuator signal actually is or is needed. This reduces wiring costs and permits a cost and time efficient deployment of the machine or plant. The field bus can be seen as an extension of the PLC backplane bus system and distributes the input/output functions. This structure draws on only a part of the optimization potential. One still finds, as ever, centralized (controller) program processing (similar to a centralized mainframe computer used in office automation).

If this centralized control structure were to be broken down and the control tasks distributed to various small devices, which can act independently, then one would have functional modules that can be separately configured, powered up and maintained.

2 Improved modularity by decentralizing control functions.

Increasing modularity is the key to success when it comes to ensuring cost efficiency when deploying and operating a machine or plant. Rigorously putting into practice just such an automation concept, has the following advantages:

- Tried and tested plant- and machine modules with built-in control functions would become available.

- High levels of modularity would ease the entire integration of a machine or plant.

- Fast and easy maintenance allowing for shorter deployment and powering up times.

  The age of the decentralized industrial automation has begun. This is clearly to be recognized by the availability of powerful, small controllers and computer networks suitable for use in industry. With these, as the later to be described decentralized controlling concept will make clear, one can create and integrate functional modules into a computer network.

3 Only two bus technologies are needed

When examining technological trends, a continuous leveling off of hierarchies within the world of automation is clearly recognizable. The often mentioned "five level CIM-pyramid model" is nowhere to be found in practical applications. On the other hand, more and more concepts are coming through which cover all automation hierarchies, based on only two bus technologies. One is Ethernet-TCP/IP and the other a real sensor/control bus such as CANopen, Profibus-DP or Interbus.

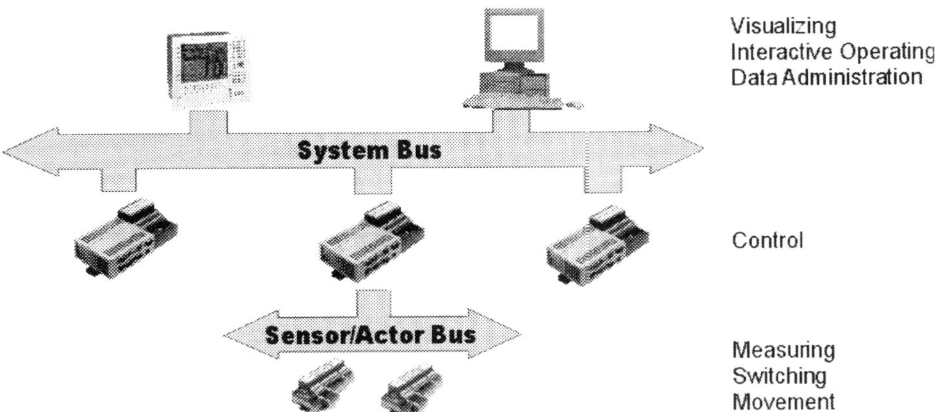

Fig. 1. two bus topology

Ethernet-TCP/IP networks the controllers (PLC, IPC, Soft-PLC) with one another, and these with the control level. The sensor/actuator bus couples the sensors and the controls, or more precisely, the intelligent periphery devices to the controller. With this concept, the functionality is distributed over three levels.

- Management level for visualizing, interactive operating, data administration and off-line storage.

- Controller level for carrying out real-time functions. The more frequently used platform being IPC with a real-time operating system, PLC and application specific microprocessor controlling.

- Sensor/actuator level in the form of process interfaces for measuring, switching and movement.

Two communication systems are needed to connect the three functional levels: the system bus for communication between the management and control levels, and the sensor/actuator bus for communication between the controller and sensor/actuator levels.

There are a number of advantages arising from the choice of Ethernet-TCP/IP as the system bus, such as high baud rates and unlimited compatibility with a large number of systems. Moreover, Ethernet-TCP/IP has already become the proven standard in different areas of operation.

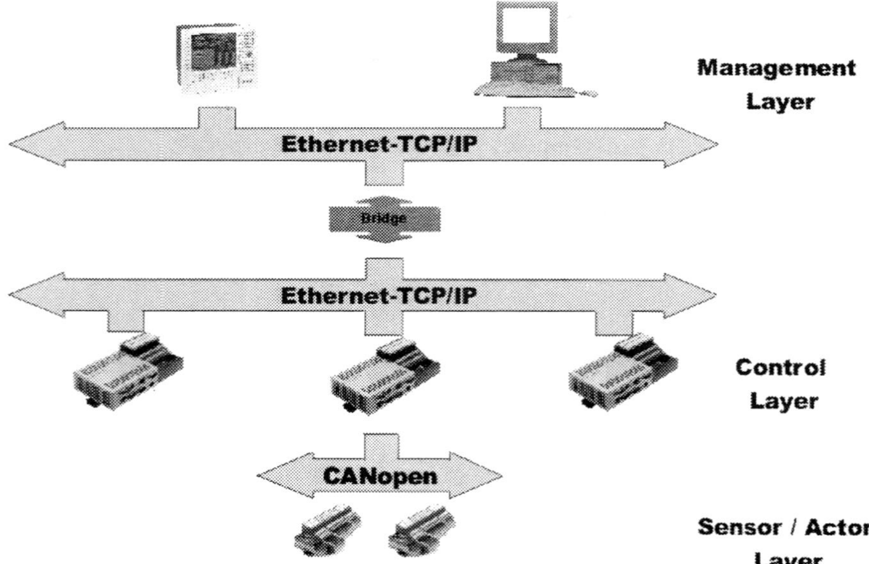

Fig. 2. topology with Ethernet-TCP/IP and CANopen.

CANopen on the other hand, is developing more and more into a powerful standardized network for the coupling of intelligent, programmable sensors and controls - either directly or by means of intelligent bus terminals.

4 Ethernet-TCP/IP: continuous communication between the management and the control level.

Ethernet-TCP/IP is already a fixed component of office networks in a great number of companies. Correspondingly, infrastructure and a large resource of knowledge in connection with concerning technology and applications are available. This know-how could be transferred without any problem, to the controller level within the manufacturing companies. Together with CANopen, this bus technology reduces engineering and maintenance work.

Various reservations concerning the use of this bus system in industrial automation have been expressed by the professional world, in respect of:

- Real-time capability (answering times, deterministic behavior),

- Robustness (immunity to interference, system suitability, maintenance)

- Availability of higher-level protocols.

To be precise, Ethernet-TCP/IP cannot be taken from an office environment and transplanted directly into an industrial environment. A number of different companies and institutions have put in a great deal of work on achieving the complete industrial suitability of Ethernet-TCP/IP, whilst keeping the one hundred percent compatibility with office systems. Although the research and results are very well documented (compare, e.g. [1], it is reasonable, to review the more important aspects once more:

4.1 Ethernet-TCP/IP is real-time capability

Ethernet-TCP/IP is not a deterministic system due to the CSMA/CD access logarithms (it cannot be made into one either without making incompatible changes to it's protocol). Added to this, there is also the latent danger of a culmination of data-collisions. Such collisions would lead to, amongst other things, intolerably long data transfer times. There are two preventative measures that can be taken:

- Dividing the system bus into "normal" and "real-time" segments; the segment separation is realized by bridges. The bridge has the task of keeping local Ethernet-data local. Thereby, local traffic in one segment cannot be seen in the other segment, and the danger of data collision is avoided.

- Introducing a load-limit in the real-time segments. By allotting a maximum network load to each station, the total load within the real-time segment is reduced. This in turn allows for certainty when predicting Ethernet-TCP/IP behavior. Complicated computer simulations have shown, that Ethernet remains stable by a network load of 10 % (= i.e. net through rate 1 Mbit/s) and that worst case transfer times of under 20 ms (mean delay: 2 ms) can be achieved. These figures can be reduced to 10 ms as well as 1.8ms, by a network load of 5 % (net through rate = 0.5 Mbit/s).

4.2 Ethernet-TCP/IP is robust

The Ethernet and TCP/IP error correction mechanisms, provide a high level of safety against malfunctioning. On the physical level, it is recommended for systems that cover large distances to be connected exclusively with optical fibers, and with hubs using shielded-twisted-pair cable (STP) to the stations.

4.3 Higher-level protocols for Ethernet-TCP/IP

Ethernet with TCP/IP is a transport protocol. It guarantees the safe transport of a bit signal from sender to receiver. Neither Ethernet nor TCP/IP issues recommendations of any kind, on how to take advantage of the bits and bytes. Therefore, the participants have to agree amongst themselves about the composition, content, and the meaning of telegrams – meaning, define a higher-level protocol (so called applications protocol). At present, the work on defining and standardizing just such higher-level protocols is well under way. The following are far advanced:

- S7-functions: access protocol over TCP/IP to the Simatic-S7,

- Maritime Information Technology Standards (MiTS),

- TCP/IP Instrument Protocol (IEEE-488 over TCP/IP),

- OPC.

4.4 OPC as applications protocol for the exchange of production data

OPC is becoming the standard in the applications level, for the exchange of production data using Ethernet-TCP/IP. OPC was originally developed for the deployment of a comprehensive interface to couple software tools with Human Machine Interfaces (HMIs), SCADAs or, manufacturing control systems to PLCs or network cards. Now it increasingly organizes the exchange of production data over and above Ethernet-TCP/IP.

The OPC standard defines in the main, a functional interface between two software systems. A productions data server (also known as OPC server) offers a method interface, which has to be operated by a data user (the OPC client). The method interface is then implemented when added onto a (quasi) standardized component model such as CORBA or COM, which organizes the communication between two software systems. By using the OPC standards on the OPC server and client side, a vendor neutral coupling between two software tools is created.

If the Remote Procedure Calls (RPC) function is supported by the component model (the possibility of calling-up of a server by a client through Ethernet-TCP/IP), there is the additional possibility of exchanging production data with OPC over and above Ethernet-TCP/IP. In the mean time, there are numerous OPC servers and clients being offered on the market, which implement OPC functions based on the Windows component model DCOM (Distributed Component Object Model). This means that process data can be used from any point within a company, via the company's own intranet or even the internet on any windows platform.

5 CANopen network: the second generation field bus

CANopen is gaining in importance as the new technological standard for networking control components at the field level. Justifiably so, CANopen can be seen as a further development of the first generation of field bus systems.

These first generation bus systems, quite often settle for a single master structure. This makes a central administrator, called Arbiter necessary, to organize data traffic on the bus. This is implemented in the classical control architecture of the controller. The very existence of an Arbiters puts up high hurdles, if decentralized independent units, designed as self-supporting (without a central controller), are to be commissioned and powered up within a network.

CANopen gets over these obstructions and opens the way for decentralized automation. To this end the central Arbiter is transferred decentrally into every single network station. This procedure is allowed for by the Arbiter mechanism CSMA/CA (Carrier Sense Multiple Access with Collision Avoidance). By using this communications oriented, prioritized collision avoidance system, each individual station can actively seek permission to transmit signals.

This turns CANopen into a multi-master network. It is now possible to create functional units from a number of CANopen stations and separately configure and commission them, without the existence of a central controller or a central Arbiter. It is therefore possible to create modules by dividing up the control functions at the field level.

6 Continuous, harmonized solutions

Weidmüller for example, offers with the products DIAcan, DIAcontrol and DIApro, a continuous, harmonized set of solutions for establishing decentrally controlled tasks. The components DIAcan and DIAcontrol control platforms are supported by open standards such as IEC 1131-3, CANopen, TCP/IP and OPC. The following modules are put to use on individual levels:

There exists on the management level, sufficient software structures for visualization, HMIs, SCADAs, production control systems like ERPs or MES. High levels of investment were needed to adapt these applications. OPC as the standard interface for the exchange of production data, also takes care, that such software packages can easily be integrated with Weidmüller's controller components.

With integrated Ethernet-TCP/IP and CANopen terminals, DIAcontrol is explicitly suitable for the control level, with the IEC 1131-3 programmable Industrial PC for switching cabinet installation. Moreover, I/O modules can be directly connected. DIAcontrol co-ordinates the workings of the functional modules networked by CANopen. In addition it implements the CANopen network manager. The complete network configuration including the applications programs and the configuration of the underlying CANopen network are stored on it's flash disk. DIAcontrol is therefore, capable of configuring the CANopen modules and monitoring their functional capability. The IPC takes on in addition the communication with higher-level systems via Ethernet-TCP/IP and exchanges production data via OPC.

Finally on the sensor/actuator level, programmable I/O modules (called DIAcan) are put to use, next to standard I/O modules. Integrated into the housing of Winbloc I/O modules, they convey the control functionality to the sensor/actuator terminal level. DIAcan fully utilizes the possibilities of the multi-master capable CANopen network. In this way, functionally independent units can be built up together with other CANopen stations. The I/O modules are optimized for short reaction times, whereby, it is possible to divide the control program into a number of tasks, prioritize these individually, and link them to an input / output.

7 Software for all necessary program- and configurations steps.

Both platforms from DIAcontrol and DIAcan are programmed and configured with the engineering tool, DIApro. It features all IEC-1131-3 program languages and contains extensive error detect functions as well as a HMI and Network Configuration for the integration of any CANopen node. The DIApro can either be adapted direct on CAN to process individual functional modules, via RS 232 on DIAcontrol for individual cells, or via TCP/IP for the whole application, depending on the degree of automation of the machine or plant. All DIApro functions can be used via a modem or internet for post powering up or long distance maintenance.

References

1. Frank J. Furrer: Ethernet-TCP/IP für die Industriautomation. Hüthig Verlag, Heidelberg, 1998. ISBN 3-7785-2641-3.

2. CiA (CAN in Automation): CiA Standards 301, 302, 401, etc. CAN in Automation e.V. International Users and Manufacturers Group, Erlangen, 1995-1998.

3. OPC Foundation: Ole for Process Control (OPC) Final Specification V2.0 OPC Foundation, P.O. Box 140524, Austin, Texas, 10/1998

Considerations on a LonWorks/IP Gateway Implementation

Stefan Soucek, Hans-Jörg Schweinzer

Institute of Computer Technology, TU Wien
{soucek,schweinzer}@ict.tuwien.ac.at

Abstract. The success of the Internet makes it a superb goal to discuss interworking aspects with fieldbus systems. LonWorks with its strict layering is especially interesting to explore within this context. The discussion of concepts for a LonWorks/IP gateway leads to three approaches exploiting different protocol mechanisms giving IP hosts access to nodes' files, network variables and application messages as well as EIA-709.1 link layer packets. An experimental implementation is based on a real-time operating system running on an embedded hardware platform featuring a powerful Motorola micro controller.

1 Introduction

The Internet has evolved to be a widely used means for information exchange and worldwide communication nowadays. Fieldbuses on the other hand are growing in acceptance and open new chances to automation in different fields. A superb goal is to enable interworking between the two worlds to exploit features of one another.

LonWorks in this respect is an especially interesting fieldbus system to investigate due to its strict layering scheme. Its wide acceptance in building and home automation also provides for promising applications of a LonWorks/IP gateway. A gateway that transparently maps between LonWorks communication objects and Internet protocols opens a new potential to remote network access and monitoring. Isolated LonWorks segments can be connected via Ethernet or other media capable of carrying IP traffic.

Today there exist solutions where a proxy collects data from fieldbus nodes and makes this information base accessible from the Internet. This approach is easy to implement but too restrictive in some areas. The intent of this paper is to present a collection of concepts that allow for a more complete access to LonWorks nodes and how they can be implemented in an embedded system.

The key functionality of a LonWorks/IP gateway comprises the manipulation of node files, the access of network variables and application messages and the processing of link layer PDUs. This is necessary in order to load and store node configuration files, to monitor and control node applications and to connect LonWorks networks over IP paths, respectively. In order to specify necessary mechanisms different strategies for interworking have to be discussed. The most promising approaches are chosen for implementation on an embedded system.

2   Approaches

A variety of different gateway implementations can be conceived, due to the multitude of possible protocol conversions. Differences that have to be considered with a LonWorks/IP gateway include address spaces, transport protocols, application layer services and application programs. Table 1 gives a summary of the most important differences between communication objects in LonWorks and the Internet protocol suite. This assembly assumes that IP is run over Ethernet.

Table 1. Protocol properties in EIA 709.1 and IP

	EIA 709.1	**IP over 10BaseT**
Application layer		
protocol	LonMark File Transfer	FTP, HTTP, SNMP, . . .
Session layer		
service	Request/Response	—
Transport layer		
reliable service	ACKD service	TCP
unreliable service	UnACKD, UnACKD repeated	UDP
Network layer		
address format	subnet/node, group, Node ID	network/host
frame format	LonTalk NPDU	IP datagram
LLC sub-layer		
encapsulation	LonTalk LPDU	IEEE 802.2 LLC

A gateway typically translates between protocols on top of the OSI layer 7, the application layer. The main application services in LonWorks are network variables, application messages and the LonMark File Transfer. A gateway should include provisions to give Internet hosts access to these services.

One approach for a gateway design is to look for Internet application protocols that provide functionality similar to LonWorks application services. The great number of protocols in the Internet protocol suite, however, creates the need to concentrate on a subset in an actual implementation. Here a trade-off decision between complexity and performance/resources has to be made, especially with respect to an embedded system implementation. Most Internet protocols are very application specific and can't be used with a LonWorks/IP gateway. Protocols that have been considered are FTP, HTTP and SNMP due to their ability to transfer files, retrieve objects and access variables, respectively.

Not all of the above are equally appropriate to use with all LonWorks application services in a gateway. HTTP ([4]), for instance, provides no mechanism to store files or objects which makes it not suitable for use with LonMark File Transfer. SNMP ([1]) is designed to retrieve and store values in MIB variables which makes it a good choice for accessing network variables. The polling nature of SNMP, however, makes it impossible for LonWorks nodes to actively send network variable updates to IP hosts. None of the mentioned protocols provides functionality needed to send or receive application messages. Therefore only FTP is considered in the implementation. A more detailed discussion can be found in [10].

Another approach for a LonWorks/IP gateway design is to implement some of the upper EIA-709.1 layers on the IP host utilizing TCP/IP protocol layer services. This requires additional communications software on IP hosts but provides best flexibility in cases where Internet protocols do not provide sufficient functionality. This approach is favored to access network variables and application messages.

3 Gateway Concepts

3.1 LT/IP Tunnel

The LonTalk/IP tunnel (LT/IP tunnel) is an approach in which EIA-709.1 link layer PDUs (LPDUs) are carried over the IP network. Speaking in OSI terms, the LonTalk[1] protocol is "tunneled" through the IP network. The two main functions of a gateway that implements this scheme are to map LonWorks addresses to corresponding IP addresses and to embed LPDUs in Internet transport.

It is sufficient to use an unreliable Internet transport service like UDP for the transport of LPDUs since LonWorks itself provides a reliable transport service. TCP with its sophisticated transmission scheme would only introduce redundant protocol overhead. Figure 1 depicts the architecture in the OSI model.

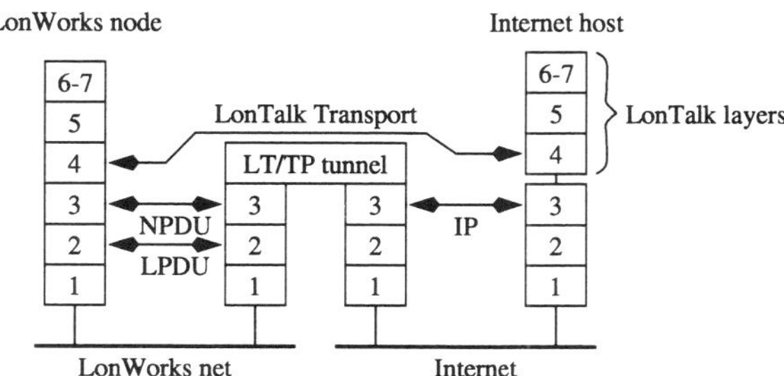

Fig. 1. LT/IP tunnel architecture

Address mapping in the gateway is based on address mapper tables (AMTs). The AMTs are assumed pre-configured in this discussion. This decouples configuration issues from pure functionality and enables the use of appropriate algorithms. The gateway has to

---

[1]LonTalk is used as a synonym for EIA-709.1 compliant protocol.

read the EIA-709.1 destination address from the LonTalk NPDU[2] and find a corresponding IP address in the AMT. For performance reasons, there exist three AMTs, due to the different addressing schemes in LonWorks: one for Node ID addressing, one for subnet/node addressing and one for group addressing. The domain ID is included in all three AMTs.

The AMTs represent forwarding tables. The gateway drops a LonTalk packet if it does not match an entry in the corresponding AMT. If the LonTalk packet is to be forwarded the gateway puts the LonTalk LPDU into a UDP datagram. The encapsulated LPDU is sent to the specified IP host using a default port address, which has to be defined for LT/IP tunnel service.

The LonTalk LPDU is considered here for tunneling because the peer IP host can be both an end-system in the LonWorks network as well as another LT/IP tunnel gateway, which forwards LonTalk packets to a remote LonWorks network. On the remote network the priority bit and channel backlog information of the LPDU is needed again for transmission. If the peer IP host is an end-system the NPDU alone is sufficient.

The described approach allows the connection of two or more LonWorks network segments over IP networks. It is also possible to connect IP hosts that implement a complete EIA-709.1 protocol stack and serve as extremely powerful "nodes". An implementation of this concept has to ensure that LPDUs are delivered in-order and in-time. This is necessary because of some EIA-709.1 protocol restrictions[3]. Especially the latter requirement is hard to meet on the global Internet as it implies the accurate measurement of packet delays.

3.2 LT Layer 7 Gateway

The idea of the LonTalk layer 7 gateway is to establish network variable and application message communication between IP hosts and LonWorks nodes. The IP host, however, does not rely on EIA-709.1 transport services it rather uses TCP to communicate with the gateway. This approach implies IP hosts have has to implement an extra communication layer that provides an EIA-709.1 application layer interface to host applications.

The protocol translations involved in this solution concern the LonTalk session layer and transport layer. End-systems use the same application layer objects, they only embed them in different transport protocols. It is interesting to note, that there is no session layer on the Internet side of the gateway. Consequently, LonTalk transport layer protocols and session layer protocols are replaced by a single Internet transport protocol on the IP side. Figure 2 illustrates this approach in the OSI model.

A way to implement this concept is to use a proxy sub-layer for the EIA-709.1 ALI on the IP host that establishes a RPC[4]-like protocol to the ALI on the gateway. The task of

---

[2][3] p. 16

[3]The limit of 15 transaction IDs and the algorithm for duplicate detection put very strict requirements on LPDU delays. Packet delays on the Internet can be very high ([5]) and can't be met by EIA-709.1 receive timers. For more discussion refer to [10].

[4]Remote Procedure Call

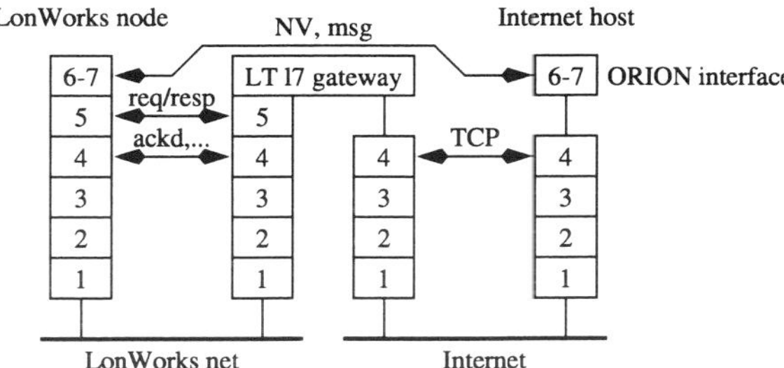

Fig. 2. LT layer 7 gateway architecture

the ALI proxy is to convey application layer function calls to the gateway, where they are executed. That means the gateway sends network variable updates on behalf of an IP host. To a LonWorks node it appears the update originates from the gateway. The gateway uses TCP for the RPC-like mechanism.

The presented scheme enables programs on IP hosts to access network variables in the LonWorks network transparently. An implementation of this concept needs mechanisms in the EIA-709.1 protocol stack on the gateway that allow the dynamic creation and deletion of network variables while IP hosts log on and off the gateway.

3.3 LonMark File Transfer/FTP Gateway

The LonMark file transfer/FTP (LM FT/FTP) gateway is a classical gateway approach where two application protocols are translated. The gateway presents FTP services to the Internet and LonMark file transfer services to the LonWorks network. The Lon-Mark file transfer ([2]) defines a simple mechanism to load and store files in a node. Its functionality can be considered a subset of FTP ([6]), which is widely used in the Internet. For example, LM FT does not support directory structure or deletion of files.

The main task of a LM FT/FTP gateway is the translation of file transfer commands. Transferred data is interpreted as a stream of bytes between an IP host and a LonWorks node and therefore needs no conversion. The gateway simply pipes the data from its FTP side to its LM FT side.

FTP commands have to be mapped to corresponding LM FT functionality. Since LM FT only defines a subset of commands the gateway itself needs to process the unsupported commands. For commands without LM FT pendants a gateway has the choice to either emulate them in an appropriate way or to return an error. Table 2 lists file transfer control commands for both protocols and illustrates how they can be mapped. In general the actions taken by the gateway can be subdivided in three categories:

- Translate to corresponding commands: This is the case with STOR, RETR and LIST. These FTP commands set up a certain transfer process, and the gateway needs to issue LonMark file transfer commands in order to initialize a corresponding transfer on the specified LonWorks node.

- Emulate commands: Only the gateway is involved with this class of FTP commands, taking appropriate actions. For instance, FTP user authentication is processed by the gateway, consulting an access control configuration file. Other commands are allowed with default settings only.

- Return an error: Commands which have no correspondence in LonMark file transfer or which make no sense to use with a gateway return an error value, telling the clients this command is not supported. All FTP commands not listed in table 2 belong to this category. For instance, the `DELE` command is not supported, because files cannot be deleted from LonWorks nodes.

Table 2. Mapping of mandatory FTP commands

FTP	LonMark file transfer
`RETR`	`FR_OPEN_TO_SEND`
`STOR`	`FR_OPEN_TO_RECEIVE`
`LIST`	`FR_DIRECTORY_LOOKUP`
`CDW`	virtual network file system
`PWD`	virtual network file system
`USER`	gateway access control
`PASS`	gateway access control
`QUIT`	kill user control thread
`TYPE`	default setting: image
`MODE`	default setting: stream
`STRU`	default: file
`NOOP`	ignored
others	error: not supported

Another important issue is addressing certain LonWorks nodes through FTP. The assignment of IP addresses for every single node is impractical due to the restricted address space in IP version 4[5]. A solution to this problem is to embed the LonWorks addressing hierarchy into a virtual network file system on the gateway.

The root of this file system is the gateway itself. This is also a good place to store configuration files. First-level subdirectories correspond to LonWorks subnets which contain subdirectories for nodes. The leaves of this directory tree are representing the nodes' file systems, which do not contain further subdirectories. The naming of subdirectories may use subnet and node numbers in order to ensure consistency. Using this convention the first file in node 11 of subnet 47, for instance, is assigned the path `'/47/11/file1'`.

The presented scheme allows the use of standard FTP clients without any modifications to load and store node files in a LonWorks network. Security issues can be handled by the gateway employing a user/password authentication scheme or other enhanced methods such as Kerberos.

---

[5] IP version 6 will remove this restriction in the future.

4 Implementation

The gateway developed to implement the described approaches is a one-box solution with a LonWorks network port and an Ethernet port. The LonWorks port is used to communicate with fieldbus nodes using the EIA-709.1 protocol. The Ethernet port connects the gateway to the LAN, where it is mainly a source of Internet traffic. This design lets you use standard TCP/IP components. Routers may be used for interconnection or firewalls for security considerations.

Figure 3 depicts the gateway architecture. It features two complete protocol stacks for both TCP/IP and EIA-709.1 communication. The LonWorks part of the gateway consists of a hardware implementation for EIA-709.1 layers 1 to 3 (LDC, [9]) and a software implementation for the upper layers (ORION Stack, [8]). The LDC Driver Interface (LDI) decouples the ORION stack from the LDCs. Applications running on the gateway can access the ORION stack through a C++ object framework that implements an object-oriented LonWorks application layer interface ([7]).

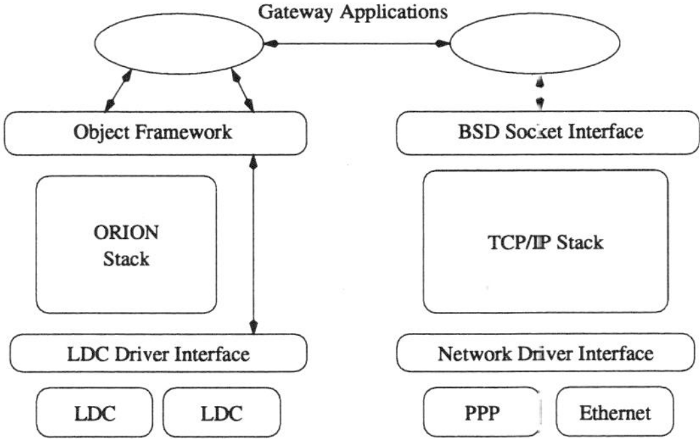

Fig. 3. Gateway architecture

Internet communication facilitates a TCP/IP stack that was ported from the Linux kernel TCP/IP implementation ([10]). The TCP/IP stack operates on top of a Network Driver Interface which abstracts the various network interfaces such as PPP or Ethernet. Applications access TCP/IP functionality through a BSD socket interface.

The gateway functionality discussed in the previous section is implemented as applications running on top of the application interfaces of the two protocol stacks. Typically, they are split in two threads communicating via some sort of IPC. Each thread runs in the context of its associated protocol stack. In cases where not the whole functionality of the protocol stack is needed, the application layer interface provides a bypass. The LT/IP tunnel for instance accesses EIA-709.1 LPDUs directly through the LDI.

The described software is running on an embedded system featuring a Motorola MPC821 microcontroller operating at 50MHz, 1 MB Flash and 4 MB EDO-RAM. The freely available real-time operating system RTEMS ([11]) was ported for this board. In order to improve portability the software uses a light-weight abstraction layer for all essential system resources such as timers, threads, semaphores or interrupt locks.

5 Conclusion

The implementation of a LonWorks/IP gateway opens a wide range of interworking possibilities between a LonWorks network and the Internet. This paper has presented several approaches how communication objects of the two protocol worlds can be translated. The implemented concepts – a LonTalk/IP tunnel, a LonTalk layer 7 gateway and a LonMark File Transfer/FTP gateway – provide IP hosts with different levels of access to LonWorks nodes and networks.

The presented implementation of the gateway was intentionally kept as simple as possible in order to focus on the principles of different approaches. All parts that rely on somehow configured parameters use configuration tables. Further discussion has to be done on appropriate configuration mechanisms for a LonWorks/IP gateway. Also, more research has to be conducted in the area of EIA-709.1 packet transmission over IP networks in order to make it work reliably over the global Internet.

References

1. J. Case and M. Fedor. *Simple Network Management Protocol (SNMP)*. ISI, May 1990. RFC-959.

2. Echelon Corporation. *File Transfer*, January 1995. LonWorks Engineering Bulletin.

3. Electronic Industry Alliance. *Control Network Specification*, March 1998. EIA STANDARD EIA-709.1.

4. R. Fielding, J. Gettys, J. Mogul, H. Frystyk, and T. Berners-Lee. *Hypertext Transfer Protocol - HTTP/1.1*, January 1997. RFC-2068.

5. V. Paxon. End-to-End Internet Packet Dynamics. In *Proc. SIGCOMM '97*, pages 139–152, September 1997.

6. J. Postel and J. Reynolds. *File Transfer Protocol (FTP)*. ISI, October 1985. RFC-959.

7. Thomas Rauscher. Object-Oriented User Layer for LonSpeed. Diplomarbeit, Technische Universität Wien, 1998.

8. Norbert Reiter. LonSpeed – Upper Layer Protocol Stack Implementation and Node Management. Diplomarbeit, Technische Universität Wien, 1998.

9. P. Rössler and D. Loy. LDC - Ein Controller für das Feldbus-Protokoll EIA-709.1. In *Proc. Austrochip 1998*, 1998. ISBN 3-901578-03-X.

10. Stefan Soucek. LonTalk/IP Gateway. Diplomarbeit, Technische Universität Wien, 1998.

11. U.S. Army Missile Command, Redstone Arsenal, Alabama 35989-5254. *Real-Time Executive for Multiprocessor Systems (RTEMS)*, May 1995. Release 3.2.x.

The Importance of Being Competent:

the Role of Competence Centres in the Fieldbus World

Thilo Sauter*, Max Felser**

* Institute of Computer Technology, Vienna University of Technology,
Gußhausstraße 27-29, A-1040 Vienna (AUSTRIA)
T. ++43-1-58801-38430 F. ++43-1-505 38 98-14
E-mail: sauter@ict.tuwien.ac.at

**Bern Institute of Engineering and Architecture, Bern University of applied sciences,
Morgartenstrasse 2c, CH-3014 Bern (SWITZERLAND)
T. ++41-31-33 55-237 F. ++41-31-33 30-625
E-mail: max.felser@hta-be.bfh.ch

Abstract. As far as their application in practice is concerned, fieldbus system are still a new technology that many potential users are not quite aware of. Competence centres can provide such users with know-how and thus help to improve the acceptance of these new concepts. This article presents an overview of the services a competence centre can provide, with a particular emphasis on training activities. We shall also describe a European project dedicated to the establishment of new competence centres in the countries of Central Europe.

1 Introduction

The development of fieldbus systems was driven by the wish to reduce the cabling in automation systems and to enhance the functionality of the installations by creating distributed and decentralised structures. For historical reasons, the creators of many fieldbus systems had actual applications or application areas in mind, and so a multitude of definitions evolved. While many potential users recognised the advantages of the emerging industrial networks, they were reluctant to use them because of their mostly proprietary nature. The great breakthrough came when the idea of open systems was born and certain fieldbus protocols were standardised. In the meantime, many national standards have been superseded by international ones like the European standards EN 50170 or the EN 50254.

Today, a user can select from a variety of (at least theoretically) interoperable products to compile the appropriate fieldbus devices for a concrete installation. With the new standards, the often inconvenient dependence on one single manufacturer is no longer an argument not to employ fieldbus systems. Yet, the concept of these industrial networks is still unfamiliar to many potential users. In particular among the small and medium-sized enterprises, there is often a lack of information and awareness preventing the use of fieldbus systems even in cases where the benefits are clearly visible for the expert. Consequently, there is a demand for competent and independent institutions that provide information as well as training for potential fieldbus users and give also practical support. This need is satisfied by competence centres.

The importance of competence centres as seed crystals for the fieldbus industry is even more pronounced in areas where the application of contemporary automation concepts is still at the beginning. This is, for example, true for the countries of Central and Eastern Europe, where the political and economical changes also called for improvements in the manufacturing processes. Among many others, the use of automation is one means to close the gap between the eastern and western parts of Europe. Therefore a transfer of fieldbus know-how is reasonable and important.

2 What is a competence centre?

It is difficult to give a concise definition of what constitutes a competence centre. A suitable approach is classification according to the services it provides, which may be different depending on the knowledge of the staff and the general policy of the centre. Therefore we can roughly differentiate between system centres, integration centres, and test centres. In practice, of course, these functions overlap, and all centres are usually also training centres for the type of service they provide (fig. 1).

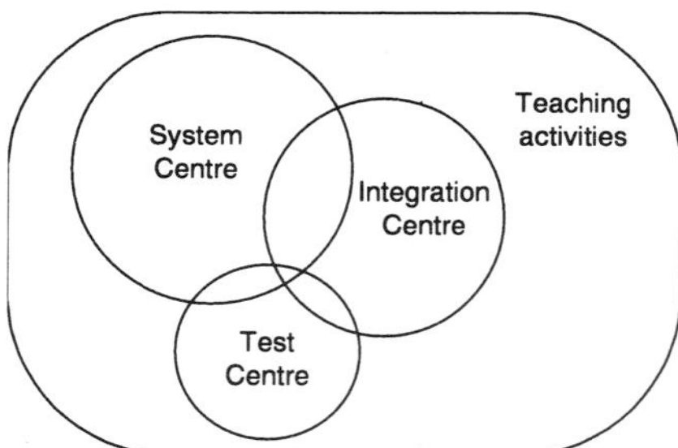

Fig. 1. Different types of competence centres

A *system centre* provides all support necessary to perform an open fieldbus system integration. This includes consulting and training as well as hotline support or experts for trouble shooting. These services require that the competence centre be independent of all manufacturers but still have good contacts to the developers of the equipment used in the installations. To achieve a good consulting quality, different types of fieldbus systems should be supported. Undoubtedly, the best knowledge is gained through own practical experience with multi-vendor systems. Appropriate and up-to-date equipment is a also must, and a large variety of devices enhances the experience and improves the services.

A fieldbus *integration centre* supplies manufacturers of automation equipment with the know-how to implement a fieldbus connection to their devices. This may be in the form of consulting, training or workshops up to the complete product development according to the needs of the manufacturer. The preconditions to provide these types of services include most functions of a system centre. However, it requires more detailed technical knowledge about the implemented fieldbus, and it is possible to be only an integration centre and specialise in one particular type of fieldbus.

An independent *test centre* checks newly developed devices for conformance to the standard and interoperability according to predefined testing rules. Devices that pass these tests get a corresponding certificate, which is an important marketing aspect. The work of a test centre, however, is sensible only if a special test procedure is specified in the standards. In the case of PROFIBUS according to EN50170, the PROFIBUS User Organisation defined these rules. An approved test centre is audited every three years by the user organisation and must fulfil precisely defined requirements.

Apart from the working areas, also the customers of the different types of competence centres and their particular needs are different. There are many system integrators that require technical support. These system integrators are mainly small and medium-sized enterprises and need short-term local help. The number of manufacturers of fieldbus devices, on the other hand, is considerably smaller. They also prefer local technical support, but normally such activities can be planned several months in advance. Finally, for a one day conformance test only a few highly specialised test centres on a continent are sufficient. Therefore a competence centre does not necessarily provide all types of services. However, a close co-operation of the different competence centres is reasonable.

3 Goals of the TRAFICC project

The name TRAFICC stands for the focal activity of a project funded by the EC: transferring European fieldbus technology to the countries of Central Europe [1]. The primary goal is to establish competence centres that are able to support the local industry. The TRAFICC consortium consists of three "technology providers" from Austria (TU Vienna), Germany (TU Munich), and Switzerland (HTA Bern), which represent already existing competence centres. On the other side, there are three "technology users" from the Czech Republic (TU Prague), Hungary (TU Budapest), and Poland (TU Wroclaw). These universities are to host the newly founded expert groups. With a total effort of some 15 man years, the project was scheduled to last 28 months. To achieve the goals of the project, three major phases were planned (fig. 2).

During the initial knowledge acquisition phase of the project, the western partners provided training courses for the CCE participants to become fieldbus experts. This "train the trainers" activity served two purposes. First, the trainees could gather both theoretical and practical experience with different fieldbus systems. Second, the practical work in the existing competence centres gave them an impression of the infrastructure required for a competence centre and the possibilities how to set up their own training programmes later on. For the sake of comparison and for a complete knowledge transfer, it was extremely important that the CCE partners visited all three competence centres of the technology providers.

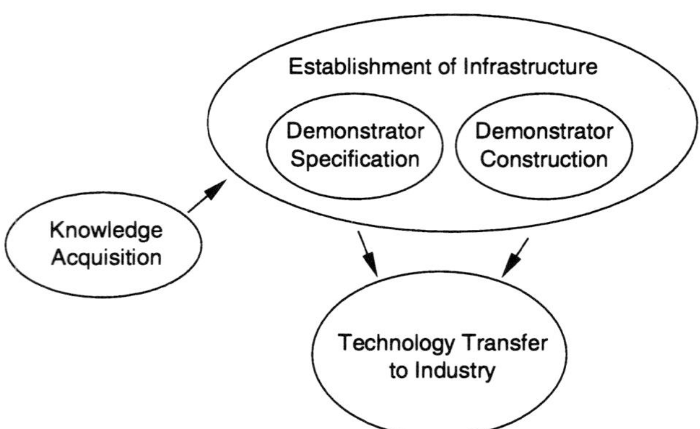

Fig. 2. Phases of the TRAFICC project

The second phase was the establishment of infrastructure in the new competence centres. This included the specification and construction of so-called "demonstrators", multi-vendor installations of fieldbus systems that provide the hardware platform for later research and education activities. The second aspect of the infrastructure was the creation of an appropriate operational environment with a proper management and – most important – a permanently employed personnel able to maintain the operability of the competence centre. The acquisition of the funding needed for a continuous employment of the experts will in most cases already require a close and persistent co-operation with the local industry. In this second phase, the western partners still provided some support, particularly regarding the set-up of the fieldbus installations.

The third phase was again a technology transfer. This time, however, it was the know-how transfer between the new competence centres and the local industry. The technology transfer was based on training courses for industrial fieldbus users, general workshops and seminars, consulting activities, but also the co-operation within the scope of research and development projects. To ease the contact with the local industry and to present the competence centres to the public, workshops were planned and held not only at the end of the project, but also as accompanying measures during the set-up phase of the competence centres. This way, the particular needs and interests of the local industry could be spotted at an early stage and taken into account.

Within the TRAFICC project, the main emphasis was laid on the fieldbus systems defined by the European standard EN 50170 and on PROFIBUS in particular. This decision was driven by the large market share of PROFIBUS in Europe and by the fact that all three technology providers have detailed knowledge about PROFIBUS. Nevertheless, the technology transfer approach adopted in the project is universal and applicable to whatever fieldbus system. In fact, the staff of a competence centre should be familiar with more than just one single fieldbus. To what extent this is possible, however, depends on the funding (and fund-raising skills) of the respective centre and thus in a way also on the readiness of the industry to support such activities.

4 Are multi-vendor systems a must?

Somehow the core of a competence centre is its fieldbus installation, which is the basis for any practical training and research activities and thus the source of knowledge and expertise for the staff. We believe that this installation should be a multi-vendor system, i.e. a compilation of a large variety of devices from as many manufacturers as possible. While this approach may be more cumbersome in the acquisition phase due to the number of companies involved, it has some distinct advantages over conventional installations furnished by a single vendor.

The first aspect is a very pragmatic and also psychological one. The key to the success of standardised fieldbus systems was the availability of devices from many different manufacturers. Since one aim of a competence centre is to promote the fieldbus concept in general, it is important to demonstrate the vendor-independence by simply showing a variety of purchasable products. On the other hand, this is not only a way to convince potential users of the advantages of fieldbus systems. It also helps the staff of the competence centre to gain experience with many different devices and to establish a broad knowledge base.

The idea behind the merely visual effect of having a multi-vendor system is of course the interoperability issue. Standards laid the grounds for interoperability, and a multi-vendor installation actually is the proof that this term is not just an empty promise. On the other hand, standards like PROFIBUS still leave many degrees of freedom for the manufacturer to decide what they want to implement and how this is to be done. Consequently, compliance with the standard does not automatically mean also full interoperability with the implementations of other manufacturers. In fact, a least common multiple of features to ensure interoperability between different devices is sometimes hard to find. These limits of fieldbus systems and their particular implementations can only be found in multi-vendor systems. Devices of one single manufacturer are much more likely to operate together without any problems.

A multi-vendor system also serves didactical purposes. It allows the trainees to work not only with different devices, but also with different software tools. From a practical viewpoint, the variety of programming environments and approaches is still a major drawback of multi-vendor systems, despite the existence of programming standards like IEC 1131. The exploration of the traps and pitfalls of interoperability may be an elucidative experience and can create an awareness of the parameters to take care of in practice.

For these reasons, we deem a multi-vendor system a prerequisite for the competence of a competence centre. As for the actual set-up, the devices are either just connected together or arranged in a meaningful model of some technical process. A simple interconnection of the devices is more flexible and allows the user to concentrate on the essentials of programming. Therefore it is ideally suitable for the preparation of small dedicated exercises with a fixed time-scale. The whole installation is easier to oversee and to maintain, it can easily be extended, and individual devices can be exchanged without great difficulties. This is the typical installation for integration and test centres.

A process model, by contrast, has a rather fixed configuration, but it is much more appealing to students and trainees and permits the programmer to cope with process control intricacies such as synchronisation and the like. Sometimes the control of the physical process is more important than the fieldbus communication principle. This is the typical installation of system centres. However, experience shows that this type of demonstrators is difficult to maintain. Very often the know-how of the installation is divided between several persons and it takes several hours (or even days) to get a complete demonstration process running again.

This is the reason why the fieldbus demonstrators constructed in the TRAFICC are of both types. The Czech partners employed the fieldbus to control a robot-based, flexible production system [2]. They also added simple demonstrators for exhibitions and exercises in the area of building automation and different fieldbusses. The Hungarian participants applied the distributed intelligence of fieldbus systems to the unusual field of medical monitoring [3]. Finally, the Polish partners stayed in the typical industrial automation area and used the fieldbus to interconnect different tooling machines [4]. A part of their installation, however, was also set up as an abstract demonstration wall without practical applications for the fieldbus devices.

5 Generic training concept

Competence centres are often located at universities or have a close relationship to such research institutions. This is for one good reason: A continued use of fieldbus systems in academic education is a guarantee for operating installations since it requires a permanent maintenance of the devices. On the other hand, students are a valuable source of man power, which is a prerequisite for experiments and tests – and finally also for new developments. Apart from marketing aspects, this is the main argument for universities to seek the support of manufacturers of fieldbus devices. So, education not only requires the availability of fieldbus installations, fieldbus installations also need their application in training courses. This symbiosis is fruitful and in our opinion very important.

Although the goals are the same, there is a fundamental difference between training courses carried out within the scope of regular academic education and those held for external participants from the industry. An ordinary academic course lasts typically one entire semester, and even if the students are not able to work all the time on the fieldbus due to space limitations, there is still the opportunity to split the actual hands-on training over a comparatively long period. This possibility does not exist in an industrial course, which lasts at maximum one week and cannot be extended beyond that time span. Apart from this obvious difference, the structure of a successful training course turned out to be more or less the same in either case. The clue is not to use one large problem the trainees have to solve during the course but to pose them a number of examples with increasing complexity. These problems may be related in that they are parts of a mosaic, but they must be solvable independently of each other. When a problem depends on a previous one, intermediate results must be at hand to help the participants get over unforeseen obstacles and to avoid an excessive loss of time. Our experience showed that only such an approach will ensure the motivation of the participants throughout the entire course, which is a necessity for positive results and the achievement of the didactical goal.

In the case of PROFIBUS, a typical training course could comprise the following topics [5, 6]:

- A theoretical introduction to the concept of fieldbus systems in general and to PROFIBUS in particular is necessary prior to the practical work. Depending on the previous knowledge of the participants, an introduction to PLC programming is often reasonable.

- PLCs are still an important class of fieldbus devices, and the programming of PLC is one of the basic skills in the fieldbus world. Some simple exercises are therefore recommended to get familiar with the programming language.

- The first fieldbus-specific examples are master-slave connections with PROFIBUS-DP, which require but correct configuration of the devices.

- Connections between PROFIBUS-FMS devices are more complicated to establish since they require appropriate object directories and communication relation lists. Master-slave connections are a little simpler to program than a data exchange between two master stations.

- The acquired skills can then be taken together to use a visualization tool, where several connections between both masters and slaves are necessary.

The complete training program is split into different lessons and exercises. The individual lessons and exercises are combined to courses and workshops tailored to the needs of the participants. The authors employed this education concept successfully many times. It proved useful not only for academic courses, where the problems can be slightly more complex due to the longer time scale, but also in the training of people from industry, and last not least in the "train the trainers" activities within the TRAFICC project. So far, the Viennese competence centre had 150 participants from industry and 300 regular students who attended the courses, while the centre in Bern trained 50 students (including postgraduates) and 80 industrial fieldbus users.

6 Conclusion and outlook

Competence centres play an important role in the ever-growing community of fieldbus users. They can serve as information bases for those who want to apply fieldbus systems and as support nodes for those who already have fieldbus systems installed.

The establishment of new competence centres was the goal of the TRAFICC project in order to initiate a technology transfer to the countries of Central Europe. This goal was fully achieved. New competence centres were founded and contacts to the local industry have been established. Now that the project is nearly finished, we can proudly state that TRAFICC was a success.

Nevertheless, the role of the competence centres will be subject to change in the near future. Originally, they could be regarded as islands in the fieldbus world. This was an appropriate position in the beginning of the fieldbus era. Meanwhile, both the fieldbus systems and their applications have become more and more complex, and it is practically impossible for one single competence centre to cover all different aspects. So what we shall see in the future is an inevitable specialisation of the competence centres together with an increasing co-operation between the individual groups. Like the fieldbus systems started to connect manufacturing cells, we will need a network of competence centres. A first and very promising step in this direction was the meeting of the PROFIBUS Competence Centres held in June 1998 in Karlsruhe. At this meeting 16 different competence centres from 13 countries from Europe and America agreed to collaborate and exchange information and experience.

Acknowledgement

This work was funded in part by the Commission of the European Communities within the scope of the INCO/Copernicus project no. INCO 960161 (TRAFICC).

References

1. K. Bender and P. Wenzel, "Transferring European Fieldbus Technology to Countries of Central Europe", *Proc. of the Symposium on Fieldbus Systems and Application Technics*, Budapest, 17.-19. Feb. 1998, pp. 3-10.

2. Z. Hanzalek, E. Schmieder, and P. Wenzel, "Creation of the Laboratory for Fieldbus-based Automation Systems", *Second IEEE Real-Time Education Workshop*, Montreal, 8. June 1997.

3. P. Várady, "Distributed Communication System in Biomedical Applications", *Proc. of the Symposium on Fieldbus Systems and Application Technics*, Budapest, 17.-19. Feb. 1998, pp. 39-44.

4. J. Reiner, J. Koch, "Distributed Control Systems for Manufacturing Cell", *Rynek Instalacyjny*, no 7, 1998.

5. M. Knizak, T. Sauter, and C. Pfeiler, "Training Engineers with a Profibus Multi-Vendor System", *Proc. of the Symposium on Fieldbus Systems and Application Technics*, Budapest, 17.-19. Feb. 1998, pp. 11-16.

6. M. Felser, J. Blatter, "Practical Experiences with PROFIBUS-DP Multiple Vendor Configuration", *Proc. of the Symposium on Fieldbus Systems and Application Technics*, Budapest, 17.-19. Feb. 1998, pp. 17-19.

Chapter 8: Research

Invited Paper presented in Plenary Session 2

Some Future Directions in Fieldbus Research and Development
Decotignie, J.-D.

Papers presented in Session 2.5

IEEE1394 in Comparison with Other Bus Systems
Stampfl, N.

On Interoperability and Intelligent Software Agents for Field Area Networks
Palensky, P.

Transmitting Voice on InterBus
Kandler, W.; Schweinzer, H.

Some Future Directions in Fieldbus Research and Development

J.-D. Decotignie

CSEM
Jaquet-Droz 1
2007 Neuchatel, Switzerland
jean-dominique.decotignie@csem.ch

Abstract. In this paper, we attempt to explore the different topics where there still exist opportunities for research and development in the fieldbus domain. Different venues are explored and the main problems are outlined. References to on-going work are given.

1 Introduction

Since the first research in the early 80's, fieldbusses have grown to maturity with numerous solutions available. Most of the problems have been solved although none of the existing solutions may claim universality. Any prediction on future developments is obviously risky. However, in this paper, we will try to look at the issues that still need to be solved as a baseline to explore future directions in the area. Such issues are speed, time bounds and predictability, distance, distribution, mobility and cabling, interoperability, cross influence between fieldbus and applications, application design methodology and cost. We shall not attempt to rank these issues as their relative importance depends on the application.

2 Time bounds and predictability

This aspect has two sides: temporal guarantees and temporal characterisation [1]. There has been a number of studies (see for example [2,3,4,5]) concerning the temporal and bandwidth guarantees that the different fieldbusses may offer. The various types of traffic, periodic and sporadic, have been analysed and the corresponding bounds for message transmission times have been derived. Results show that strict bounds can be guaranteed in absence of errors and for bounded traffic. For most networks among the popular ones, the guaranteed deadlines are longer than what is required by many applications. In many cases, the assumptions taken in the calculations do not hold in case of alarms or even cannot be enforced for normal traffic even in absence of errors. It is for instance easy to show that CAN will not operate properly when a station is the source of an unbounded traffic. Profibus also suffers from traffic limitations [6]. To overcome, these limitations several schemes have been proposed for CAN [7], Profibus [6] and P-Net [8]. It is useful to note that there exist networks, such as some variants of CSMA, that do not suffer from the above mentioned restrictions [4]. Work still needs to be done to assess the behaviour in case of transmission errors. Analyses of CAN and Profibus also show that is impossible to control message transmission jitter with the existing protocol. Low jitter is sometimes required by the applications (for example, motion control). Jitter control is also useful to enhance predictability as it leads to simpler solutions. There is a need to study new schemes to improve this aspect.

Concerning the second aspect, many of the available solutions do not incorporate any notion of time in the communication stack. In fact, at our best knowledge, FIP and the forthcoming IEC standard are the only fieldbusses that include this aspect. For the others, at best, priorities may be defined to enforce deadlines. For a number of applications, it is important to know the age of the received information either in absolute terms or relative to others [1]. This may be solved by time stamps based upon synchronised clocks. However, such solutions are heavy and thus in opposition with the general fieldbus approach. Theoretical and practical solutions have been designed in the research community [9] and need to be incorporated in the existing solutions.

Opportunities exist to improve jitter, analyse the behaviour in presence of transmission errors and add temporal characterisation of the transferred information..

3 Mobility and cabling

Mobility is sometimes required and may also reduce cabling costs. There have been some attempts to provide mobility, through wireless radio and infrared links, to fieldbus devices [10, 11]. The IEC fieldbus and LON both have a wireless physical layer. An Esprit project (MOFDI) is still under way on the subject (http://www.mofdi.com/). However, work still needs to be done to take into account the special characteristics of these media such as higher error rate and turnaround time. Simple and efficient access protocols to reduce device consumption are also needed as some applications require up to 10 years of operation on a single battery.

Another way to reduce cabling costs is to reuse existing cables. Signal transmission on power lines has been offered for example by LON. However, transmission speeds are limited to some kilobits per second. Recent developments show that transmission speeds in the megabit range may be reached (see for example http://www.intelogis.com/). Its application to fieldbusses may be studied.

4 Distributed applications

When a fieldbus is available as a communication means between a number of processing units, it is tempting to implement a fully distributed application. A lot of results are available on distributed systems, distributed algorithms and dependability issues [12]. They give the basic building blocks and means to construct a distributed system. However, application distribution is still performed manually on an ad-hoc basic. Automatic or assisted distribution taking into account the available resources, the fieldbus temporal characteristics and the application timing constraints would be more than useful to design engineers. Although some research has been done to help distribution either by looking at the impact of the network on the application [13, 14], there is still a lot to be found to allow for automatic or at least systematic distribution of the application over the various nodes of a fieldbus.

5 Speed and distance

If applications are often distributed or decentralized, with the advent of very powerful computers at low cost, it is tempting to implement a whole control application such as a CNC or a robot controller on a single personal computer. The fieldbus is then only used for input acquisition and output distribution. Existing solutions are often too slow, due to transmission speed, protocol overhead or intrinsic limitations, to be used. New fieldbusses with higher speed and better efficiency are need. Solutions based on Ethernet (IEEE802.3) [4,16] or Firewire (IEEE1394) [17] are currently being explored. If designed properly, one can expect an improvement of an order of magnitude in the performances.

6 Cross influence between fieldbus and application, design methodology

By introducing a delay, often variable, in the transmission of information, the network impacts the application. In addition, simultaneous sampling readily available when inputs and outputs are local is not granted when they are on the other side of a fieldbus. These parameters directly affect the application design. Their influence has been theoretically evaluated for applications such as control loops [19, 14]. This gives useful guidelines on which fieldbus is adequate and on the best way to exploit it [18]. In the future, we may expect that this research will be extended to other types of applications and that the results will lead to application design methodologies that take into account the communication system.

7 Interoperability

In the recent years, a lot of effort has been put on achieving interoperability between the fieldbus devices provided by various vendors which is still a large problem for users. This has been done through the definition of profiles, function blocks [20], device description languages or intelligent sensors and actuators behavior definition. Research work still needs to be done to bring together these ideas into a more formal framework with simpler mechanisms and better handling of temporal issues.

8 Cost

Personal computers have made their way in industrial automation as a mean to reduce costs. Along this line, it is tempting to reuse the input and output networks, USB or Firewire, readily available in each PC as fieldbusses. There is obviously some work to be done to make them adequate for this use. Suited data link and application layers should be designed. However, the very low cost of these solutions makes them very attractive as substitutes of existing solutions for a number of applications.

9 Conclusion

This paper has tried to provide some insights of some possible future reseach and developments in the fieldbus domain. One may raise other issues such as the availability of a single international standard which seems to be a mirage. We hope that this paper has shown that room exists for a number of opportunities.

References

1. H. Kopetz, K. Kim, "Temporal Uncertainties in interactions among Real-Time Objects", 9th symposium on reliable distributed systems, Huntsville, USA, Oct. 9-11 ,1990, pp.165-174.

2. K. Tindell et al., Analysing Real-Time Communications : Controller Area Network", proc. of the IEEE Real-Time System Symposium, pp.259-63, 1994.

3. E. Tovar, F. Vasques, "Guaranteeing Real-Time Message Deadlines in Profibus Networks", proc. of the 10th Euromicro Workshop on Real-Time Systems, Berlin, 1998.

4. G. Le Lann, N.Rivierre, "Real-Time Communications over Broadcast networks: the CSMA-DCR and the DOD-CSMA-CD Protocols", INRIA research report 1863, 1993.

5. G. Agrawal et al., "Guaranteeing synchronous message deadlines with the timed token medium access protocol", IEEE Trans. On Computers, vol.43, pp.327-39, 1994.

6. M. Li, "Real-Time Communication in an Industrial Network- Profibus", PhD Thesis, Swiss Federal Institute of Technology, Lausanne, 1996.

7. G. Cena, A. Valenzano, "An improved CAN fieldbus for Industrial Applications", IEEE Trans. On Industrial Electronics, vol. 44 (4), pp.553-64, 1997.

8. E. Tovar et al., "Adding Local Priority-based dispatching Mechanisms to P-Net Networks: A fixed Priority Approach", Proc. of the 11th Euromicro Conf. On Real-Time Systems, York, June 9-11, 1999, pp. 175-84.

9. J.-D. Decotignie, P. Prasad, "Spatio-Temporal Constraints in Fieldbus: Requirements and Current Solutions", 19th IFAC/IFIP Workshop on Real-Time Programming, Isle of Reichnau, June 22-24, pp.9-14, 1994.

312

10. P. Morel, A. Croisier and J.-D. Decotignie, *Wireless Extension of a Factory Instrumentation Fieldbus (FIP)*, In Proceedings of the 4th International Conference on Control, Automation, Robotics and Vision (ICARCV'96), Westin Stamford, Singapore, pp. 2282-2286, 1996.

11. I. Izikowitz, M. Solvie, "The OLFCHA project, Industrial Needs for Time-Critical Wireless Communication and Wireless Data Transmission and Application Layer Support for Time-Critical Communication", Informatik Aktuell / Euro-ARCH'93, 1993.

12. S. Mullender, "Distributed Systems", ACM Press, New-York, 1993.

13. Yeqiong Song et al., "Validation of distributed real time systems thanks to performance evaluation of their physical architecture", Proc. CESA'96

14. M. Törngren, "Modelling and Design of Distributed Real-Time Control Applications", PhD Thesis, Royal Institute of Technology, Stockholm, Sweden, 1995.

15. T. Kouthon, J.-D. Decotignie, "Improving Time Performances of Distributed PLC Applications", ETFA'96, 5th IEEE International Conference on Emerging Technologies and Factory Automation, Kauai, Hawaii, November 18-21, 1996, pp.656-62.

16. Synergetic Micro Systems inc, "Industrial Ethernet", white paper, 1999 (available at http://www.IndustrialEthernet.com/)

17. L.Ruiz et al, "Using Firewire as Industrial Network", 19th Int. Conf. Of the Chilean Computer Science Society, Talca, Chile, Nov. 8-13, 1999.

18. S. Koppenhoefer et al., "Fieldbus Based Integrated Communication and Control Systems", Proc. CESA'96, Lille, France, July 9-12, 1996, pp.1316-21.

19. Y. Halevi, A. Ray, "Integrated Communication and Control Systems: Part I - Analysis", J. of Dynamic Systems, Measurement, and Control 110 (12), pp.367-373, 1988.

20. IEC TC65C, "Function Blocks for Industrial-Process Measurement and Control Systems - Part 1: Architecture", committee draft 65/240/CD, June 4, 1999.

IEEE1394 in Comparison with Other Bus Systems

Norbert Stampfl

Vienna University of Technology
Institute of Computer Technology
Gusshausstrasse 27-29
A-1040 Vienna

Abstract. IEEE1394, also known as FireWire™, is a relatively new high performance serial bus system. It was originally designed to connect multimedia devices, but due to its properties it can also be used for other applications, e.g. for industrial automation. In order to consider the use of IEEE1394 in combination with or instead of traditional and established communication systems, it is necessary to have comparison values. This paper shall point out these pros and cons of IEEE1394 by comparing it with field area networks and Ethernet.

1 Introduction

Every year new communication systems are presented and find more or less acceptance by the developers and users. In 1995 the formerly proprietary FireWire™ became an international standard [1] named IEEE1394. For that reason, IEEE1394 is an open system and several devices from different manufactures, especially multimedia devices like cameras, VCRs, etc. are already available on the market.

The key property of IEEE1394 is its high data rates of up to 400Mbps together with hard real time transmission capability. Therefore it is interesting to compare IEEE1394 with high speed bus systems as well as with time deterministic communication systems which are usually used at lower levels of industrial automation, e.g. field area networks. It makes no sense to compare it with each system, therefore a lot of publications already exist [2, 3]. In this paper only the properties of a representative subset are compared.

Another reason for the very diverse systems which are considered in this paper is the infiltration of the lower level of the industrial CIM pyramid [4] by high speed bus systems, especially Ethernet. In the past, the assumption was justifiable, that, at the top of the pyramid, the office field, data rates are high, but real-time constrains are less important, while this changes as one descends the pyramid. At the bottom level, called field area level, sensors and actuators need little amount of, but time deterministic data. As devices like sensors become more and more intelligent and complex, the data rate requirements also increase at this level. Furthermore, image processing for quality checks, process supervision, automatic sorting, etc. and hence the need to transmit uncompressed images becomes more and more important in industrial automation. One approach to meet these requirements is the use of Ethernet in this field, which is already realized by several companies [5]. But Ethernet has the disadvantage of its non realtime data transmission capability. There exist solutions to solve this problem by using switch technology and Fast Ethernet, but a more promising approach would be the use of a system that supports all these features generically, e.g. IEEE1394.

2 Comparison

Before starting to compare communication systems, we have to define what properties are worth considering. Concerning the OSI architecture, layers have to be compared, but systems very often do not support all 7 protocol layers. IEEE1394 only covers layer 1 and layer 2. This is similar to Ethernet and a lot of other bus systems like CAN and BITBUS.

2.1 Topology and medium

The first self-evident criteria of communication systems are topology and the variety of media. The original IEEE1394-1995 standard supported all topologies except loops. The new improved specification IEEE1394a also allows loops, whereby one connection in the loop is deactivated automatically by the system. This is a big advantage over other systems since most field area networks allow only a bus structure, and even Ethernet does not support closed loops. With loops it is possible to make the system tolerant to cable faults because in such a case the second path is used automatically. Several systems reach more flexible topologies by using special nodes like bridges, routers (LON, Ethernet) or multiport-masters (P-Net), which is not possible for IEEE1394 yet because of the lack of bridge definition.

The number of devices per system varies form 127 (PROFIBUS) up to 32385 (LON) using repeaters and even more nodes when using bridges or routers. With IEEE1394 just 63 nodes are allowed per bus because of addressing limitations. The standard allows the interconnection of up to 1024 busses, but due to the lack of standardized bridging devices this is not possible yet. Another serious limitation of IEEE1394 is the low maximum distance of the network. The cable length between two nodes must not exceed 4.5 m, and a maximum of 17 devices are allowed to be connected in series. This gives a maximum distance of 72 m. Most other bus systems can cover distances up to several km (Ethernet, LON, FIP, Profibus) and therefore exceed this distance. Only BITBUS (30) and CAN (40m) at their maximum speeds are heavily restricted. To overcome this limitation, plastic optical fibers (POFs) are used to connect IEEE1394 nodes. Here a maximum distance of 70 m is possible between two devices. The drawback of this solution is, that power cannot be delivered over POFs, whereas IEEE1394 supports power over the standard copper cable.

Concerning several media, IEEE1394 already supports copper cable, with or without galvanic isolation, POFs and Infrared (at 100Mbps). Other media like radio (LON, Ethernet) or power line (LON, EIB) are not supported.

2.2 Data rates and packet size

IEEE1394 supports different speeds of 100/200/400 Mbps even concurrently on the same network. This is a big difference to field area networks where the maximum speed is about 12 Mbps (Profibus) or 2 Mbps (FIP, BITBUS). Also 10Mbps Ethernet is left behind, just Fast Ethernet approaches this data rates.

The packet size in field area networks is quite small, as most of the time time short messages like sensor data or actuator command have to be sent. This ranges from 8 Bytes (CAN) up to 246 Bytes user data for Profibus. Ethernet, which was developed for the exchange of big amounts of data, e.g. files, has a maximum payload of 1500 Byte. The maximum payload of IEEE1394 depends on the used transmission speed. At 400 Mbps the maximum data payload is 2048 Bytes.

2.3 Bus access control and timing constrains

Bus access control primarily depends on the use of single- (ASI, BITBUS, Interbus, Profibus-DP) or multi-master systems (CAN, LON, P-Net, Ethernet). For the latter the bus arbitration method has to be considered in detail. While LON, CAN and Ethernet use CSMA or some of its variations, P-Net and Profibus use token passing to gain access to the bus. This leads to completely different behavior concerning deterministic transmission times and therefore real-time properties. CSMA does not grant any transmission time limit, just some protocols (LON) support different levels of priority. Other attempts to reach real-time capabilities are to use special higher protocols or switching technology (Ethernet in combination with Fast Ethernet).

IEEE1394 devices have two possibilities to send data, either in asynchronous or isochronous manner. Asynchronous means that the node requests the bus, and if granted, gets a acknowledgement whether the transmission succeeded or not. Therefore this is well suited for control information. It is possible to determine the maximum time a node has to wait for the bus because of a special algorithm that blocks a node to send one more request before all other nodes had the chance to get access to the bus. If only asynchronous transaction are used this maximum delay occurs when all nodes of the network want to send a packet with maximum size.

The situation changes completely in case of the use of isochronous data transfer. If devices require guaranteed bandwidth, e.g. for audio or video data, they can allocate the desired bandwidth from the system. Afterwards the node can place a data packet of the appropriate size every 125 µs. So the maximum time delay for getting access to the bus is approximately twice the cycle time of 125 µs (which considers that a packet can be placed at the beginning or end of the cycle). As isochronous data have the highest priority and can use up to 80% of the bus bandwidth, the transmission time of asynchronous data increases up to a factor of 5 (1/5 of the bandwidth left).

It has to be pointed out that no token is necessary, therefore it cannot be lost, duplicated, etc., no rigid time slots are wasted if a device does not use its allocated bandwidth and that this works properly even in case of full bus load. Systems which use native CSMA (Ethernet) can handle a maximum bus traffic of approximately 35%, otherwise too many collisions occur.

Real-time capable are further CAN, P-Net, Profibus-FMS and all single master bus systems but at much lower data-rates.

A side effect of the isochronous data transmission is that a global time exists in an IEEE1394 network. The local clocks are updated every 125 µs. This easily allows time stamping of time-critical data as well as synchronized actions between different devices, which is very attractive for industrial applications, e.g. axis control, common trigger conditions, time stamps for data validity, etc.

The only event that can avoid this deterministic timing model is a bus reset, which occurs every time a device is added or removed from the bus. As this is a legal bus state, and if the system is not protected from changes in topology, the time needed for a bus reset has to be considered for the real-time systems. The time span from a bus reset to the first user data transmission could take up to 1s. As bus reset is a global bus event, all nodes and transactions are affected.

2.4 Decentralization and addressing

IEEE1394 is a fully decentralized system as there is no distinction between master and slaves on the bus. This is also the case for Ethernet and some field area networks like CAN and LON. Fully decentralized means that the desired application is distributed over the whole network, and, in the ideal case, no device plays a dominant role in the system.

The addressing model is another important point for communication systems and there are a lot of different techniques in use. Some systems like CAN use Identifiers in order to label different kinds of information packets and broadcast them over the network. Every node interested in certain information can take it from the bus by filtering the packets by their identifier. Other systems like LON, P-Net and Ethernet use addresses for each node. So not the information is labeled but the destination is specified. For Ethernet these are the well known MAC-Addresses. Within the node the information is channeled by means of higher protocol layers, e.g. by the concept of network variables at LON.

Within IEEE1394 the addressing model depends on the kind of data exchange. Isochronous data are labeled by IDs (so-called channels) and broadcasted over the whole network. So every device that is interested in it can forward the information to the application. For asynchronous data exchange, the situation is completely different. As the information is dedicated to a certain node, the destination address (the node ID) is sent within the packet. As IEEE1394 supports automatic address assignment, which means no DIP-switches or address configuration, it has to be considered that this node ID may change after adding or removing devices from the bus. Therefore, after such an event, every device has to look for its former communication partner and find its, possibly changed, node ID. In order to do this, every node has a so-called Global Unique ID (GUID) and configuration ROM, which remain constant, even during bus resets. The GUID identifies the used IEEE1394 chip, while the configuration ROM provides information about the used higher protocols, subunits, manufacturer information, etc. Furthermore the IEEE1394 system can be seen as a distributed memory architecture with 64 bit address space. The highest 16 bits are used to address the device, while the lower 48 bits remain for the addressing within the device. So data is explicitly read or written from or to dedicated memory spaces. As shared resources are also placed at certain addresses, IEEE1394 supports special atomic functions called Lock-Transactions to access these resources.

3 IEEE1394 in combination with field area networks

Field area networks are already established technologies at the industrial automation field level (Profibus, BITBUS, Interbus, P-NET, CAN, etc.) and for home and building automation and management (EIB, LON). They are well suitable for the exchange of small amounts of data, like sensor values or actuator commands, but not for data-intensive transmission necessary for audio or video applications. As this becomes more and more important for all fields, some attempts were made to transmit such data over field area networks [6, 7, 8]. Although this is possible under certain conditions, the low bandwidth avoids high quality and further improvement of such applications.

On the other hand, there are the high speed systems like IEEE1394 and Ethernet which are well suitable for the transmission of multimedia data, but too expensive to equip simple sensors and actors with connectors and communication chip-sets. Furthermore the packet overhead of such systems leads to an inefficient exchange of small data packets. Therefore the combination of both field area networks and high speed systems would be a very appropriate solution. As already described, Ethernet more and more enters that region and even tends to displace field area network. But Ethernet is not real-time capable and therefore suitable for audio- and video transmission only with some restrictions. IEEE1394, in contrast, was developed for such applications, so the combination of field area networks and IEEE1394 seems to be very promising. In industrial applications, an IEEE1394 system can be used as high speed, real-time capable backbone for field area networks (Fig. 1). Uncompressed data from cameras

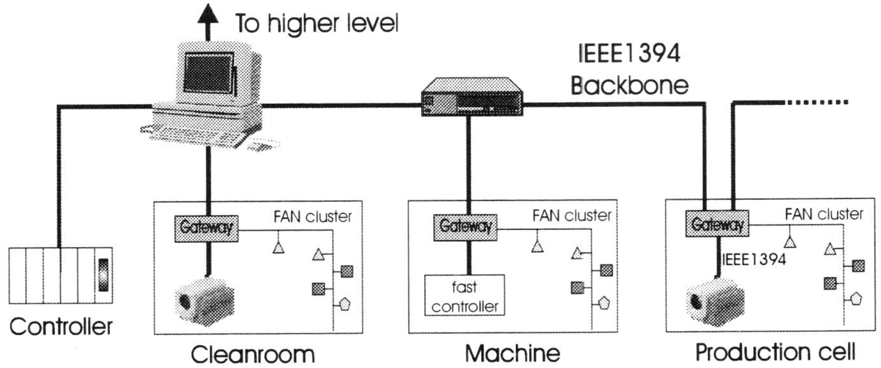

Fig. 1. IEEE1394 as backbone for field area networks

can be transmitted next to real-time data shared by different field area network clusters. What is needed is an interconnection device, e.g. a gateway or router between the different systems. Other devices like fast control systems e.g. to synchronize robot axes [9] can be connected directly to the IEEE1394 bus.

Also in the home and building environment this combination is very attractive. As more and more consumer electronics companies equip their devices with IEEE1394, the system will be available in most future TV-Sets, PCs, etc. An IEEE1394 gateway would bring the data from a sensor connected to a field area network to the TV screen, or a motion-detector would switch on the IEEE1394-camera and the video-recorder to monitor rooms, or bring the camera picture taken at the front-door to the TV screen next to the user.

4 Conclusion

At the moment IEEE1394 is the only system which supports high data rates as well as real-time capability, free topology and hot plug&play. The system is not only suitable for home-multimedia-systems, but also worth considering for applications in industrial automation. However, for applications that need hundreds of nodes sending small amounts of data and for distances of more than some hundred meters, field area networks are more suitable. So the combination of IEEE1394 and field area network could be a very promising solution to meet the requirements of many applications in industrial automation.

References

1. IEEE Standard for a High Performance Serial Bus, IEEE 1996, ISBN 1-55937-583-3

2. P-Net in Comparison with other Fieldbus Systems, Martin Manninger, Dietmar Dietrich

3. Feldbussysteme im Vergleich, Robert Busse, Pflaum 1996, ISBN 3-7905-0722-9

4. LON Technology, Dietrich, Loy, Schweinzer, Hüthig Verlag, ISBN 3-7785-2481-6

5. Das Netz ist die Steuerung, M.Buchwitz, M.Jetter, Elektronik 8/1999

6. Voice using LonWorks Technology, Hans-Jörg Schweinzer, Tagungsbeitrag FieldComms, Boston; June 1997

7. LonWorks basierende Standbild- und Sprachübertragung, Stampfl N., Posta R., in: Dietrich D., Schweinzer H. (Hrsg.), Feldbustechnik in Forschung, Entwicklung und Anwendung, Springer, Wien NewYork, 1997, S.184-199

8. Sprachübertragung über CAN, M. Kugelmayer, M. Mayer, W. Zimmermann, Elektronik 8/1999

9. Roboterachsen digital synchronisiert, Elektronik 12/1999

On Interoperability and Intelligent Software Agents for Field Area Networks

Peter Palensky

Institute of Computer Technology
Vienna University of Technology
Gusshausstr. 27-29/384
A-1040 Vienna, Austria

Abstract. Intelligent software agents are software programs that are currently used to solve complex problems in the domains of network management or information retrieval. The usage of agent technology for field area network would offer a lot of new and enhanced applications and would be "the second chapter" of distributed applications. The missing link to the commercial exploitation of agent technology on field area networks is a standard for agent communication. This paper gives an overview and takes the first steps towards standard agent communication for field area networks.

1 Special properties of field area networks

The large variety of field area networks (FANs) makes a general description of their characteristics almost impossible. But there are still some basic properties that are common to most of them.

The first one is that FANs usually communicate via a serial communication channel. This channel can be for instance a twisted pair wire, a radio channel or a fiber optical cable. Based on this physical channel the members of the FAN have to arbitrate and share this common communication means. Only one of the FAN nodes is allowed to talk, while the other may only listen. There is of course the possibility of setting up separate segments, connected via routers or bridges, but this problem then still exists within one segment or via one router. This communication channel is usually relatively slow. The low bandwidth of the FAN communication channels results in an increased robustness, which is more important for FAN applications than a high throughput [2].

Another property is that FANs tend to keep computation in the field. Modern field area networks consist of a number of small nodes that usually incorporate various means for sensor/actor interaction with the environment and a microprocessor. This microprocessor is able to execute parts of a distributed application. A temperature sensor node for instance operates the sensor and its analogue digital converter (ADC). The other nodes that need the sensed value do not have to care about the timing of the ADC or the linearization or calibration of the sensor. They get the sensed value without caring about if the ADC is 8 bit or 12 bit wide, what is its offset and so on. The other nodes receive the temperature value in degrees Fahrenheit or Celsius already formatted in the right and understandable way. Most of these microprocessors are 8 bit (or less often 16 bit) controllers with very low computational power and memory.

The microprocessor has usually two main tasks. The first one is the handling of the FAN communication protocol. Sometimes the lower layers are handled by hardware circuits but usually most of the layers are implemented in software. The second task is the application itself. These applications can reach from simple light switch program to more complicated thermostat controllers. The programming languages for these controllers are either assembler or some dialects of high level programming languages like ANSI-C or Pascal [4, 3]. These high level programming languages for FANs are designed especially for real time applications and hardware oriented programming.

The last common property for FANs that will be mentioned here is their message oriented communication. Measuring values or commands are communicated via short messages. FANs are usually not designed for continuos data streams like needed for video cameras or audio transmission. Other types of networks are more suitable for these kinds of application [1].

2 Interoperability for Field Area Networks

Communication of field area networks is based on communication protocols. These protocols are usually using a layered architecture. The lower layers are more concerned with the physical properties of the communication system, while the upper layers tend to be more abstract. The ISO-OSI seven layer model is a well known example for a layered protocol suite and is used as reference for all kinds of layered protocols [14]

The communication protocols describe the rules how to transmit data. The structure of packets and frames, how to establish a connection and other aspects of data transmission are defined in the protocol specifications. But following these specifications is not sufficient for real *understanding* of information. Two network partners that want to exchange information need to settle on certain rules that exceed ordinary protocol conformance [2]. Interoperability guidelines define how to encode and format physical values that have to be transmitted over the network. They define the syntax and the semantics of communication objects and provide functional profiles for the components of distributed applications.

Without these interoperability guidelines a packet could be transmitted in a proper way (means the bits are in the right order etc.), but the recipient would not know if that packet represents a temperature value or a command to turn on the light. The interoperability guidelines define the structure of the application layer interface of the field area network nodes. With these guidelines it is possible to identify, distinguish and assign transmitted data and its functionality.

Various types of field area networks have interoperability guidelines for HVAC-systems, lighting, sunblinds or fire alarms. These profiles are to be configured and installed by a network management tool that knows the their purpose and functionality.

LonMark, the interoperability association for LonWorks, for instance defines standard network variable types (SNVTs) and functional profiles (LonMark profiles) [11, 10]. EIBA, the EIB association, defines EIB interworking standards (EIS) and object interworking standards (ObIS) for the European Installation Bus (EIB) [12]. Both interoperability guidelines define types for physical values like temperature, speed or humidity. The profiles describe how to use this types to create an interoperable application layer interface. This paper uses LonWorks and EIB as examples for FANs.

3 Intelligent Software Agents

Intelligent Software-Agents (or short agents) are autonomous and usually distributed software programs that communicate and cooperate with other agents and humans. They try to reach a given goal without any further interaction with their creator (another agent or a human) by using sensors, actuators and communication. Agents are widely used in the Internet for information retrieval and filtering. [9] describes the properties of agent and its differences to ordinary software programs as being autonomous, goal-oriented, learning, flexible and others.

One of the most important abilities of agents is to cooperate. Usually cooperation is based on communication, and therefore agents have to take part in a kind of communication network. Agents use agent communication languages (ACLs) to exchange data, queries and other kinds of information via a network. Well known examples for ACLs are the knowledge query and manipulation language (KQML) [5, 6] and the FIPA (foundation for intelligent physical agents) ACL [8]. These ACLs are implemented for various platforms and networks like the world of TCP/IP and Java. This paper focuses on KQML which offers the mechanisms for establishing and routing messages without the need of a specified language of the actual content. It is just more "wrapping" protocol than a protocol.

4 Intelligent Agents on Field Area Networks

The convergence of agent technology and field area networks can be done in two ways. The first approach is that agents are just used as a metaphor for component- or object oriented development.

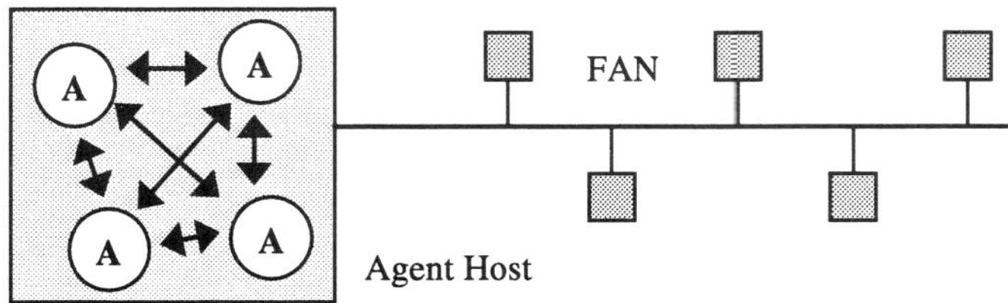

Fig. 1. Agents on an agent host

An application is split up into individual processes that are executed on an agent-host like depicted in Fig. 1. The FAN is simply the sensor- and actuator interface for the agent society, the inter-agent communication is done on the agent-host. The advantage compared to a monolithic program is its structure, the agents can be simple and robust program components that are more easy to program and maintain.

The second way, and that is the one that is investigated here, results in distributing the agents in the FAN (Fig 2.). They are executed on the FAN nodes and communicate via the FAN, therefore the ACL has to be mapped onto the FAN protocol.

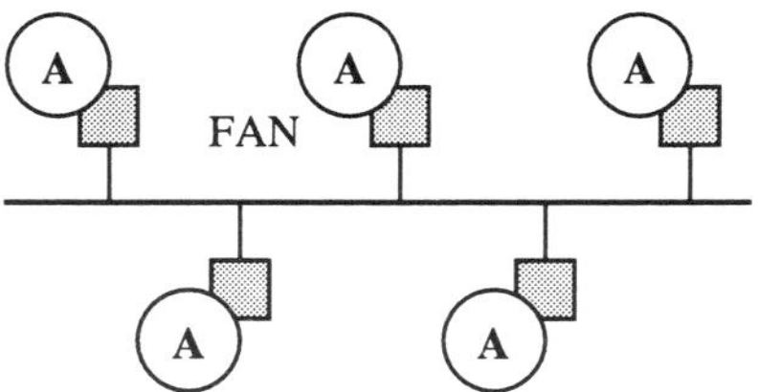

Fig. 2. Agents on FAN nodes

This creates a couple of problems. FAN-nodes are probably to weak for agents and the bandwidth is not suitable to transfer large messages, databases or knowledebases.

Special types of agents are required. They have to be compact, robust and simple. They have to get along with a minimum of communication, preferably based on messages. So called "reactive" agents meet these requirements [15]. They can not be called intelligent on their own, but a multi agent system (MAS) built up with reactive agent shows intelligent behavior [16]. With these agents it is possible to realize efficient problem solvers and complex algorithms although only primitive nodes are used [17]. The functionality of the network is modeled with agent structures, that can be developed and simulated on workstations using suitable simulation environments.

5 FAN-Interoperability for Agents

We suppose a field area network that connects agent-hosting nodes. What is missing is a data-type for agent communication and a functional profile for agents.

A KQML message has the following format [7]:

```
(<speech act type>
      :content <statement/speech act type>
      :sender <name>
      :receiver <name>
      :language <text>
      :ontology <text>)
```

One proposal could be to define a FAN-type for every speech act type. The other possibility is to define one FAN type for all KQML messages and to distinguish the different speech acts inside the type. In the case of LonMark SNVTs the latter would be the right choice, in case of EIBA EIS the first one. EIS9 for instance is suitable to contain various physical values like temperatures, voltages or speeds. The same can be done for the KQML speech acts.

The sender and receiver address section of the KQML message can be abandoned when it is ported to a FAN packet, because this information is then already included. There is no existing standard for the content language even for existing Internet agents. There are approaches to use KIF (knowledge interchanging format), but this ASCII-encoded language would enlarge our packets too much. A more FAN-specialized approach is necessary where every bit of a message is used. A reasonable approach would be to map KIF symbols on numerical tokens. The ontology is domain and application-specific. The language and ontology fields in the KQML message can be sufficiently covered by a couple of bits.

So it is possible to port KQML messages to FAN variable types. These types can then be used for functional profiles. The application layer profile for an agent would look quite simple (Fig. 3).

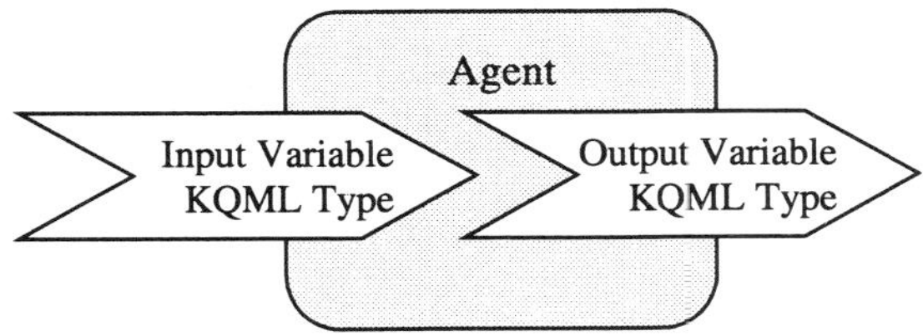

Fig. 3. The functional profile of an agent

One input and one output variable cover the needs for agent communication. This profile should only be used for agent specific aspects of the distributed application like knowledge exchange, cooperation and learning. Ordinary values or commands should not be transferred via this KQML interface but use the ordinary variables. Otherwise it would be a simple protocol tunneling and all nodes would just consist of these two variables.

So an agent enhanced FAN node is an ordinary node with a couple of communication objects in his profile, including one agent object.

6 Conclusion

This paper showed that intelligent agents can be used for FAN applications. They help to make a complex and distributed system more structured and more powerful. It is possible to implement and use them now, without any standards and agreements on communication means. The goal should be to find a standard which can be used for future applications for interworking and cross vendor communication.

References

1. IEEE Standard for a High Performance Serial Bus, IEEE Std. 1294-1995, ISBN 1-55937-583-3, IEEE Computer Society, 1995

2. D. Dietrich, D. Loy and H.-J. Schweinzer: "LON-Technologie", Hüthig Verlag, Heidelberg, Germany, 1998

3. "Process Pascal Users Manual", Version 2.0, Proces-Data A/S, Silkeborg, Denmark, 1991

4. "Neuron-C Programmers Guide", Revision 4, Echelon Corporation, Palo Alto, USA, 1995

5. Tim Finin, Yannis Labrou and James Mayfield: "Software Agents, chapter KQML as an agent communication language", MIT Press, 1997

6. Yannis Labrou and Tim Finin: "A semantics approach for KQML -- a general purpose communication language for software agents", Third International Conference on Information and Knowledge Management (CIKM'94), 1994

7. James Mayfield, Yannis Labrou, and Tim Finin: "Evaluation of KQML as an Agent Communication Language", in Intelligent Agents Volume II - Proceedings of the 1995 Workshop on Agent Theories, Architectures, and Languages. M. Wooldridge, J. P. Müller and M. Tambe (eds). Lecture Notes in Artificial Intelligence, Springer-Verlag, 1996

8. Donald Steiner: "Die FIPA-Initiative für Agenten-Standardisierung", in: "it+ti", Vol. 4/98, ISSN 0944-2774, Munich 1998

9. S. Franklin and A. Graesser: "Is it an agent, or just a program? A taxonomy for autonomous agents", in Jörg P. Müller, Michael J. Wooldridge and Nicholas R. Jennings (editors): "Lecture Notes in Artificial Intelligence 1193, Intelligent Agents III, Proceeding of ECAI'96 Workshop", ISBN 3-540-62507-0, Springer, 1997

10. LonMark Application Layer Interoperability Guidelines, Version 3, LonMark Interoperability Association, USA, 1996

11. LonMark Layer 1-6 Interoperability Guidelines, Version 1.3, LonMark Interoperability Association, USA, 1994

12. EIBA Handbook for Development, Issue EIB 2.21, 12/95

13. O. Etzioni, S. Hanks, T. Jiang, R. M. Karp, O. Madani and O. Waarts: "Efficient Information Planning on the Internet", Proceedings FOCS-96, 1996

14. A. S. Tanenbaum: "Computer Networks", Second Edition, Prentice Hall Inc., 1998

15. Darryl N. Davis: "Reactive and Motivational Agents: Towards a Collective Minder", in Jörg P. Müller, Michael J. Wooldridge and Nicholas R. Jennings (editors): "Lecture Notes in Artificial Intelligence 1193, Intelligent Agents III, Proceeding of ECAI'96 Workshop", ISBN 3-540-62507-0, Springer, 1997

16. R. A. Brooks: "Intelligence without representation", in Artificial Intelligence Vol. 47, pp139-159, 1991

17. P. Palensky and M. Gordeev: "Demand Side Management by using distributed artificial intelligence and fieldbus technology", IARB99 Intelligent and Responsive Buildings Conference, Brugge, Belgium, 1999

Transmitting Voice on INTERBUS

W. Kandler, H. Schweinzer

Institute of Electrical Measurement and Circuit Technology (E354),
University of Technology, Vienna
Gusshausstrasse 25, A-1040 Vienna, Austria

Abstract. This paper presents requirements and solutions for speech communication via INTERBUS. Using an appropriate data compression algorithm the data rate of speech does not exceed rates of conventional INTERBUS applications. By using standard INTERBUS slaves the cycle time increases according to the number of INTERBUS Voice Stations. By connecting several speech stations the system is soon overloaded. In a more effective solution a specified data area of the total frame of INTERBUS is used. This data area is called "Shared Data" channel and is alternately used by all speech stations. In this way few flexible connections can be established without overloading the system.

1 Motivation

Speech transmission with fieldbus systems is required in many applications. There are already some examples in the area of building automation like elevator control, fire alarm systems, network installation and maintenance. By using INTERBUS speech communication can also be useful for installation, testing and tuning of manufacturing plants and other industrial applications. Installation costs can be reduced by using the same already installed transport media for instance to combine alert or loudspeaker systems with INTERBUS.

Two alternative methods for speech communication are considered: one based on standard INTERBUS slave modules and another one based on Shared Data modules. A Shared Data channel is an area of the total frame protocol of INTERBUS which can be used by several modules. Speech channels have first to be established and afterwards they can be used bidirectionally. This kind of communication over Shared Data is also useful for enabling flexible tests with real time data or visualization. Temporary uni- or bidirectional test connections can be managed in that way. This paper presents an approach for speech communication on INTERBUS and discusses the requirements necessary for this purpose. Furthermore the need of transfer capacity in relation to the cycle time of INTERBUS is examined.

2 Introduction

INTERBUS is defined as a digital, serial communication system for transmitting data between control systems and devices for the entire field of industrial sensors and actuators. The single master concept, the total frame protocol and the ring structure makes the protocol very efficient. The cycle time is fixed and predictable [6].

Simple and complex devices are often combined in the same network and therefore different classes of data are necessary. The INTERBUS protocol is very suitable for data that has to be updated in a cyclic manner in real time. Process data normally need this form of update. On the other hand parameters are only loaded once by using PCP (Peripherals Communication Protocol). A time slot of the total frame is reserved for transmitting parameters gradually. In this way cycle time is not increased.

Digitized voice is real time data with a special feature: the trunk between two parties has not to be held all the time. By using standard IB slaves to transmit voice, transmitting and receiving data of each speech station is updated at every cycle. The data area of the total frame of each IB slave has to be large enough to guarantee the required channel capacity. In this way all connections can be held at the same time but this is only necessary in very rare special applications. A temporary real time connection among only few stations at the same time does not entirely make use of the transfer capacity provided by the standard solution. In this situation communication via a Shared Data channel is a very promising approach.

To transmit voice over INTERBUS, an appropriate data compression algorithm has to be employed to digitize and compress the voice signal. The resulting data stream has to be transported in a suitable way over INTERBUS to the receiving node. There the inverse function has to be applied to reconstruct the voice signal. Some constraints must be met like cycle time, delay time and size of the channel.

3 Basics of Digitized Voice Transmission

Channel bandwidth and the resolution of the quantized signal are important parameters of the transmitted signal. Both should be as small as possible to avoid an unnecessary transfer load of the system.

3.1 Bit Rate of ADPCM Voice Signals

A PCM-system (Pulse Code Modulation) quantizes the speech signal directly. The voice signal is sampled at a rate of 8 kHz and each sample is quantized with 12 bit resolution. A companding filter converts the digital signal with a resolution of 12 bit into one with 8 bit. Primarily the digital signal requires a bit rate of 64 kbit/s. By using ADPCM (Adaptive Differential Pulse Code Modulation) only the difference between the speech signal and a prediction of the speech signal are transmitted. The original signal with 64 kbit/s will be transmitted in a reduced form as a 16 kbit/s ADPCM signal. This bit rate is the basis for the following performance analysis [4, 7].

3.2 Requirements for Packet Voice Networks

The end-to-end delay requirements for a voice packet is about 100ms - 200ms and the optimal packet length in a packet voice network is between 300 and 700 information bits [2]. By using PCM technologies, longer packets (about 700 bits) can be used.

To guarantee high quality of voice in case of packet loss, the packet size must be smaller than 32 Bytes containing 16ms of speech data. A degradation of the speech quality is practically not assumed with speech loss rates less then 0.5% and package lengths smaller than 32 Bytes. The lost packet has to be replaced by a zero-sequence or the last packet. Losses of small package lengths (<32 Bytes) lead to crack noises, while losses of longer packages (>128 Bytes (>64ms)) lead to losses of syllables or words [2].

4 Principle of Transmission

To compensate for varying delays of the transmission, buffers have to be used. Send and receive buffers are serial FIFO registers (First In First Out). INTERBUS is used as a transmitting channel in between (Fig. 1). To transmit a continuous bit stream,

- the communication channel throughput must be higher than the data rate of the bit stream and

- the buffers on both sides must be large enough to store the digitized voice signal when the propagation delay time is at a maximum [4].

A part of the sending buffer is transmitted freeing up space to be filled with new data. On the other side, the receiving buffer has to wait until the entire propagation delay time before emptying to guarantee a continuous data stream.

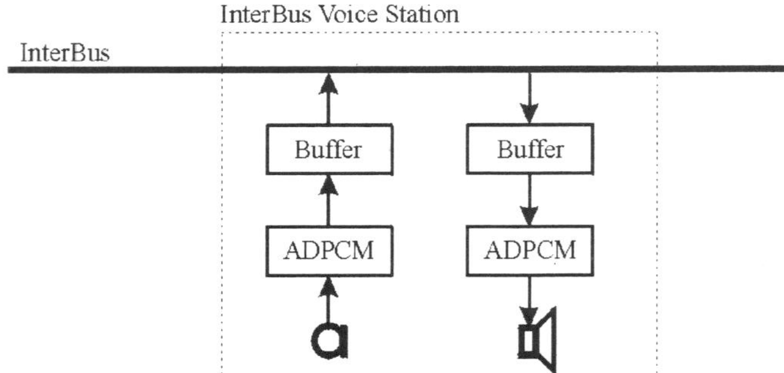

Fig. 1. Principle of the Transmission

5 Implementation on Base of Standard INTERBUS Slaves

In INTERBUS systems with 90% extent of utilization the cycle time is about 15ms. In order to achieve an appropriate data transmission rate for communication by using ADPCM, 30 bytes per cycle are necessary. Each INTERBUS Voice station approximately consumes 6.5% of the transfer capacity. The cycle time and the corresponding size of the total frame are shorter in INTERBUS rings with fewer I/Os. This causes also a reduced number of bytes for ADPCM, e.g. 10 bytes with 5ms cycle time. An other way to reduce the load of INTERBUS is to use the 2 Mbit/s version of INTERBUS. The increase of the required transfer capacity with increasing numbers of users is still a substantial disadvantage.

6 Implementation on Base of the Shared Data Channel

6.1 Principle of the Shared Data Channel

Shared Data is a defined data area of the total frame protocol which can be accessed by different stations alternately in accordance to the protocol [5]. The last station in the INTERBUS ring represents this data area physically with its shift register. All stations participating in this communication are situated on the return cable of the INTERBUS (Fig. 2). Thus it is ensured that the first piece of data in the cycle is always the Shared Data.

The size of the SD area is indicated to all stations during initialization. Additionally each SD module has to transfer individual parameters like name and type to all modules in the ring within the initialization phase.

A token, a token count and an additional priority bit manages the access to the SD area. An additional transfer code enables a variety of functions such as switching or connection indication. Thus a overhead of two bytes per cycle is necessary.

In order to guarantee the servicing of real time connections in parallel to low priority background activities within the SD channel, an additional high priority level can be used.

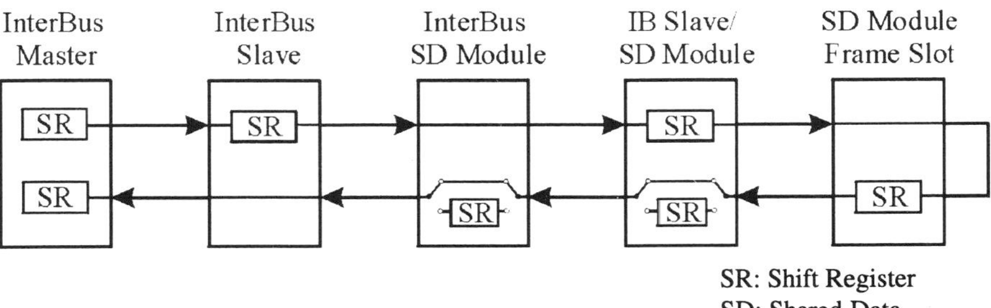

Fig. 2. Principle of the Shared Data Channel

6.2 Token and Priority Levels

When the sending buffers are empty the token is passed to the next station. A token count guarantees the sending station to get the token before complete filling of the buffer. This is managed by a priority bit, which is set when the token counter turns zero.

6.3 Establishing a Connection

To establish a connection in a simple but flexible way without an additional switching station, a decentralized switching algorithm will be used. Before establishing the connection, the speech station must check the holding of voice channels. Because a fixed transfer code is assigned to each voice channel, an allocated channel can be determined by listening to any channel activity.

If a free voice channel does exist, the calling speech station fetches the token (1st Byte). Having the token it allocates the free channel and uses a specialized transfer code (switching) together with the transfer sub code TSC1 (connection establishment), TSC2 (associated speech channel number), TSC3 (identifier of the desired station) and TSC4 (identifier of the own station) to transmit the request for establishing the connection. Further on, the transfer code of the allocated speech channel (TSC2) is used for communication. When the called party picks up and gets the token, the connection is established by sending the first speech data. The transfer code of the connection is used as long as the connection is dismounted by hang up which implies a corresponding telegram.

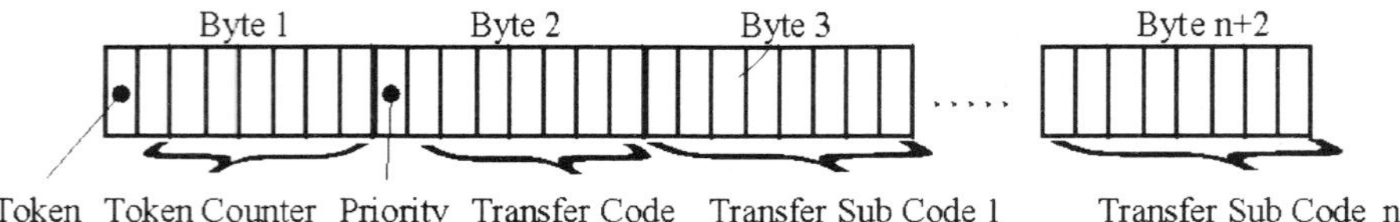

Fig. 3. Protocol of the Establishment of the Connection

6.4 Bidirectional Communication

Establishing and dismounting of a speech channel uses unidirectional multicast or broadcast transfers. However, transfer of speech data is performed bidirectionally. After the connection is established local buffering of speech data (FIFO) at both parties is started. Every time a station is receiving data it also transmits its speech data from its FIFO. If there are not enough speech data in the FIFO, blank bytes are transferred instead of. This method has a synchronizing effect.

The station which first gets a predefined number of speech data bytes initiates the transfer fetching the token. After having received the token it places the transfer code of the speech channel (TSC2) and its speech data on the SD channel. The second module replies with its speech data in the following output bytes immediately.

6.5 Configuration of the Shared Data Voice Connection

The possible channel capacity of the SD Channel does not only depend on the number of used data bytes in the Shared Data area but also on the total number of INTERBUS users and their size in the total frame. Therefore it is necessary to configure the size of the Shared Data channel.

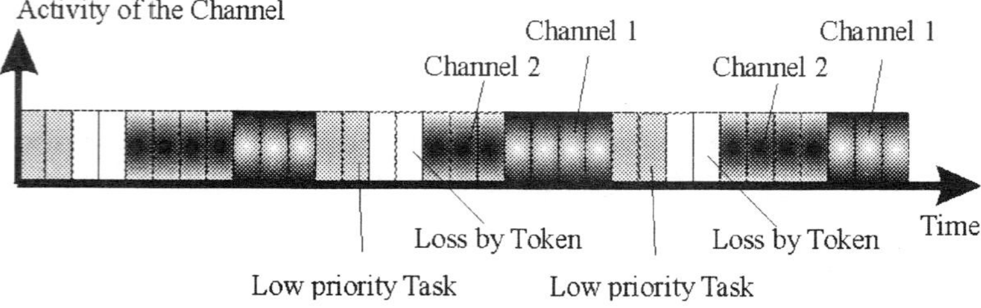

Fig. 4. Activity of the Shared Data Channel

Circling of the token is a loss of channel capacity which has to be limited. In order to prevent circling of the token, it should remain at the same station as long as possible. On the other hand, this results in a larger delay of the speech signal and in the need of larger transmitting and receiving FIFOs. Thus token holding time and delay have to be well balanced.

Fig. 4 shows a typical example of the alternate usage of the SD channel by two speech communications. Each slot characterizes an IB cycle. Between speech communications low priority tasks can be executed. Each time channel 1 and 2 sends again, the token has to cycle at least for one time. In this way, transmitting capacity is lost.

Table 1. Characteristics of the Shared Data channel varying the channel size, cycle time of IB: 10ms, assumed delay time of speech data: 100ms

Size of the SD channel	32 Byte	64 Byte	128 Byte
Channels for voice in parallel	1	2	4
Capacity of the SD channel (Bytes/second)	3200	6400	12800
Overhead of the SD channel	5,94 %	2,97%	1,48%
Lost capacity caused by circles of the token	10%	10%	10%
Remaining capacity on SD channel	21,5%	24,4%	26,0%
Shared Data in % of total frame data at IB	10,4%	20,83%	41,7%
Packet length at one cycle	16ms	32ms	64ms

Tab.1 shows characteristics of the Shared Data channel for three different sizes of the channel. The consequences of various channel sizes can be seen. The number of the transferred packets of speech data per second is defined by a given delay time of 100 ms. From this delay also results a remaining capacity on the SD channel which can be used for low priority tasks.

Table 2. Characteristics of the Shared Data channel by changing the cycle time, size of SD channel: 64 Byte, assumed delay time of speech data: 100ms

Cycle time	5ms	10ms	15ms
Load of IB referred to maximum load	30%	60%	90%
Channels for voice in parallel	4	2	1
Capacity of the SD channel (Bytes/second)	12800	6400	4266.7
Overhead of the SD channel	2,97%	2,97%	2,97%
Lost capacity caused by circles of the token	5%	10%	15%
Remaining capacity on SD channel	29,4%	24,4%	35,0%
Shared Data in % of total frame data at IB	49,2%	21,69%	13,9%
IB data including SD data per cycle	130 Bytes	295 Bytes	460 Bytes
Cycles per second	200	100	66,67

A packet is transferred in one or more cycles. The overhead is caused by the two header bytes. A basic assumption in Tab. 1 (and Tab. 2) is the use of a "hardware bridge" [5]. If no hardware bridge is used the loss caused by cycles of the token becomes larger reducing the remaining capacity.

Table 2 shows characteristics of a Shared Data channel of given length (64 Bytes) with different cycle times. This assumption reflects the situation of a typical hardware implementation of an SD module. While the size of the Shared Data channel enables only one speech channel by full loading of INTERBUS, typically two speech channels can be used in parallel at medium bus loads.

7 Conclusion

Utilizing an appropriate compression, the data rate of the speech signal may remain slow against the process data stream in many cases. In this paper two ways of implementing speech communication over INTERBUS have been broadly discussed. Having permanent connections to IB slaves, this standard solution does not fit well for speech communication. By expansion of the facilities of INTERBUS with the Shared Data channel it is possible to integrate an almost unlimited number of speech stations. However, similar to normal telecommunication systems the number of parallel connections is limited.

References

1. Alfredo Baginski, Martin Müller: INTERBUS -S: Grundlagen und Praxis, Hüthig Verlag Heidelberg, 1994.

2. Minoli, D.: Optimal Packet Length for Packet Voice Communication, IEEE Transactions on communications, Vol. 27 (1979), No. 3, pp. 607-611.

3. Klaus Sauer: Integration von Sprach- und Datenkommunikation in lokalen Netzen, 50. Bericht über verkehrstheoretische Arbeiten, Universität Stuttgart, Institut für Nachrichtenvermittlung und Datenverarbeitung, 1990.

4. Hans-Jörg Schweinzer: LonVoice – Transmitting Control Data and Voice by using the LonTalk Protokoll, in: D. Dietrich, H. Schweinzer (Hrsg.), Feldbustechnik in Forschung, Entwicklung und Anwendung, Springer, Wien NewYork, pp.167-175, 1997.

5. Herbert Schweinzer, Wilfried Kandler: Shared Data on Interbus, in: D. Dietrich, P. Neumann, H. Schweinzer (Hrsg.), Fieldbus Technology, Springer, Wien NewYork, pp. 22-29, 1999.

6. CENELEC, prEN 50254 Volume 2: High Efficiency Communication Subsystem for Small Data Packages, 1997.

7. John Bellamy: Digital Telephony, John Wiley & Sons, Inc., 1991.

Chapter 9: Real-Time Aspects

Simulation of Communication Systems in Industrial Area

Detlef Bäuerle, Wolfgang Popp

Martin-Luther-Universität Halle-Wittenberg
Fachbereich Ingenieurwissenschaften
Lehrstuhl für Elektrotechnik
eMail: {detlef.baeuerle|wolfgang.popp}@iw.uni-halle.de

Abstract. The large assortment of networks used in industrial area (e. g. field-buses, LANs and telephone systems) calls for computer based tools for modeling and simulating their performance. This paper presents such a tool, which is implemented in a communication-protocol independant structure to determine the usability and efficiency of different networks. Furthermore it allows not only evaluating the improvements of classical factory communication systems, but also analysing the introduction of complete new systems for factory applications.

1 The Method

Looking for the best communication system for an intended application, a first choice can be done on the basis of technical criteria (like maximum extension, desired redundancy, etc.) by the use of checklists as described in [1] or with the help of expert systems like [2]. For the final assessment however it's necessary to get detailed information about the waiting time of all telegrams, their loss rate, the probability of losing telegrams and the capacity under certain load in individual situations. Tools like [3], based on Markovian arrival processes, are specialized on communication systems in wide-area networks with thousands of stations attached. Solutions with the help of petri-nets presented in [4] have to be developed independently for each system. This chapter will describe a method allowing the simulation of different communication systems in the same way.

1.1 Description of the Process

First we have to find the essential, communication relevant, characteristics of the industrial process. Normally process control is done by PLCs, PCs and/or intelligent sensors and actuators, with centralized or distributed intelligence. Looking for a description independent of the respective implementation of the network, a better abstraction is necessary. Regarding a *running* process in the first step there are only two interaction points between process and network: communication input from sensors and output to actuators via telegrams. Controllers are only recipients and senders of these telegrams, generate no *additional* traffic and therefore can be ignored. The hand over of a telegram from the controller over the network to an actuator can be regarded as a request for a command as well, for it represents an expected or rather required reaction on a certain behaviour or event of the process. Then in the view of the communication system there is no difference between sending and receiving telegrams.

They are just requests to the network, waiting in a queue, describing the process by their chronological sequence. These two abstraction steps are shown in figure 1.

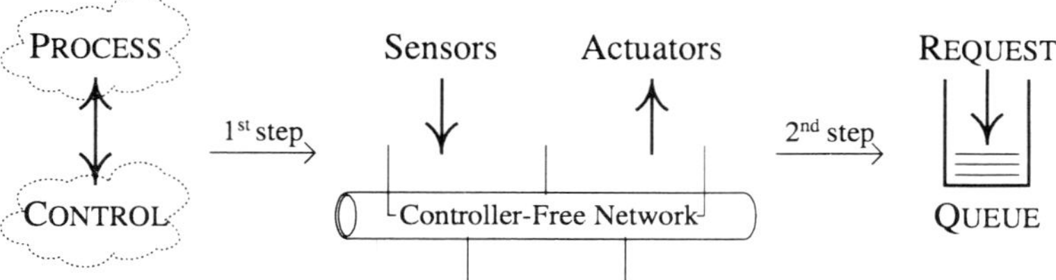

Fig. 1. Abstraction Steps in Process Modeling

The queue as interface to the communication system only needs information about data length (as the payload of the telegram) and priority of all the requests (to enable the network to act in an optimized way). An identifier is necessary to distinguish between the different subscribers. For analysis, the time of entrance and the requested delivery time should be included, too.

1.2 Description of the Network

Next we look for a way of describing the networks so that they can be regarded as exchangeable modules serving the queue. Examing their behaviour, following three actions are found to be common:

1. searching the queue for the telegram to be sent,

2. acting on the queue (taking the telegram for transport or generating a copy of the former one as a dummy, if there is no new information (as shown in the example in section 2.2); deleting double telegrams if there are no buffers),

3. waiting for the service time (time for establishing a link, transporting the telegram and disconnecting).

All communication systems can now be described in this unique way. Special features, like master-slave behaviour, token passing and binary exponential backoff after collisions, are encapsulated. This allows the use of the same analysing tool for all simulated networks.

1.3 Description of the Analysis

To get a statement about the quality of service, a supervisor analysing the queue is necessary. This analyser controls the waiting time in the queue, the losses of telegrams and the idle time of the service. Therefore the difference between service and entrance time is calculated, the number of double telegrams is accumulated and the dummies and cycled tokens are recorded. These informations can be statistically evaluated.

2 The Tool

As the main elements of the method are *abstraction*, *encapsulation* and *modularity*, the best way of implementing a simulation tool based on this method should be object-oriented [5]. This chapter presents the different classes of requests and services developed in our research project until now and shows their relationship.

2.1 Classes of Requests

Most of the literature of queueing theory starts with the *simple case of stream*, the Poisson stream. Most of the simulation tools use it (e. g. as input for a petri-net [4]) or are based on it [3]. Therefore a class with an exponential distribution of interarrival times is implemented by a transformation of an uniform deviate.

Despite of the longstanding use in research, the real world hardly behaves in a Poisson manner as [6] discovered. To enlarge the practical use of our tool, a module with the opportunity of a mathematical description of the interarrival time is added as a quite adequate solution for manufactory automation (like assembly-line production). With small deviations superimposing the calculated (deterministic) times, the network can be tested for its robustness against short delays in this mathematical description.

Of course not all kinds of processes can be simulated with these modules or a combination of them. Therefore an interface module for the input of external data is implemented. This allows the integration of data recorded in real industrial processes and results of process simulation tools like MATLAB/SIMULINK.

2.2 Classes of Services

For the current research on the efficiency and usability of different networks, following communication systems are implemented: AS-INTERFACE, CAN, INTERBUS, PROFIBUS (based on common layer 2), IEEE 802.3 and ATM (Asynchronous Transfer Mode with constant, variable and available bit rate). Because of the object-oriented programming capability and the new Standard Template Library STL (with the predefined container `<queue>` and algorithms like `find()`, `for_each()` and `erase()` as described in [7]), C++ was chosen for all development.

As an example, the steps of simulating INTERBUS are explained in detail:

1. **Searching the queue:** Every place in the summation-frame telegram is designated to a specific slave. Therefore the module is looking for a request for transmission of data from this slave (`queue.find(ActualSlave)`).

2. **Acting on the queue:** Normally the specific request is taken out of the queue (`queue.erase(ActualSlave)`). As INTERBUS transports data even if there's no new request of the respective slave, the module has to be able to generate a dummy with the information sent in the previous frame and take this one for the next frame. If the transmission rate is not fast enough, double requests should be deleted and reported.

3. **Waiting for the service time:** For the time of transmission, the module has to do nothing, because there can be no collisions using INTERBUS. Sensors and actuators continue adding requests in the queue.

These three steps are repeated until end of simulation.

2.3 Object-Oriented Presentation

With the object-oriented design of the tool, the integration of further types of request and any kind of communication system is possible. Figure 2 shows the current object model of the implemented classes in UML (*Unified Modeling Language* [8]) with their public (+), protected (#) and private (-) members and their relationship between each other.

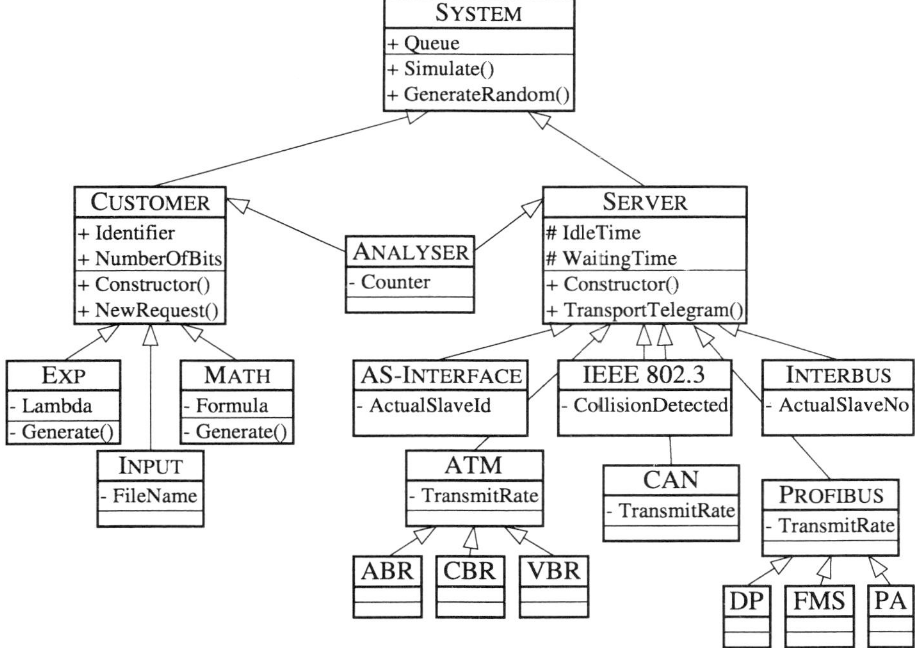

Fig. 2. Presentation of the Implementation (Class Diagram)

As the queue is in use by all objects, it is part of the over all SYSTEM-superclass. The common functionality of the requests is integrated in the CUSTOMER-class, the super-class of all the communication systems is the SERVER-class. Variants of PROFIBUS can be regarded as chields of a common class, the same is done for the different variants of ATM.

3 The Results

The interchangeability of simulation objects allows the comparison of the usability of different communication systems for the networking of a specific process. This chapter will bring an automated cleaning facility as an example and shows some results.

3.1 Exemplary Process

The specification of an automated cleaning facility for bottles in a brewery can be done in five parts as shown in figure 3. The flow of bottles is illustrated as ⇨, the symbol for a sensor is ↓, ↑ stands for an actuator.

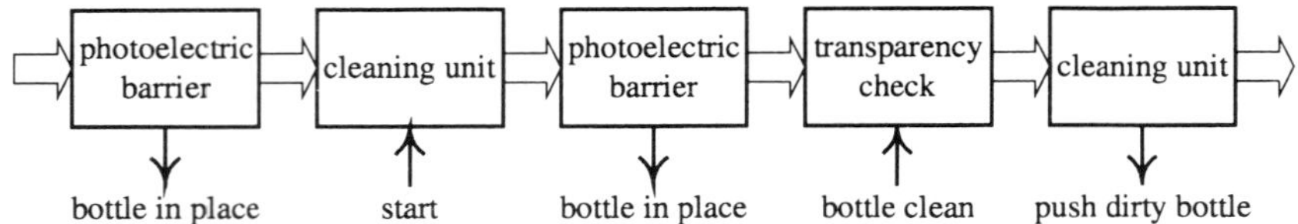

Fig. 3. Parts of an Automated Cleaning Facility

For the simulation of a facility with an input of 60,000 bottles per hour, this means one bottle every 60 ms, the following modules are used: as ten bottles can cleaned simultaneously, ten sensors inspect the presence of the bottles at the beginning of the process. This functionality was simulated with ten requests coming every 500 – 600 ms. Then an actuator opens the water valve for the cleaning unit, this means one telegram every 600 ms. Ten sensors control the exit and ten others examine the cleanness (both simulated with telegrams every 600 ms with a deviation of 50 ms). At the end a switch divides clean and dirty ones, this can be described by a Poisson process. Figure 4 shows the interarrival time between subsequent requests of this process, simulated with 32 modules (thirty sensors, one actuator and one switch). This will be the input of the queue to be served by the network.

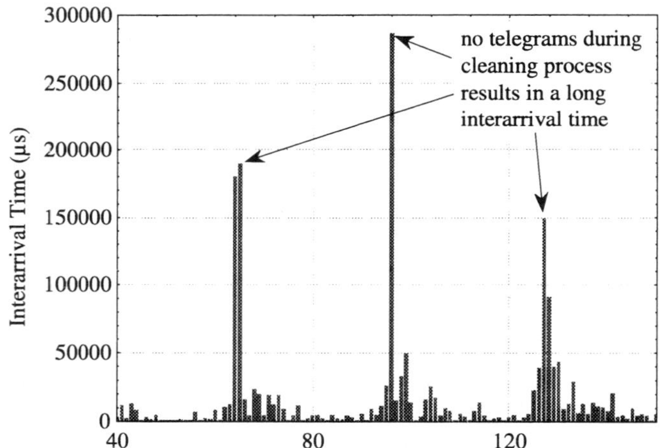

Fig. 4. Interarrival Time between Requests (No. 40-150)

3.2 Comparison of INTERBUS and CAN

For the communication of all sensors and actuators in this example two systems are compared: CAN and INTERBUS, both with a transmission rate of 500 kbps. Using CAN for this problem results in a good solution as figure 5 shows: nearly 99% of all requests can be served instantly, the longest waiting time is (in this simulation) 144 µs.

Unfortunately, this time can be much larger under bad circumstances, e. g. all sensors and actuators send at the same moment. Then the telegram with the lowest priority has to wait until all other 31 requests are served. With a telegram length of about 55 bits (frame and one byte payload including stuff bits, acknowledge and interframe space),

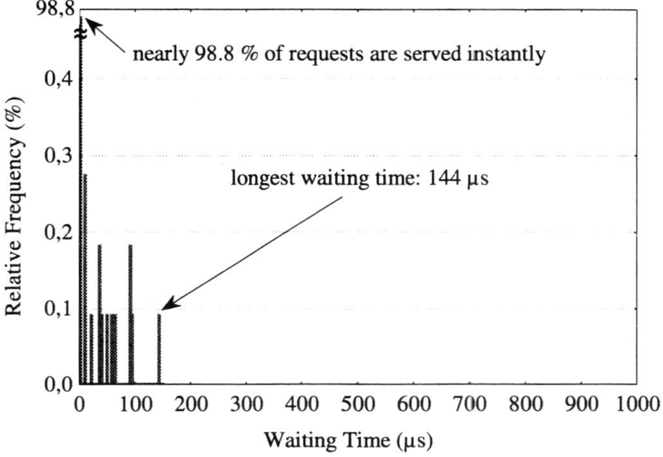

Fig. 5. Waiting Time Distribution Using CAN

the waiting time will be 3.41 ms in the worst case. Only with a long (nearly infinite) simulation time, this result can be determined. Therefore it's necessary to calculate the probability of such a situation mathematically.

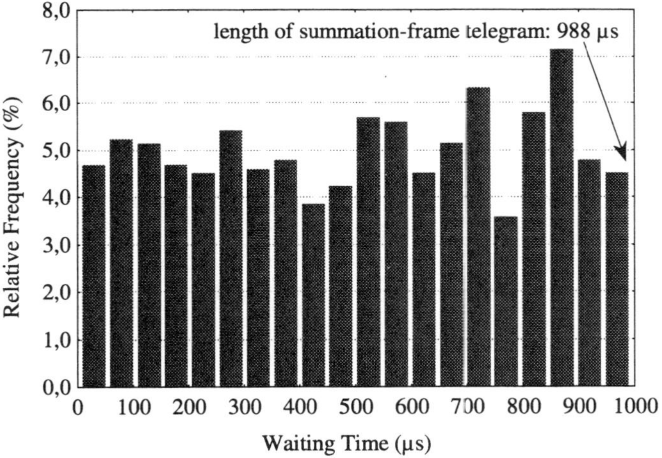

Fig. 6. Waiting Time Distribution Using INTERBUS

The INTERBUS-technology with its summation-frame protocol guarantees the longest waiting time not to be larger than (in this example) 988 µs. But figure 6 shows, that the waiting time is equaly distributed, telegrams are hardly served immediatly. The final decision between these two systems has to be made by the user.

3.3 Examination of Idle Time

Both INTERBUS and CAN don't work to their capacity. Using the simulation tool, it's possible to say that for less than 1% of all requests, subsequent slots in the summation-frame telegram are used, but for 38% the idle time of INTERBUS is longer than 10 ms.

This result is shown in figure 7. It would be more economic if less relevant data like reports, new recipes or even telephone calls could be transported in this time. The common fieldbus systems are not able to do this.

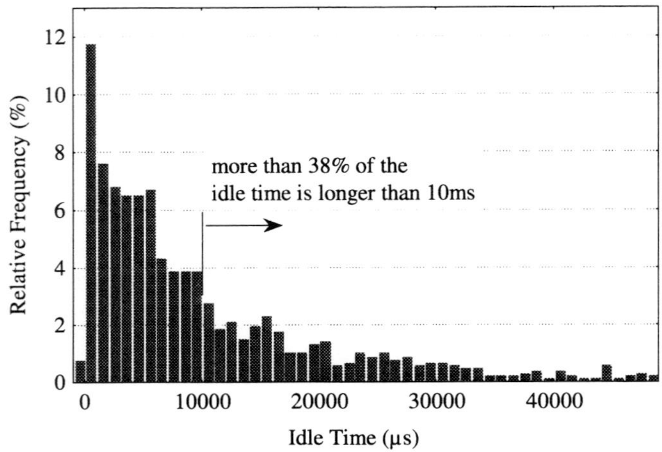

Fig. 7. Idle Time Distribution of INTERBUS

3.4 Present Work

Present work deals with validating the simulated data. Therefore the simulation will be compared with recorded network traffic of a large system equipment sponsored by the DFG (German Research Community) and the Land Sachsen-Anhalt, with PLCs and communication systems like PROFIBUS DP, FMS and PA, INTERBUS, AS-INTERFACE, CAN, TCP/IP and ATM. Furthermore the possibilities of transporting all the different requests with one fieldbus with extended functionality (using its idle time) are examined and the results are compared with the possibilities of integrating ATM. For the examination of other process models of real industrial plants, there are contacts with industrial partners.

References

1. WERNER KRIESEL ET AL.: Bustechnologien für die Automation: Vernetzung, Auswahl und Anwendung von Kommunikationssystemen. – Heidelberg: Hüthig, 1998.

2. ANONYMOUS: *BUS-choice: Bedienungsanleitung.* – Reutlingen: Steinbeis-Transferzentrum Sensoren und Systeme für die Automatisierung.

3. M. SCZITTNICK AND C. WYSOCKI: *MACOM: A Tool for Evaluating Communication Systems.* – 9. ITG/GI-Fachtagung MMB'97, Kurzbeiträge und Toolbeschreibungen – Informatik in Freiberg, Band 4, 1997.

4. JÖRG KIEFER: *Methodische Partitionierung und Parametrierung von Feldbussen.* – Düsseldorf: VDI-Verlag, 1996.

5. GRADY BOOCH: *Object-Oriented Analysis and Design with Applications.* – 2nd ed. – Redwood City: Benjamin/Cummings, 1994.

6. V. PAXSON AND S. FLOYD: *Wide-Area Traffic: The Failure of Poisson Modeling.* – Proc. SIGCOMM '94 Conf., ACM, p. 257-268, 1995.

7. BJARNE STROUSTRUP: *The C++ Programming Language.* – 3rd ed. – Reading: Addison-Wesley, 1997.

8. UNIFIED MODELING LANGUAGE: *UML Resource Center.* – UML v1.3 Draft – `http://www.rational.com/uml/index.jtmpl`.

A Flexible Time-Triggered Communication System Based on the Controller Area Network: Experimental Results

Luís Almeida, José A. Fonseca, Pedro Fonseca
{lda, jaf, pf} @ua.pt

Universidade de Aveiro, DET, UI-INESC
Campo Universitário
3810 Aveiro, Portugal

Abstract. Controller Area Networks are normally used in event-triggered communication systems. These are known for not supporting composability in respect with the system temporal behavior. When using a time-triggered communication paradigm, such composability is achieved. However, common time-triggered systems rely on static message scheduling which compromises system flexibility. This paper proposes the use of centralized scheduling together with a planning scheduler to achieve flexible time-triggered communication on a CAN system. A model is deduced to allow the prediction of message production delays for a particular implementation using low processing power micro-controllers (Philips 80C592). Experimental results shown confirm the validity of the model. These results are also in agreement with a model for the scheduler execution time presented by the authors in previous works concerning the use of the planning scheduler in FIP-like networks.

1 Introduction

Several authors have stressed the growing importance of flexibility in modern real-time systems, particularly those used in industry [1, 2, 3]. Only flexible real-time systems will be capable of operating in dynamic environments, i.e., those where a complete environment specification is not possible and/or where operational requirements may change during system lifetime. However, flexibility in real-time systems must be achieved without jeopardizing the understandability and predictability of the system temporal behavior, allowing safe and predictable upgrades with negligible down-time. Two important properties to support such a flexibility are composability (the subsystems' properties remain valid upon integration in the whole system) and scalability (the extendibility of the system is not limited by some predefined upper limit).

The communication system has a central role in determining either the composability and scalability of distributed real-time systems. In particular concerning composability, Kopetz [3] stresses that the event-triggered communication paradigm does not support composability with respect to the temporal behavior while, on the other hand, the time-triggered paradigm does. However, typical time-triggered communication systems use static message scheduling and this strongly limits system flexibility.

Controller Area Network [4] is a communication standard which was originally develo-ped for distributed automotive control systems. However, its acceptance has expanded to other application areas including industrial systems at the field level, as a fieldbus. This fact justifies the authors' proposal to use CAN to build a flexible time-triggered communication system particularly meant for field level industrial applications.

This paper extends the material presented in a previous work-in-progress paper [5] proposing the use of a CAN network in an untypical centralized scheduling architecture. The CAN native distributed arbitration is still used to reduce the communication overhead. The central scheduler uses a planning scheduler [8] to combine flexibility with temporal predictability and low runtime overhead. This paper focuses on the experimental results obtained on a home-made CAN system (CANivete) [9] based on low processing power microcontrollers (Philips 80C592).

2 Time-triggered communication in CAN

The motivation to perform time-triggered communication over CAN is to achieve composability with respect to the temporal behavior of the communication system [3]. Among several possibilities, two antagonistic paradigms have been considered: that of the Time-Triggered Protocol (TTP) [10, 3] and that of the Factory Instrumentation Protocol (FIP) [11, 12]. The former can be briefly characterized by a fully distributed MAC scheme based on global synchronization of all nodes and by a very low communication overhead. The latter uses centralized access control and has a higher communication overhead.

A negative aspect of both of these paradigms is that they are based on static table scheduling thus compromising the system operational flexibility. Once the system is running, changes in the set of messages that have to be broadcast, such as adding a new message or changing the phase or period of an existing one, normally require to stop system operation and to load the new table(s).

The FIP-like apporach follows the Producer-Distributor-Consumers (PDC) communications model [13] with a centralized distributor. The authors' proposal follows this approach, in a different way than suggested in [12], inheriting the following advantages:

- Implicit nodes' synchronization using the regular transmission of master messages;
- Robust autonomous control of the communication system;
- Simple global management due to the communication system centralized control.

The effect of the most common disadvantages can be reduced by using appropriate techniques:

- Operational flexibility can be improved replacing the static table-based scheduler with a planning scheduler [8];
- The communication overhead can be drastically reduced when compared to the FIP-like approach by taking advantage of the CAN-native distributed arbitration.

3 The communication protocol

As mentioned above, the proposed communication system relies on a centralized planning scheduler which has been developed by the authors to improve flexibility in

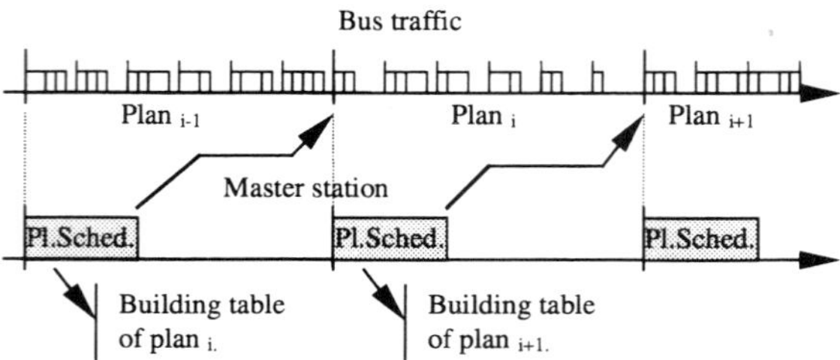

Fig. 1. The planning scheduler.

FIP-like networks [6, 7, 8]. The name planning scheduler comes from the fact that the scheduling table is rebuilt every fixed duration period of time called a plan. Changes to the message set can be easily accommodated in the next plan thus improving system operational flexibility (fig. 1).

Each plan is divided into a fixed number of constant duration elementary cycles (Ec). All periods and deadlines of messages are expressed as integer multiples of the Ec duration. Consequently, within each of these cycles, each message can be transmitted only once. The Ec duration determines the temporal resolution of the communication system.

In this implementation of centralized scheduling in CAN, the master node sends a special control message with a particular identifier, the master message, in the beginning of each Ec. This message indicates which data messages must be transmitted during that particular Ec. The order by which the data messages are transmitted is left for the CAN native distributed arbitration to sort out (fig. 2). This scheme allows to reduce the overhead of FIP-like centralized arbitration to just one extra message per Ec. As an example, using a transmission rate of 125 Kbit/s, an Ec duration of 8.9ms and a master message with the maximum data size of 8 bytes yields a communication overhead of just 11.7%. This value decreases proportionally with an increase in the transmission rate. It is called centralized scheduling and not arbitration, because it is the scheduling activity that is centralized while the arbitration is still distributed.

The data messages to be produced in each Ec are codified in the data field of the master message. Each bit has a direct correspondence with a given data message and thus, using the maximum data length of 8 bytes yields a maximum of 64 different data messages (fig. 3). A simple modification to the protocol can be done if more than 64 different data messages are required. In that case the master sends two or more master messages in the beginning of the Ec.

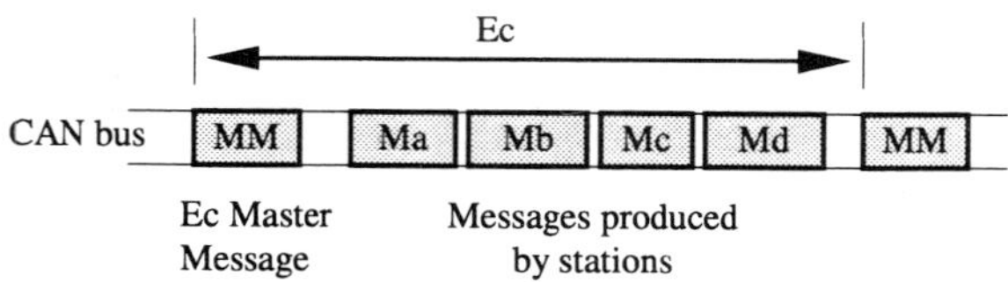

Fig. 2. Messages transmitted during one Ec.

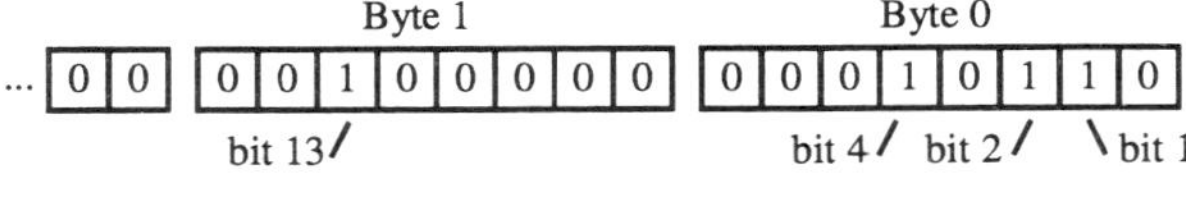

Fig. 3. Codification of messages to be produced in one Ec.

Upon reception of the master message each node scans its local table to see whether it is the producer or consumer of any of the messages to be broadcast in that Ec. After this table scanning each node prepares itself to either produce and/or consume the appropriate messages.

Two important issues are: the messages' production delays and the planning scheduler execution time. The former has to be controlled so that the messages specified to be broadcast in a given Ec always fit within the limits of that Ec. The latter must also be controlled so that after doing changes in the message set the scheduler is still able to build any plan within the plan duration. For both cases appropriate models are used, based on parameters which values are taken from experimental observations.

In order to obtain more predictable production delays for the data messages broadcast in one Ec, the scanning of the messages' table was separated from the actual messages' production. Therefore, the Ec was split in two windows, the scanning window, during which the master message is decoded and the table scanning is done, and the production window, during which the data messages are produced. The duration of both windows is fixed at configuration.

During the scanning window the bus is idle in what concerns the transmission of time-triggered messages. If required, this window can be used by any node to transmit sporadic messages with probabilistic delay guarantees. This effectively allows to raise the bus utilization factor without jeopardizing the properties of the time-triggered communication system provided that the transmission of sporadic messages does not extend beyond such window.

4 Experimental setup

An experimental setup was built using four CANivete boards, one for the master

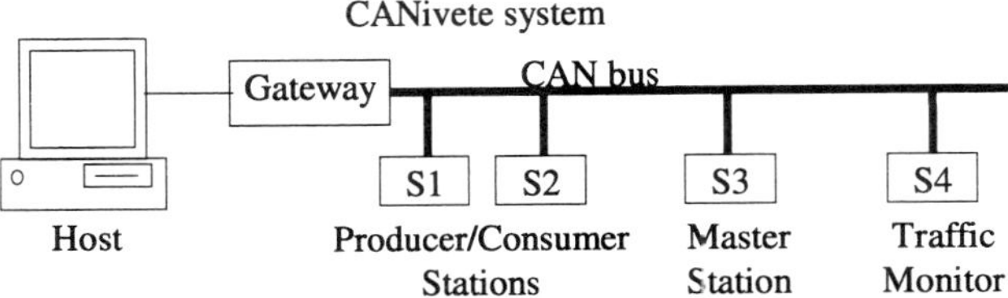

Fig. 4. Experimental setup.

station, two for variables' producer/consumer stations and one for monitoring (fig. 4). The bus transmission rate was 125 Kbit/s and the Philips 80C592 controllers were clocked at 11.06MHz. The Ec duration was set to 8.9ms and the plan duration to 20 Ec's, i.e. 178ms. The time measurements were carried out using an 8-bit timer of the controller with a resolution of 35µs. The message set used is described in table 1.

Table 1. Message set used in the experiments.

Message ID	Period (# of Ec's)	DLC (bytes)	Max. # of bits	Tx delay (µs)
0x101	1	1	63	513
0x102	2	2	73	594
0x103	3	2	73	594
0x104	4	1	63	513
0x105	5	1	63	513
0x106	6	4	92	749
0x107	7	4	92	749
0x108	8	4	92	749
0x109	9	4	92	749

5 Messages' production delays

In [5] some preliminary measurements were carried out with this system. It was shown that the processing and queuing delays at the microcontroller were larger than the transmission delays of the messages on the bus. In those measurements no scanning window as referred in section 3 was used and the production of data messages was started as soon as they were found in the local table. This resulted in a production delay which was difficult to predict because it included the time to scan part of the table. Using the scanning window with a fixed duration (ScWin) the production delay of message i ($t_{p,i}$) becomes the sum of just this fixed value and the queuing (t_q) and transmission (t_x) delays of all preceding messages in the same Ec, including message i. The resulting production delay can be modeled by a linear expression of the message data sizes (DLC) with a constant offset (1). Of course, this applies to the worst-case, only, when all the data messages of one Ec are to be produced by the same node and, consequently, are all serialized.

$$t_{p,i} = ScWin + \sum_{j=1}^{i} \left(t_{q,j} + t_{x,j} \right) = ScWin + \sum_{j=1}^{i} \left(K_1 + K_2 * DLC_j \right) \tag{1}$$

Figure 5 shows the histograms of production delays during each Ec, measured by the monitor station, with just one node producing the whole message set (left) and the same message set being produced by two different nodes (right). In this latter case the queuing delays of some messages can be overlapped with the transmission delays of other messages leading to shorter production delays. Figure 6 shows the differences in the production delays of consecutive messages in the same two situations as in figure 5

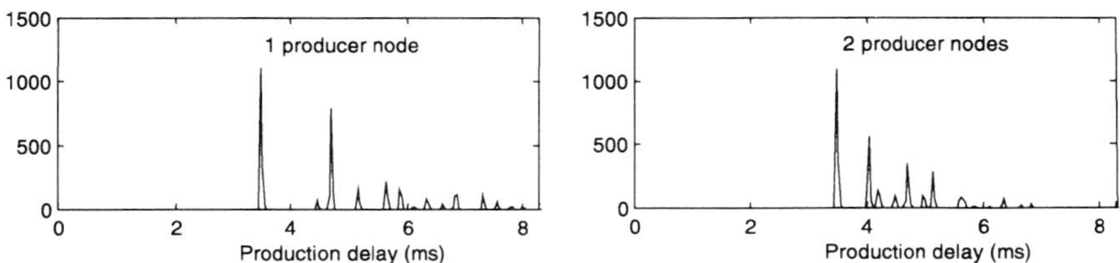

Fig. 5. Histograms of data messages' production delays.

(left and right respectively). The values in figure 6 (left) correspond to the queuing and transmission delays of the messages. The first spike corresponds to the 1-byte messages, $(t_q + t_x)_{,max} = 1015\mu s$, the second to the 2-byte messages, $(t_q + t_x)_{,max} = 1260\mu s$, and the third to the 4-byte messages, $(t_q + t_x)_{,max} = 1715\mu s$ (all values measured with a resolution of $35\mu s$). These values are in clear accordance with the model in (1) resulting in $770\mu s$ for K_1 and $245\mu s$/byte for K_2. Notice that these values are implementation dependent.

The model expressed in (1) is used by the scheduler to decide how many data messages fit within one Ec. Using the values of K_1 and K_2 deduced from the maximum production delays allows the scheduler to have a high degree of confidence that the messages scheduled for production in a given Ec will fit within the Ec boundaries, whichever are the nodes that produce them.

6 Scheduler execution time

The operational flexibility of the whole system is, essentially, based on the planning scheduler running on the master node. This flexibility means that it is possible to introduce changes to the message set during system operation. These changes normally cause the scheduler execution time to change, too. To assure continuous proper operation of the system it is important to guarantee that the scheduler execution time will never grow beyond the plan duration. In some cases it might be desirable to restrict its maximum value even more so that there is always some slack for the master CPU to execute other tasks.

Therefore, a model has been developed by the authors [7] [8] to assist the acceptance of changes to the message set in the master CPU for FIP-like networks. Notice, also, that there are other factors that have to be considered to accept such changes, e.g. schedulability guarantees, but that is not being discussed at this point.

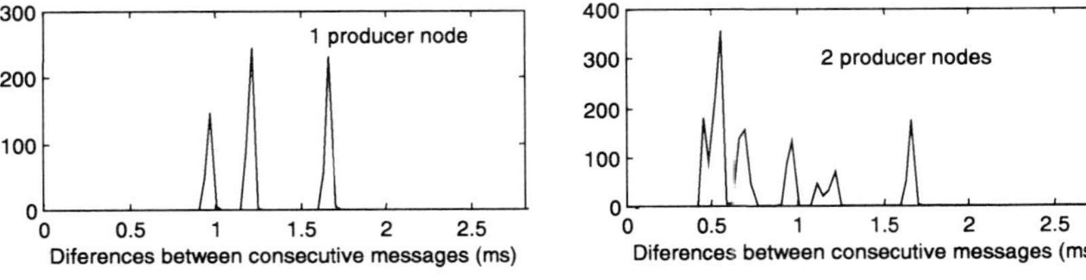

Fig. 6. Histograms of time differences between consecutive messages.

Two main experiments were carried out concerning the scheduler execution time. These experiments aimed at verifying the validity of the model for this implementation of the planning scheduler. In the first one, the plan duration (W) was varied from 1 Ec to 50 Ec's. As expected, increasing W caused the scheduler execution time (t_{sched}) to increase linearly with an offset (2).

$$t_{sched,av} = K_{3,av} + K_{4,av} * W \qquad t_{sched,wc} < K_{3,wc} + K_{4,wc} * W \qquad (2)$$

In the average case, the experimental data matches this linear trend ($K_{3,av}$= 3.31ms and $K_{4,av}$= 2.74ms/Ec) with an error lower than 1% (except for the first data point, W=1Ec). In the worst-case, the experimental data is upper bounded by a similar linear trend, as expect from the model, with a similar slope and a higher offset ($K_{3,wc}$= 13.23ms and $K_{4,wc}$= 2.76ms/Ec).

Figure 7 shows the value of t_{sched} relative to the plan duration. This ratio represents the computational load imposed by the planning scheduler. With W=1 Ec (close to a dynamic scheduler) such computational load is very high, near 88% in the worst-case. Since the dispatcher imposed a constant load of 11%, the CPU was operating very close to the limit (99% load!). However, with a plan duration increased to 20 Ec's, the maximal computational load required by the scheduler decreased to 33% (44% including the dispatcher). The remaining CPU capacity can be used by the scheduler, e.g. to add new messages to the set, or by other tasks.

The second experiment concerned the addition of messages with equal period (P_i=10 Ec's) while the plan duration was kept constant (W=20Ec's). The basic message set used is described in table 2. The new messages were added with identifiers 0x101 to 0x109 and had a DLC of 2 bytes. The results for the worst-case are depicted in figure 8. Two situations were used: a) the messages were added in phase and b) the added messages were distributed along the phase space so that they would never occur simultaneously.

Notice, again, that the results are in accordance with the trends predicted by (3). For in-phase harmonic messages, $t_{sched,wc}$ is upper bounded by a quadratic function of N, the number of messages, while for phase-distributed messages it is upper bounded by a

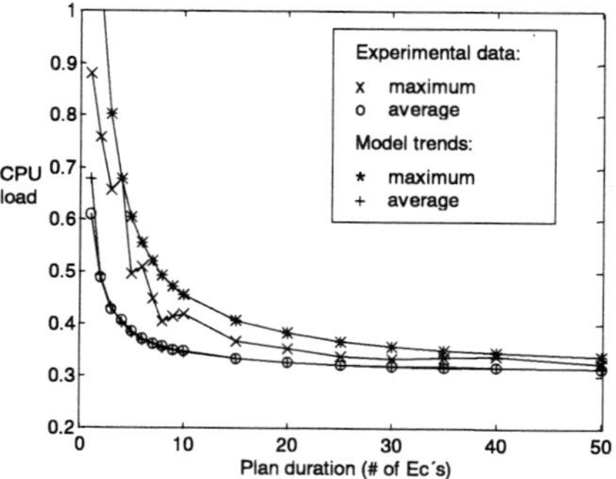

Fig. 7. CPU load imposed by the execution
of the planning scheduler.

Table 2. Basic message set used in the experiment of t_{sched} *vs* # of messages.

Message ID	Period (# of Ec's)	DLC (bytes)	Max. # of bits	Tx delay (μs)
0x10a	10	2	73	594
0x115	21	4	73	749
0x117	23	4	73	749

linear function of N.

$$in\_phase:\quad t_{sched,wc} < K_5 + K_6 * N + K_7 * N^2$$
$$dist\_phase:\quad t_{sched,wc} < K_8 + K_9 * N$$

$$(3)$$

The values obtained for the constants are: K_5=9.13ms, K_6=0.59ms/mess. and K_7=0.174ms/mess^2 (messages added in-phase) and K_8=6.49ms and K_9=1.76ms/mess (messages added with distributed phase).

Both experiments show that the referred model is still valid for this implementation of the planning scheduler. Therefore, after quantifying the respective basic time parameters, it is possible to use the model to determine whether a given change to the message set does not cause the scheduler to exceed its time budget. More experiments are currently being carried out in order to quantify those parameters.

7 Conclusions

The results of the practical experiments show that the proposed communication system is feasible. This system uses widespread low processing power microcontrollers (Philips 80C592) and yields the robustness and composability typical of time-triggered systems with autonomous control. Besides, it allows for flexible operation in the sense that changes to the message set can be accomplished without any down time, during normal system operation. Also notice that the protocol is fully implemented in S/W (written in "C") and thus, it is highly portable in the sense that it can be easily implemented on any CAN-based distributed infrastructure.

However, the low processing capabilities of the microcontrollers impose a limitation on the bus utilization factor. Using more powerful processors will cause a reduction in the

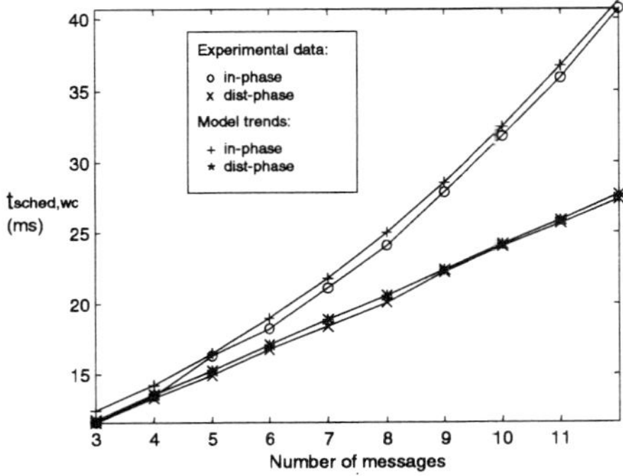

Fig. 8. Scheduler execution time vs. number of messages.

messages' queuing delays and processing overheads that will allow a more efficient bus usage. Also, making use of CAN-controllers with multiple transmission buffers, such as the Intel 82527, will contribute to improve the overall bus utilization.

The model presented for the messages' production delays produces results which are in accordance to those obtained in the experiments. Also, the trends expected from the model previously developed by the authors for the scheduler execution time have been verified by the experiments described in this paper. More experiments are currently being carried out to quantify the model basic time parameters so that it can be effectively used to predict the scheduler execution time with any message set.

References

1. Sha, L. Editorial: Industrial Computing. *IEEE Computer*, **27(1)**, 1994.
2. Stankovic, J.A. *et al.*. Strategic Directions in Real-Time and Embedded Systems. *ACM Computing Surveys*, **28(4)**: 751-763, 1996.
3. Kopetz, H. *Real-Time Systems Design Principles for Distributed Embedded Applications*. Kluwer Academic Publishers, 1997.
4. ISO, Int. Standard Organization. *Road Vehicles - Low speed serial data communication - Part 2: Low speed Controller Area Network*. ISO11519-2, 1994.
5. Almeida, L,. J.A. Fonseca and P. Fonseca. Flexible Time-Triggered Communication on the Controller Area Network. *Proc. of Work-In-Progress session of RTSS'98 (IEEE Real-Time Systems Symposium)*, Madrid, Spain, 1998.
6. Pasadas R., L. Almeida and J.A. Fonseca. A Proposal to Improve Flexibility in Real-Time Fieldbus Networks. *Proceedings of SICICA'97 (Int. Symp. on Intelligent Components and Instruments for Control Applications)*, Annecy, France, 1997.
7. Almeida, L., R. Pasadas and J.A. Fonseca. Using the Planning Scheduler in Real-Time Fieldbuses: Theoretical Model for Run-Time Overhead. *Proc. of WFCS'97 (IEEE Int. Work. on Factory Communication Systems)*, Barcelona, Spain, 1997.
8. Almeida, L. and J.A. Fonseca. The Planning Scheduler: Compromising between Operational Flexibility and Run-Time Overhead. *Proc. of INCOM'98 (IFAC Int. Symposium on Information Control in Manufacturing)*, Nancy/Metz, France, 1998.
9. Fonseca, P. *et al.*. A Dynamically Reconfigurable CAN system. *Proc. of ICC'98 (5th Int. CAN Conference)*, San Jose, USA. CAN in Automation-CiA,1998.
10. Kopetz, H. and G. Grünsteidl. TTP - A Protocol for Fault-Tolerant Real-Time Systems. *IEEE Computer*, **27(1)**, 1994.
11. CENELEC. *General Purpose Field Communication System*. EN 50170 Vol. 3/3. CENELEC, 1996.
12. Peraldi, M.A. and J.D. Decotignie. Combining Real-Time Features of Local Area Networks FIP and CAN. *Proc. of ICC'95 (2nd Int. CAN Conference)*. CiA, 1995.
13. Thomesse, J.-P. Time and Industrial Local Area Networks. *Proc. of COMPEURO'93*. Paris, 1993.

Analysis of Switched Ethernet Networks with Different Topologies Used in Automation Systems

S. Rüping[1], E. Vonnahme[1], J. Jasperneite[2]

[1]Heinz Nixdorf Institut
University of Paderborn
System and Circuit Technology
Paderborn, Germany

[2]Phoenix Contact GmbH
Concept Development
Blomberg, Germany

Abstract. Ethernet is the most famous type of communication network used for office applications. Currently, networks used on field level and on office level are different due to their specific requirements. Especially for real-time systems, the deterministic behavior of the network is very important. Now Switched Ethernet seems to be applicable also on field level, because collisions can be avoided and priorities can be used. In this paper Switched Ethernet networks with different topologies are analyzed concerning the transmission delays. Therefore a typical master-slave scenario of an automation system is investigated.

1 Introduction

Today nearly every computer is connected to a network. The most famous type used for local area networks (LAN) is Ethernet [2]. Since 1975, when Ethernet was introduced by Xerox, a steady development has been done up to the newest Gigabit Ethernet. At the moment research is carried out for Terabit Ethernet.

The large number of Ethernet networks worldwide leads to the following consequences. First, many hard- and software components are available. Second, the components are inexpensive due to the high number of sold units. And third, there are many people, that have high knowledge about Ethernet. These consequences combined with the fact, that the transmission rates become higher and higher, make it sensible to think about Ethernet in automation systems even at field level.

One of the main problems of applying Ethernet to field level is based on the medium access control protocol CSMA/CD (Carrier Sense Multiple Access / Collision Detection), which cannot guarantee a maximum delay time for data transmission. A solution for this seems to be the Switched Ethernet technology [1, 3, 6]. In this paper a theoretical analysis of Switched Ethernet networks applied to field level tasks is presented. The main point of interest is the analysis of different topologies, which can be used to build these networks.

2 Switched Ethernet Networks

Ethernet networks based on twisted pair connections usually have a star topology with a central Hub. A Hub is a multiport repeater that forwards the data received by a port to all other ports. If the network is enlarged, up to 2 Hubs can be added. But nevertheless, the whole network is a so-called shared medium, which means only a single node can send data at a time. If the number of nodes becomes too high, a router or a bridge can be used to split the network into different collision domains [4]. The problems with router or bridges are, that they are relatively expensive and often have only a few ports.

A Switch is an intelligent Hub, that can read and process the destination address of the incoming data and send it only to the required out-ports. Figure 1 shows the difference between a network based on Hubs and on Switches.

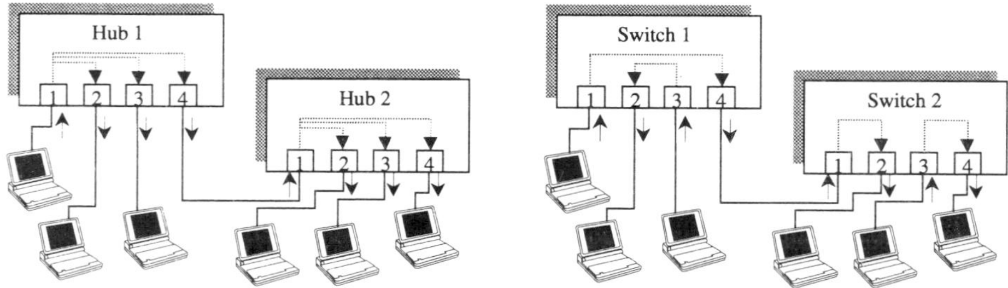

Fig. 1. Difference between Hub and Switch technology

In the Hub network the data, which is sent by a node, is received by all other nodes. In the Switch network, the data is sent only to the destination nodes, which means several nodes can send data at a time.

3 Requirements of Automation Systems

The requirements of automation networks are different to the requirements of computer LANs. On field level data packets are often small (only a few Bits or Bytes), and the latency and overall delay time must be very short. Especially for real time systems the network has to guarantee a deterministic behavior [5].

In order to investigate the suitability of Switched Ethernet in automation applications, the following system will be analyzed.

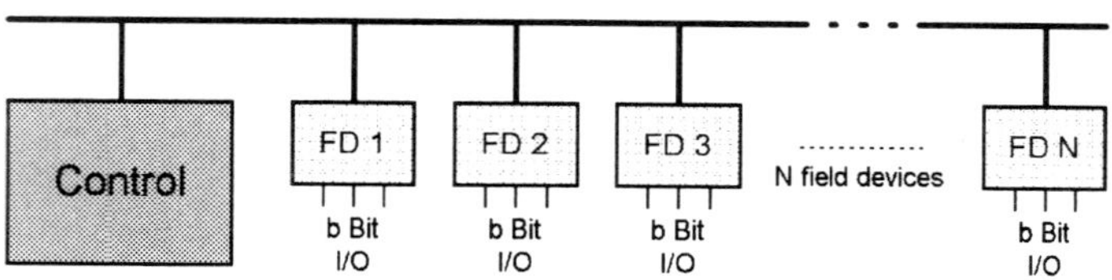

Fig. 2. Example of an Automation System

The master of the system is the node "Control" (e.g. a CNC or PC node). Besides this there are N field devices, which are connected to the master by a network. Each field device has b bits input/output data. The system runs cyclic. A cycle begins with transmitting individual data from each field device to the master (sensor data). After that the master calculates new actuator data for all field devices and transmits it. It is important, that the network has a small and deterministic latency and delay time in order to guarantee the defined cycle time.

4 Calculations for different Topologies

With Switched Ethernet systems can be built, that realize a completely deterministic behavior. In order to do so, two things are important. First, each device in the network must have it's own switch. Second, the switches must work in the so-called Store-and-Forward mode. This mode makes it possible, that a switch, which receives multiple data for just one output-port, can store the data and send it one after the other. If now every device has it's own switch, there are only point-to-point connections in the system. Therefore no collisions on the medium can occur. By using priorities or other mechanisms to order the sequence of packets in the switches, the system will be deterministic [6].

Figure 3 shows three different topologies, that can be used to connect the nodes of the example system. Each connection between two nodes enables full duplex communication. For all following calculations Fast Ethernet with 100 Mbits/sec is used. The transmission of data from the master to the field devices can be done on two different ways. The first is, to build a large packet, in which the data for all devices is included, and to send it as a multicast. Therefore each field device has to know, at which position in the packet the right data is. In the following, this is called the multicast-solution. The second is, to send individual packets to each field device. This will be called the unicast-solution.

All calculations have been done for simple protocols only, which are SNAP (Subnetwork Access Protocol) or SNAP with UDP (User Datagram Protocol) [1]. Based on the rules, how packets for these protocols have to look like, the theoretical minimum cycle times depending on the number of field devices and the number of input/output bits per device are calculated. This is done for the three presented topologies. The time, that is needed to process the data in the control unit and the field devices is assumed to be zero.

For the unicast solution the following formulas can be given. The time that is necessary to transmit a packet is:

$$t_f = \max\{h + b, 512\} \cdot t_b + t_g \tag{1}$$

h is the number of header bits, b is the number of data bits. A frame is not allowed to be smaller than 512 bits, therefore the maximum must be taken. t_b is the bit time, t_g is the interframe gap time.

For the line topology a minimum cycle time t_c can be calculated as:

$$t_c = 2(N+1)t_s + 2(N+2)t_f \tag{2}$$

N is the number of field devices, t_s is the switch delay time.

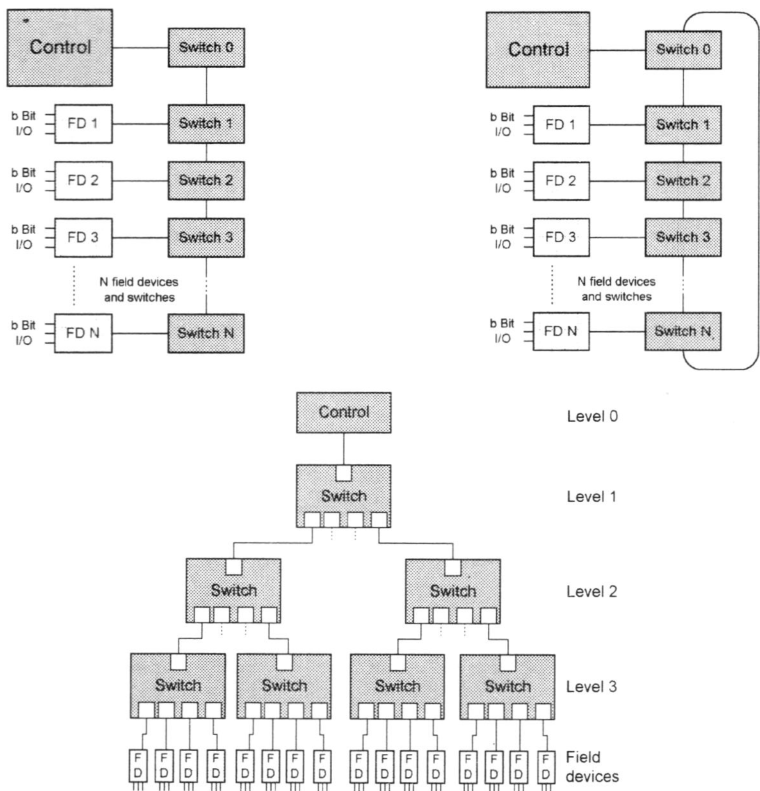

Fig. 3. Line Topology (top left), Ring Topology (top right) and Tree Topology (bottom)

For the ring topology the minimum cycle time is:

$$t_c = (N+3)t_s + 2(N+2)t_f \tag{3}$$

And the same for the tree topology, where E is the number of switch levels:

$$t_c = 2Et_s + (2E+N+1)t_f \tag{4}$$

For the multicast solution, the packet that is sent from the master to the field devices has the following transmission time:

$$t_{fm} = \max\{h+Nb;512\} \cdot t_b + t_g \tag{5}$$

The answers from the field devices still need the same time as before:

$$t_{fa} = \max\{h + b; 512\} \cdot t_b + t_g \tag{6}$$

Again the minimum cycle time for the topologies are calculated:

Minimum cycle time for the line topology:

$$t_c = 2(N+1)t_S + (N+2)(t_{fm} + t_{fa}) \tag{7}$$

Minimum cycle time for the ring topology:

$$t_c =$$
$$\max\{t_1 + (N+1)t_S + (N+2)t_{fa}; (N+1)t_S + (N+2)t_{fm} + t_S + 2t_{fa}\} \tag{8}$$
$$+(N-1)t_{fa}$$

Minimum cycle time for the tree topology

$$t_c = 2 E t_S + (E+1)t_{fm} + (E+N)t_{fa} \tag{9}$$

5 Presentation of the results

In order to present the calculated results, the following figures show plots of a comparison between the topologies. For the number of field devices the minimum cycle times are calculated and shown as a graph.

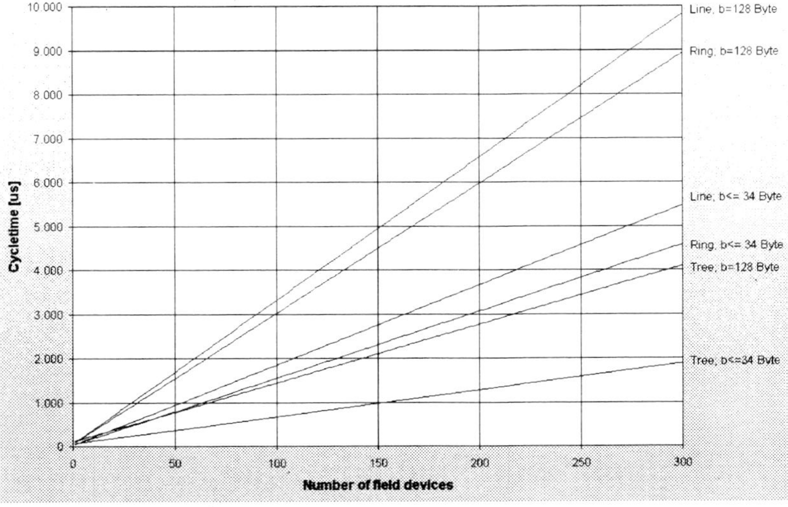

Fig. 4. Comparison of the topologies for the unicast solution

Figure 4 and 5 show the results of the topology comparison, when the unicast-solution is used (figure 4) and when the multicast-solution is chosen (figure 5). It can be seen, that the usage of unicasts is much faster (e.g. line topology, 300 devices, b=32: about 5.400 µs) than the multicast solution (about 34.000 µs for the same parameters). This is a result that has not been expected at the beginning. The main reason for it is the usage of store-and-forward switches. A packet has to be received completely before it can be transmitted to the next switch. If the packet is large and the number of switches is high, the system becomes slow. Optimization of the longest data-path by reordering the packets cannot be untertaken when using the multicast-solution.

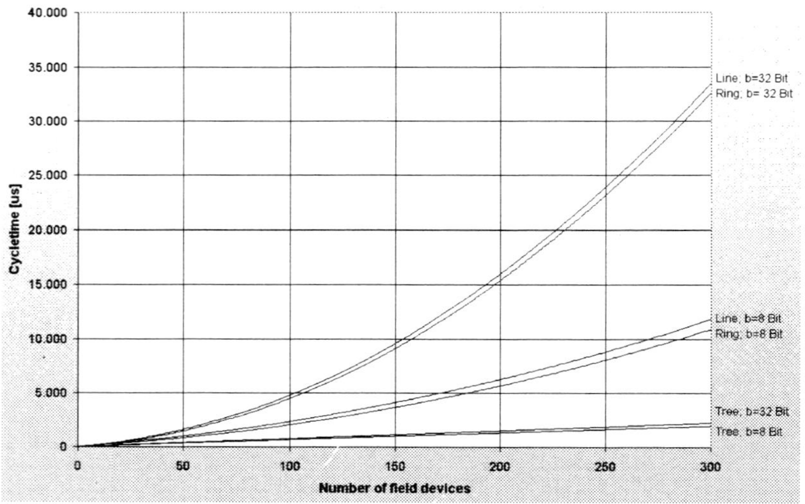

Fig. 5. Comparison of the topologies for the multicast solution

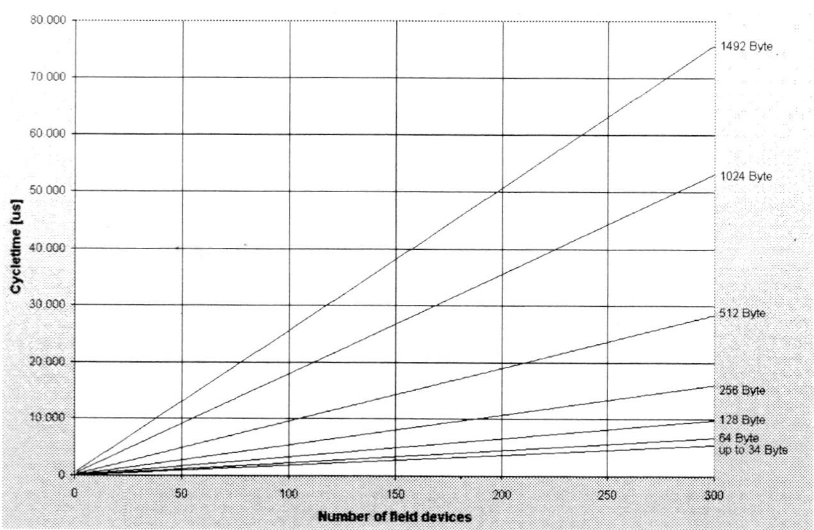

Fig. 6. Influence of the number of bits per device (line topology)

As another example for visualized results, the figures 6, 7 and 8 present the minimum cycle time for the line topology when the number of data bits per device (figure 6), the type of protocol (figure 7) and the switch delay is a parameter. All numbers are based on the unicast-solution.

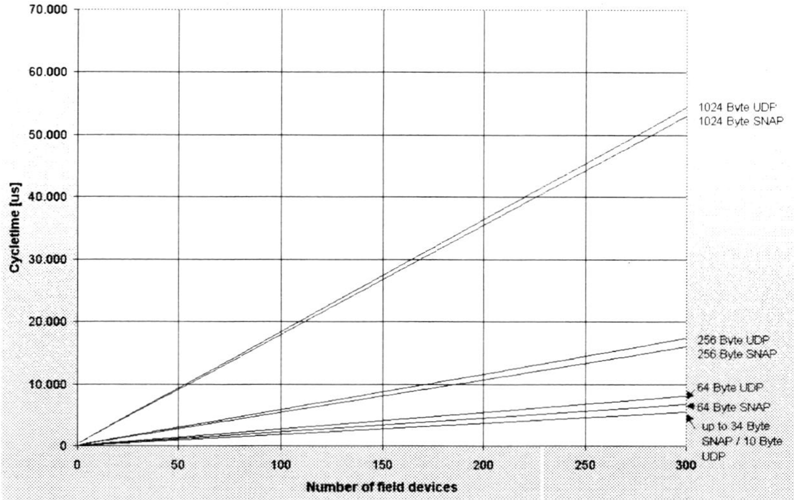

Fig. 7. Influence of the protocol (line topology)

The cycle time increases linearity with the I/O data size of the field devices, but if the size is smaller than 34 Bytes with SNAP, the frame must be padded to be at least 64 Bytes long. The type of the analyzed protocols has only little effect on the cycle time, connection oriented protocols like TCP would have a greater influence.

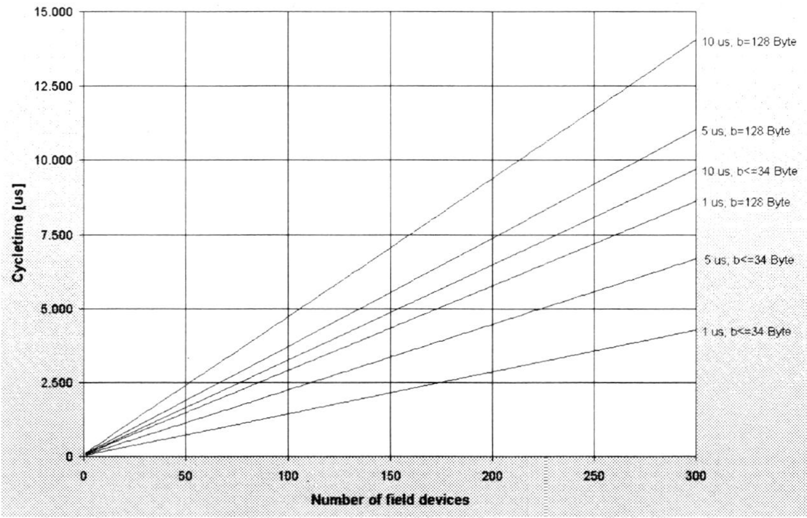

Fig. 8. Influence of the switch delay (line topology)

6 Conclusions

In this paper different topologies of Switched Ethernet networks applied to automation systems are analyzed. The theoretical minimum cycle time for a simple master-slave communication model is calculated based on line, ring and tree topology. Further parameters are the switch delay, the protocol type, the number of field devices and the I/O data size. Concerning the data transmission from the master to the field devices, two different concepts are investigated: the unicast- and the multicast-solution.

The results show, that the unicast-solution performs better than the multicast-solution. The tree topology achieves shorter cycle times, than line and ring topology. Another advantage of the tree topology is, that the number of required switches is smaller.

The ring topology performs better than the line topology. Especially the redundancy in case of an error is important for many applications. The problem is, that a special routing strategy and redundancy management is necessary.

Contact: rueping@hni.uni-paderborn.de

References

1. Held, G., "Ethernet Networks: Design, Implementation, Operation, Management", Wiley, New York, 1996

2. IEEE Standard 802.3, "Carrier Sense Multiple Access with Collision Detection (CSMA/CD) Access Method and Physical Layer Specifications", 1998

3. Johnson, H., "Fast Ethernet: Dawn of a new Network", Prentice Hall, New Jersey, 1996

4. Perlman, R., "Interconnections: Bridges und Routers", Addison Wesley, Bonn, 1993

5. Schnell, G., "Bussysteme in der Automatisierungstechnik", Vieweg, Braunschweig, 1996

6. IEEE Standard 802.1

Analysis of the Worst-Case Real Token Rotation Time in PROFIBUS Networks[1]

Eduardo Tovar[‡], Francisco Vasques[†]

[‡] Polytechnic Institute of Porto (ISEP-IPP), Portugal, e-mail: emt@dei.isep.ipp.pt
[†] University of Porto (FEUP), Portugal, e-mail: vasques@fe.up.pt

Abstract. This paper provides a comprehensive study on how to support time critical distributed applications using PROFIBUS. We show that, despite the absence of synchronous bandwidth allocation, it is possible to guarantee real-time behaviour for the high-priority traffic in PROFIBUS networks. The main contribution of this paper is to give a methodology for the setting of the T_{TR} parameter, by proper analysis of the worst-case real token rotation time (T_{RR}).

1 Introduction

Communication networks aimed at the interconnection of sensors, actuators and controllers are commonly known as fieldbus networks. In the past, fieldbus scope was dominated by vendor specific solutions, which were mostly restricted to specific application areas. Moreover, the concepts behind each proposed network were highly dependent on the manufacturer of the automation system, each one with different technical implementations and also claiming to fulfil different application requirements, or the same requirements with different technical solutions [1].

In the recent years, more and more standardised fieldbuses are accepted supporting the open system communication concept and thus having a vendor independent communication. Controller Area Network (CAN) [2] and PROcess FIeld BUS (PROFIBUS) [3] are two of the most popular fieldbuses, which are defined as international standards, respectively by ISO and CENELEC. Though these two protocols were originally conceived for different purposes, they have significantly evolved so that they can be used in distributed computer controlled systems connecting remote sensors and actuators. CAN, in fact, was primarily conceived for automotive applications, to solve cabling problems found in some kind of vehicles. However, due to its interesting features, it is also being considered in the automated manufacturing and process control environments [4]. PROFIBUS and its communication stack, instead, were created as a communication support in automated factory environments. Its great flexibility however was achieved at the cost of a reduced network responsiveness, which means that PROFIBUS is not suited for high-speed real-time data exchanges. To overcome this problem, in the past few years a new upper layer protocol have been developed for PROFIBUS, called decentralised periphery (DP) [5]. Such development makes PROFIBUS particularly suited for use as a control network

---

[1] This work was partially supported by ISEP, FLAD, DEMEGI/FEUP and FCT.

in those applications where a number of small-sized data variables have to be exchanged at a high rate. Considering the general characteristics of CAN and PROFIBUS, several studies have shown that CAN is more suited for small-size event-driven distributed systems while PROFIBUS gives better performances when used in medium-to-large systems, whose interactions are mostly based on polling schemes [6].

The remainder of the paper is organised as follows: in section 2 we briefly describe the PROFIBUS MAC mechanisms. In section 3, we introduce some basic concepts concerning the real-time characteristics of the PROFIBUS protocol, which were previously presented in [7,8]. In section 4, we give an accurate bound for the worst-case token rotation time based on the evaluation maximum token lateness in each master station. Such worst-case token rotation time is the basis for the setting of the T_{TR} parameter, which is the main contribution of this paper. Finally, in section 5 we draw some conclusions.

2 The PROFIBUS Implementation of the Timed Token Protocol

The PROFIBUS MAC mechanism is based on a token passing procedure, used by master stations to grant the bus access to each one of them, and a master-slave procedure used by master stations to communicate with slave stations. The PROFIBUS token passing procedure uses a simplified version of the timed token protocol [9]. One of the PROFIBUS MAC main functions is the control of the token cycle time, which will now be briefly explained.

After receiving the token, the measurement of the token rotation time begins. This measurement expires at the next token arrival and results in the real token rotation time (T_{RR}). A target token rotation time (T_{TR}) must be defined in a PROFIBUS network. The value of this parameter is common to all masters, and is used as follows. When a station receives the token, the token holding time (T_{TH}) timer is given the value corresponding to the difference, if positive, between T_{TR} and T_{RR}. PROFIBUS defines two categories of messages: high priority and low priority. These two categories of messages use two independent outgoing queues. If at the arrival, the token is late, that is, the real token rotation time (T_{RR}) is greater than the target rotation time (T_{TR}), the master station may execute, at most, one high priority message cycle. Otherwise, the master station may execute high priority message cycles while $T_{TH} > 0$. T_{TH} is always tested at the beginning of the message cycle execution. This means that once a message cycle is started it is always completed, including any required retries, even if T_{TH} expires during the execution. We denote this occurrence as a T_{TH} overrun. The low priority message cycles are executed if there are no high priority messages pending, and while $T_{TH} > 0$ (also evaluated at the start of the message cycle execution, thus leading to a possible T_{TH} overrun).

3 Timing Requirements in PROFIBUS Networks

As previously described, and as far as there are high priority message transfers pending, a master station is guaranteed to transmit, at least, one high priority message per token arrival (no matter if there is enough token holding time left). We define the upper bound between two consecutive token arrivals to a particular master station k as T^k_{cycle}. Based on the knowledge of the upper bound for the token cycle time, we will derive a methodology for setting the T_{TR} parameter in order to guarantee the real-time requirements of PROFIBUS messages.

3.1 Network and Message Models

We consider a bus topology containing n master stations. A special frame (the token) circulates around the logical ring formed by the masters. We denote the logical ring latency (token walk time, including node latency delay, media propagation delay, etc.) as τ. Message cycles generated at run-time in the system are either high priority or low priority messages. In each master k we consider nh^k high priority message streams. A message stream corresponds to a temporal sequence of message cycles related, for instance, with the reading of a process sensor. We denote the ith ($i = 1, 2, \ldots nh^k$) high priority stream associated to a master k as Sh_i^k. A high priority message cycle stream Sh_i^k is characterised as:

$$Sh_i^k = \left(Ch_i^k, Dh_i^k\right) \tag{1}$$

Ch_i^k is the maximum amount of time required to perform a high priority message cycle of stream i belonging to a master k. Dh_i^k is the message cycle relative deadline, which is the maximum amount of time that may elapse between the transmission request and the reception of the related response, both at the application process level. We consider that, in the worst-case, the deadline can be seen as the minimum inter-arrival time (MIT) between two consecutive message requests in the same stream.

As we do not intend to guarantee deadlines for the low priority traffic, a low priority message stream Sl_i^k is merely characterised as follows:

$$Sl_i^k = \left(Cl_i^k\right) \tag{2}$$

where Cl_i^k is the maximum amount of time required to perform a message cycle of a low priority stream i belonging to a master k.

3.2 Queuing Delay for a PROFIBUS Message Cycle

Contrarily to other timed token based networks, in PROFIBUS we can not use analysis similar to those proposed in [10,11], since in PROFIBUS protocol there is no synchronous bandwidth[2] to be allocated in each master. However, it is possible to guarantee a real-time behaviour for the high priority traffic using the PROFIBUS protocol. In [7,8] the authors analyse and propose two different approaches. One of the approaches considers a worst-case scenario with only one high priority message cycle processed per token visit. As already mentioned, in PROFIBUS it is guaranteed that

---

even if a master receives a late token, it will still be able to execute one high priority message cycle. Then, if there are m messages pending in the outgoing queue, in the worst-case it will take m token visits to execute all those high priority messages.

It is obvious that the queuing delay (Q^k) very much depends on how the outgoing high priority queue is implemented. In PROFIBUS the majority of the implementations use a First-Come-First-Served (FCFS) queue. In this paper we assume this type of queue implementation. For the queuing delay analysis, it is important to note that, in master k, the maximum number of pending messages will be nh^k, corresponding to one message per each Sh_i^k stream. This assumption results from the fact that if there are two messages of the same stream pending in the high-priority outgoing queue, then a deadline for that message stream was missed.

It is now clear that, if we assume that message deadlines are not missed (thus the maximum number of high priority pending messages is nh^k), the upper bound for the message queuing delay in a master k is:

$$Q^k = nh^k \times T_{cycle}^k \tag{3}$$

3.3 Pre-Run-Time Schedulability Condition

The worst-case end-to-end communication delay (E) can be expressed as follows:

$$E = g + Q + C + d \tag{4}$$

In (4), g represents the worst-case generation delay for the master application task to generate and queue that specific read request. The term Q corresponds to the worst-case delay for that request to gain access to the communications device after being queued. The term C corresponds not only to the worst-case for transmitting the request, but also to the time needed to receive the response from the slave (including processing at the slave side, slave turnaround time, propagation delay and other network latencies). Finally, d represents the delivery delay, that is, the time needed to process the response before finally delivering it to the destination task, which in the case of PROFIBUS is in the same host processor as the sending task.

It is now possible to re-write the end-to-end communication delay for a high priority PROFIBUS message stream as follows:

$$E_{Sh_i^k} = g_{Sh_i^k} + nh^k \times T_{cycle}^k + Ch_i^k + d_{Sh_i^k} \tag{5}$$

Thus, a set of PROFIBUS high priority message streams guaranteedly satisfies its deadlines if the following pre-run-time condition is verified:

$$Dh_i^k \geq E_{Sh_i^k}, \ \forall_{master\ k,\ stream\ Sh_i^k} \tag{6}$$

that is, if all high priority message stream deadlines are greater than or equal to their worst-case end-to-end communication delay (as defined in (5)). We need now to evaluate T_{cycle}^k. As the authors shown in [12], T_{cycle}^k is a function of T_{TR}. Thus, having defined T^k_{cycle}, it will then be possible to set the T_{TR} parameter such as the pre-run-time schedulability condition (6) is verified, otherwise the high priority traffic is not schedulable, in the sense that deadlines can be missed.

4 Analysis of the Worst-Case Real Token Rotation Time

4.1 Evaluation of the Token Cycle Time

In PROFIBUS, the real token rotation time (T^k_{RR}) will always be smaller than T_{TR}, except when one or more masters in the logical ring induce the token to be late. Two reasons may justify a late token at a master k.

1. As once a message cycle is started, it is always completed, even if T^k_{TH} has expired during its execution, a late token may be delivered to the following stations. We define this occurrence as an *overrun* of the T^k_{TH}.

2. If a master receives a late token, it will still be able to send one high priority message. This may further increase the token lateness.

Based on these two conditions and on the following three theorems (the interested reader is referred to [12] for full treatment),

- *Theorem 1* - In PROFIBUS networks, if the master holding the token releases it before the T^k_{TH} expiration, then, the following master in the logical ring will receive an early token.

- *Theorem 2* - In a PROFIBUS network, in a specific token cycle, only one overrun will contribute to the token lateness.

- *Theorem 3* - If the token holding time in a PROFIBUS master k is greater than $T_{TR} - \tau$, all the following masters up to master $k - 1$ (modulo n) will receive a late token.

T^k_{cycle} is defined as follows:

$$T^k_{cycle} = T_{TR} + T^k_{del} \tag{7}$$

where T^k_{del} is the worst-case token lateness.

Such worst-case token lateness may result not only from an overrun in the master k but also from one occurring in one of the following masters ($k + 1$ until $k - 1$) (modulo n). Using fig. 1 as an illustrative example, assume that master 1 do not overruns its T^1_{TH} (since the last message to be transmitted starts before the end of T^1_{TH}). Then, master 2 may use its available token holding time and produce an overrun.

If this overrun is longer than the sum of the maximum overrun in station 1 with the longest high priority message cycle of station 2, then this would led to a higher value for T^1_{del} (fig. 2).

364

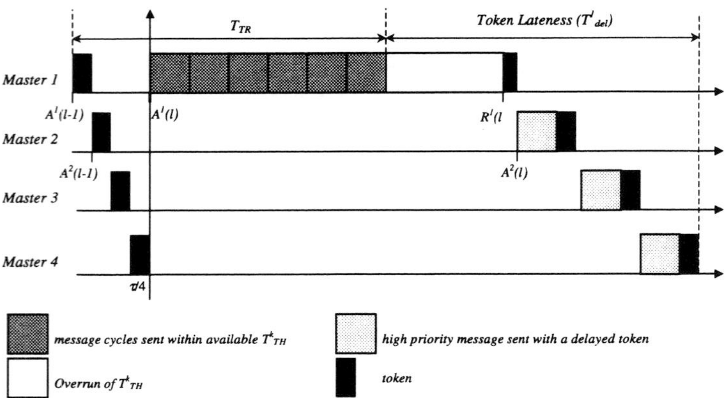

Fig. 1. Illustrative example for theorem 3

Similarly, if the maximum overrun in master 3 is longer than the sum of both an overrun in station 2 and the longest high priority message cycle of station 3, then this would led to a higher value for T^{l}_{del}. Note that by theorems 2 and 3, for the token lateness evaluation in master k, we should only consider one overrun in station j (j ranging from k to $k - 1$ (modulo n)) and one high priority message cycle per each station whose address is between j and station $k - 1$ (modulo n).

Basically, we can conclude that the T^{k}_{cycle} will depend on which master produce the worst-case overrun and on its relative positioning in the logical ring.

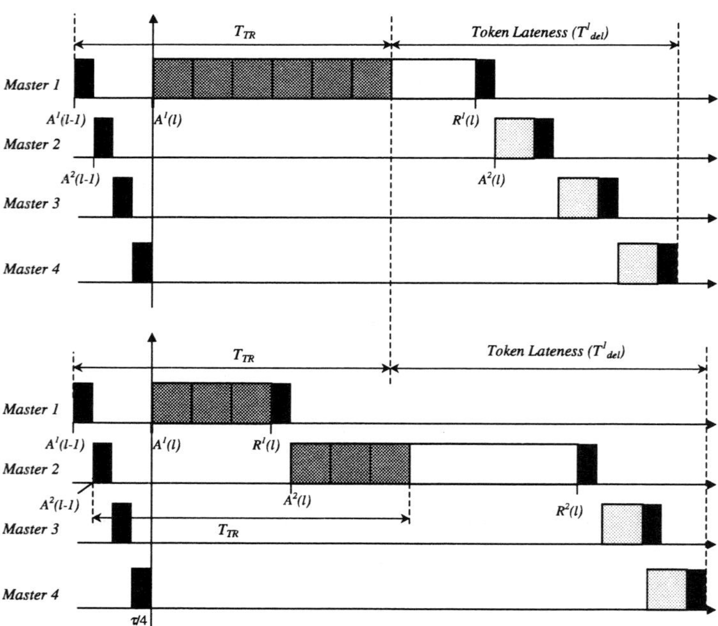

Fig. 2. Comparison between two overrun situations

For the token lateness (T^k_{del}) evaluation, we introduce the following master parameters: Ω^k, Φ^k and Ψ^k. Ω^k is the longest high priority message cycle transfer requested by a station k:

$$\Omega^k = \max_{i=1,..,nh^k} \left\{ Ch^k_i \right\} \tag{8}$$

Φ^k is the longest low priority message cycle transfer requested by a station k:

$$\Phi^k = \max_{i=1,..,nl^k} \left\{ Cl^k_i \right\} \tag{9}$$

Finally, Ψ^k is the longest message cycle transfer requested by a station k (including both types of message cycles):

$$\Psi^k = \max\left\{ \Omega^k, \Phi^k \right\} \tag{10}$$

Using the outlined analysis, we can thus define the maximum token lateness in a PROFIBUS master station k (T^k_{del}) as:

$$T^k_{del} = \max_{j\in\phi_1}\left\{ \Psi^j + \sum_{i\in\phi_2}\Omega^i \right\} \tag{11}$$

where ϕ_1 is defined as

$$\phi_1 = \begin{cases} k..n, & \text{if } k=1 \\ k..n, 1..k-1, & \text{if } k>1 \end{cases} \tag{12}$$

and ϕ_2 as

$$\phi_2 = \begin{cases} j+1..n, & \text{if } k=1 \\ j+1..n, 1..k-1, & \text{if } k>1 \end{cases} \tag{13}$$

4.2 Setting the T_{TR} Parameter

Using the Tkcycle definition (7), the pre-run-time schedulability condition (6) ca be re-written as follows:

$$Dh^k_i \geq g_{Sh^k_i} + nh^k \times \left(T_{TR} + T^k_{del} \right) + Ch^k_i + d_{Sh^k_i}, \forall_{master\ k,\ stream\ Sh^k_i} \tag{14}$$

Therefore, we may have an expression for setting the T_{TR} parameter:

$$0 \leq T_{TR} \leq \frac{Dh^k_i - Ch^k_i - \delta_{Sh^k_i}}{nh^k} - T^k_{del}, \forall_{master\ k,\ stream\ Sh^k_i} \tag{15}$$

where δ aggregates both the generation and delivery delays of message stream.

5 Conclusions and Discussion

In this paper we have provided a comprehensive study on how to use PROFIBUS field bus networks to support real-time communication. The major contribution of this paper is to provide a methodology for the setting of the Target Token Rotation Time, in order to guarantee communication real-time behaviour using the PROFIBUS protocol.

One possibility to reduce the worst-case response time relies on the implementation of a priority-based queue at the application process level. The PROFIBUS protocol uses First-Come-First-Served (FCFS) outgoing queues at the communication stack. Thus, as future work, we are to evaluate the use of a priority-based queue for the high-priority message requests at the application process level, limiting the stack outgoing communication queue to one pending request.

References

1. Cardoso, A., Tovar, E.: "Industrial Communication Networks: Issues on Heterogeneity and Internetworking", Proceedings of the 6th International Conference on Flexible Automation and Intelligent Manufacturing, FAIM'96, May 1996, Atlanta, USA, pp. 139-148.

2. ISO 11898, "Road Vehicle - Interchange of Digital Information - Controller Area Network (CAN) for High-Speed Communication", 1st Edition, ISO, 1993.

3. EN 50170, "General Purpose Field Communication System", European Standard, CENELEC, 1996, Vol. 2/3.

4. Zuberi, K., Shin, K.G.: "Real-Time Decentralised Control with CAN", Proceedings of the IEEE Conference on Emerging Technologies and Factory Automation, November 1996, pp. 93-99.

5. DIN 19245, "PROFIBUS-DP - Process Field Bus Decentralised Periphery (DP) - Part 3", Draft Standard DIN 19245, issue 1994.

6. Cena, G., Demartini, C., Valenzano, A.: "On the Performances of two Popular Fieldbuses", Proceedings of the 2nd IEEE International Workshop on Factory Communication Systems, WFCS'97, October 1997, Barcelona, Spain, pp. 177-186.

7. Tovar, E., Vasques, F.: "Guaranteeing Real-Time Message Deadlines in Profibus Networks", Proceedings of the 10th Euromicro Workshop on Real-time Systems, Berlin, Germany, IEEE Press, 1998, pp. 79-86.

8. Tovar E., Vasques, F.: "Setting Target Rotation Time in Profibus Based Real-Time Applications", Proceedings of the 15th IFAC Workshop on Distributed Computer Control Systems (DDCS'98), Como, Italy, 1998, pp. 1-6.

9. Grow, R.: "A Timed Token Protocol for Local Area Networks", Proceedings of Electro'82, May 1982, Token Access Protocols, Paper 17/3..

10. Montuschi, P., Ciminiera, L., Valenzano, A.: "Time Characteristics of IEEE802.4 Token Bus Protocol", IEE Proceedings, January 1992, 139 (1), pp. 81-87.

11. Malcolm, N., Zhao, W.: "Guaranteeing Synchronous Messages with Arbitrary Deadline Constraints in a FDDI Network", Technical Report, Department of Computer Science, Texas A&M University, March 1993.

12. Tovar E., Vasques, F.: "Cycle Time Properties of the PROFIBUS Timed Token Protocol", September 1998, to appear in Computer Communications, Elsevier Science.

Chapter 10: Realisations

Papers presented in Session 1.6

Powerline Communication in der Feldebene
Beikirch, H.

Design of Bus Media Redundancy in CAN
Rufino, J.; Veríssimo, P.; Arroz, G.

Programmable Process Interfaces for Smart Bus Nodes
Beikirch, H.; Voß, M.

Eine Feldbusarchitektur mit realzeitfähiger Fehlerkorrektur
Erdner, T.

Further Papers

Safety-Oriented INTERBUS - INTERBUS Safety
Meyer-Gräfe, K.

MultiPort RAM: a PC-Based Implementation of the Fieldbus Concept
Lobachov, M.; Sauter, T., Plagemann, B.

Fieldbus Physical Layer in the Application
Fetzer, U.

Powerline Communication in der Feldebene

Helmut Beikirch

Fachbereich Elektrotechnik und Informationstechnik
Universität Rostock, D-18051 Rostock
e-mail: beikirch@e-technik1.uni-rostock.de
phone: +49 381 498 3514, fax: +49 381 498 3608

Abstract. Derzeitige Entwicklungswege in der Feldbustechnik beinhalten den Ausbau neuer Interfacestrukturen zur Vergrößerung der Anwendungsbreite bestimmter Netzwerke. Dabei werden andere als die Standardübertragungsprinzipien genutzt. (beispielsweise drahtlose Interfaces oder Powerline-Funktionalität). Allein das Problem der Powerline Communication für zeitkritische Feldbusnetzwerke umzusetzen, bringt erhebliche Probleme mit sich. Auf Erfahrungen und umfangreichen Untersuchungen basierend werden Powerline Prinzipien differenziert dargestellt und wesentliche Probleme diskutiert. Konkrete Lösungen wurden für das Netzwerk CAN entworfen und getestet. In einer Übersicht wird die Spannweite industrieller Lösungen für PLC-ähnliche Strukturen funktionell bewertet.

Abstract. Present a consequential standardization becomes generally accepted for fieldbus networks. Each of the available network solutions is shaped by a typical communication standard, that optimally employable is for special application areas. Existing industrial solutions establish themselves with it more and more strongly and hardly still appear it new networks. By optimization of the network structures and expansion of the application width innovations are advanced. The realizations of new designs of the physical interface of field bus networks contribute to it particularly. With the aim, through new circuits the original bus node interfaces to expand, the application-width of the respective field bus is rising. For the data transfer the field bus networks use different physical media. To the expansion of the utilization some research projects concentrates of the bus integrated power supply. Important technical and economic effects through simultaneous transfer of information and energy over the same wire can be achieved.

1 Einführung

Bei Feldbusnetzwerken setzt sich derzeitlich eine konsequente Standardisierung durch. Jede der verfügbaren Netzwerklösungen wird durch einen typischen Kommunikationsstandard geprägt, der für ganz bestimmte Applikationsbereiche optimal einsetzbar ist. Damit etablieren sich vorhandene industrielle Lösungen immer stärker und es erscheinen kaum noch neue Netzwerke. Innovationen werden durch Optimierung der Netzwerkstrukturen und Erweiterung der Anwendungsbreite vorangetrieben. Dazu tragen besonders die Ausprägungen neuer Entwürfe des physikalischen Interfaces (Schicht 1 des OSI-Modells) von Feldbusnetzwerken bei.

Mit dem Ziel, die ursprünglichen Busknoten-Interfaces durch neue Schaltungen zu erweitern, steigt die Anwendungsbreite des jeweiligen Feldbusses.

Neben der eigentlichen Datenübertragung, die als Hauptfunktion über verschiedene physikalische Medien (Draht, optische Übertragung, drahtlose Übertragung über Licht oder Funk) erfolgen kann, konzentrieren sich eine Reihe von Forschungs- und Entwicklungsarbeiten auf die *busintegrierte Stromversorgung*. Mit einer gemeinsamen Übertragung von Information und Energie über die gleiche Leitung (meist Zweidrahtleitung) lassen sich bedeutende technische und wirtschaftliche Effekte erzielen.

2 Powerline-Konzeptionen

Für die Art und Weise, Versorgungsleitungen mit zur Datenübertragung zu nutzen, hat sich immer mehr der Begriff *Powerline Communication (PLC)* durchgesetzt.

In der Feldebene von Automatisierungsstrukturen geht es vorrangig um die Übertragung prozeßrelevanter Informationen (Meß-, Stell- und Regelgrößen) und erst in zweiter Linie um die Übertragung umfangreicher Informationspakete. Deshalb sollte grundsätzlich zwischen zwei globalen Powerline-Konzeptionen unterschieden werden:

a) Powerline Communication for Intranet Protocol (PLC-IP)
   - reiner, relativ zeitunkritischer Datentransfer;
   - Reaktionszeiten wenig kalkulierbar;
   - Übertragung globaler umfangreicher Informationen (besonders File-Transfers);
   - große Bedeutung für Realisierung von Internet Protocol-Transfers.

b) Powerline Communication for Fieldbus Protocol (PLC-FP)
   - zeitkritischer Datentransfer;
   - Echtzeitfähigkeit zum großen Teil gefordert;
   - Übertragung kurzer Prozeßinformationen;
   - deterministischer Informationsaustausch;
   - Realisierung optimaler Feldbusprotokolle.

Diese konzentrierte Darstellung von Besonderheiten und Differenzen beider Konzeptionen zeigen, daß man nicht von *Powerline Communication* global sprechen kann. Es gibt sehr deutliche Unterschiede, die sich in der Realisierung von Transceivern (Hardware, Schaltungen, Modulatoren/Demodulatoren) und Zugriffsorganisationen (MAC - Medium Access Control) niederschlagen.

In beiden Ansätzen steht die Forderung nach möglichst großer Datenübertragungsrate im Fordergrund. Physikalische Gesetzmäßigkeiten begrenzen die Forderung recht schnell, da man keine reine Datenleitung benutzt, sondern eine bereits stromführende Leitung mitbenutzt. Das bedeutet, man muß sich mit dem Übertragungskonzept nach der zu Verfügung stehenden Leitung (z.B. 230V/400V Endverbraucherversorgung der Energiewirtschaft) richten. Die sehr unterschiedlichen Leitungseigenschaften und fast unkalkulierbaren Lastzustände beeinflussen die Übertragungsbedingungen extrem.

Ebenfalls steht die Störfestigkeit und Störabstrahlung als Hauptproblem dieser Technologie, die sehr stark von dem verwendeten Energieträger (Leitungseigenschaften, Belastung) abhängt. Konkrete Meßergebnisse zu diesen bneiden Größen sind i.a. nur für spezifizierte Einsatzbedingungen ermittelbar. Praktisch nutzbare Trägerfrequenzen mit entsprechend niedrigen Signalpegeln sind derzeitig in Europa nur eingeschränkt verfügbar (Chenelec-Band). Neue erweiterte Frequenzbereiche (150kHz ... 20MHz) stehen zur Diskussion.

3 Powerline-Verfahren

Zur Übertragung von Daten auf Stromversorgungsleitungen muß grundsätzlich auf ein Modulationsverfahren zurückgegriffen werden. Unterschiedliche Verfahrenswege verweisen auf verschiedene Anwendungsgebiete. Powerline-Verfahren kann man nach folgender Systematik unterscheiden:

a) *Breitbandige Modulationsverfahren*
Mehrfrequenzverfahren/Vielfrequenzverfahren, Bandspreizverfahren (z.B. DSSS - Direct Sequence Spread Spectrum, FHSS - Frequency Hopping Spread Spectrum, Chirp)

b) *Schmalbandige Modulationsverfahren*
Ein- bzw. Zwei-Trägerverfahren (ASK-, PSK- bzw. FSK-Modulation), Basisbandverfahren (Direktmodulation des Bitstroms)

Einige interessante industrielle Verfahren von PLC-Systeme des Feldbereichs sind in Tabelle 1 zu einer Übersicht zusammmengestellt. Die gegenübergestellten Daten vergleichen einige wesentliche Parameter und Eigenschaften der Systeme Profibus-PA (Process Automation), Interbus-Loop, AS-Interface (Actuator Sensor-Interface), LON (Local Operating Network), EIB (European Installation Bus), M-Bus (Meter-Bus) und Hart (Highway Adressable Remote Transducer). Zum Vergleich ist in der letzten Zeile der Stand der eigenen Arbeiten an CAN-PLC-Interfaces mit aufgenommen worden. Bemerkenswert ist die die Variationsbreite bei Bitraten und Leitungslängen, die bei genauerer Betrachtung immer inverse Proportionalitäten beschreibt. Bereits bei den dargestellten Systemen sind Einschränkungen in der Echtzeitfähigkeit bzw. Signaldeterministik zu erkennen. Mit dem aufgenommene HART-System wird ein typischer Vertreter der Begründung von digitalen Feldbussystemen vorgestellt, der ursprünglich das 4-20mA Einheitssignal zur analogen Meßsignalübertragung und zur Stromversorgung des Meßumformers nutzte.

Seit einiger Zeit wurden auch zum extrem zeitkritischen Feldbussystem CAN (Controller Area Network) Untersuchungen angestellt und Lösungswege eines PLC-Konzepts (neuer Physical Layer) entworfen. Als praktisch nutzbare Ergebnisse wurde ein Single Carrier-Verfahren und ein Base Band-Verfahren entworfen und weiterentwickelt. Die Ergebnisse sind verschiedentlich veröffentlicht worden. An neuen störfesteren Verfahren (Multiple Carrier-Verfahren) wird z.Zt. gearbeitet. Das Grundprinzip dieses Verfahrens liegt in der gezielten (flexiblen) Zuordnung einiger weniger Trägerfrequenzen, die eine codierte Bitinformation übertragen.

Tabelle 1. PLC-Funktion ausgewählter serielle Bussysteme der Feldbusebene
(BPSK - Binary Phase Shift Keying; SFSK - Spread Frequency Shift Keying; PTP - Point-to-Point; * CAN-PLC Single Carrier Verfahren, **CAN-PLC Base Band-Verfahren)

Bussystem	Datenübertragungsrate	Modulationsform	Leitungsanforderungen	Norm des PLC-Layer	Hersteller, typ. Entwickler/ Provider
Profibus-PA	31,25kbps	geformte I-U-Impulsmodulation, Manch.-II-Cod.	Zweidraht, geschirmt, bis 1000m	IEC 1158-2, DIN EN 61158-2, ISA S50.2	u.a. Siemes, PNO
Interbus-Loop	500kbps	geformte U-Impulsmod., Manch.-II-Code	Zweidraht, geschirmt, 100m Ring (10m Segm.)	DIN E 19 258, EN 50 254	Phoenix Contact
AS-Interface	167kbps	$\cos^2$-Spannungsimpulse, Manch.-II-Code	Zweidraht, ungeschirmt, bis 100m	IEC 62026, EN 50295	u.a. Siemens
LON	2/4,8/78kbps	Spreizband; BPSK-Modulation	Netzkabel/Zweidraht, verschied. Mod.verf.	EN 50065-1 (IEC-Entwurf)	Echelon
EIB	1,2/9,6 kbps	SFSK-Mod.; geformte U-Impulsmodulation	Zweidraht, gesch., bis 1000m	EN 50 090	viele Hersteller; EIBA
M-Bus	0,3 ... 9,6kbps	Master-Slave: U-Impulse, Slave-Master: Stromtastung	Zweidraht (ungesch.), < 1000m; auch drahtloses Verfahren mittels ASK	EN 1434-3, IEC 870-5	Texas Instruments, Uni/GH Paderborn
HART	1,2kbps (PTP), 19,2kbps (Bus)	FSK 1,2/2,2kHz	Zweidraht, 3000m PTP (100m Bus)	nach Bell 202-Standard	Rosemount
CAN	10 ... 30kbps*, bis 250kbps**	ASK-Sinus-Trägerm.*, geformte I-U-Impulsmodulation**	Zweidraht, bis 1000m*/ 100m**	(könnte eine ISO-DIS werden)	Uni Rostock/ Uni Magdeburg

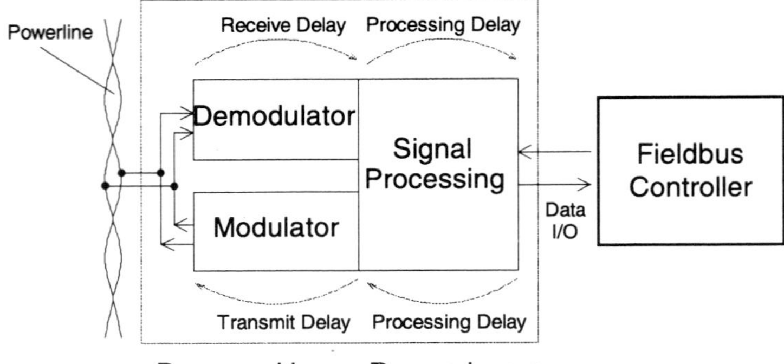

Abb. 1. Struktur eines PLC-Receivers bei Verzögerungeinflüssen

Auch bei diesen Arbeiten steht die Problematik der Verzögerungen durch die Leitung, durch die Modulation und die Signalverarbeitung als Hauptrestriktion. Dies ist besonder schwerwiegend bei Frames, die von mehreren Busknoten generiert werden. Das betrifft bei CAN besonders die bitweise Arbitrierung und das Quittierungsbit (Acknowledge). Abb. 1 zeigt das Verzögerungsprinzip mit den beteiligten Komponenten und Abb. 2 modelliert die daraus resultierende Verzögerungskette. Da die Leitungslänge bei CAN ohnehin nicht sehr lang gewählt werden kann (bitratenabhängig), ist der Leitungseinfluß gegenüber den anderen Verzögerungen relativ gering.

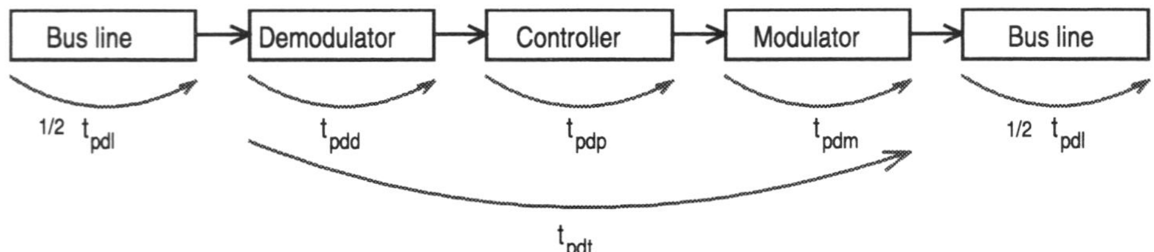

Abb. 2. Timingmodell eines PLC-Receivers bei Verzögerungeinflüssen

Verzögerungszeiten:
t_{pdl} - Delay time Bus line; t_{pdd} - Delay time Demodulator; t_{pdp} - Delay time Processing; t_{pdm} - Delay time Modulator; t_{pdt} - Delay time Transceiver

Berechnung der max. Zeitverzögerung durch das Modulationsprinzip erfolgt mit:

$$t_{pd}\,\text{max} = t_{bitseg} \leq t_{pdt} + t_{pdl} \qquad (1)$$

(t_{pdmax} - maximale Zeitverzögerung; t_{bitseg} - min. Bitzeitlänge für sichere Biterkennung)

Berücksichtigt man bei den Entwurfsarbeiten die Einflüsse der Signalverzögerungen, so stehen sich sich zwei gegensätzliche Prinzipien gegenüber.

a) *Breitbandige Modulation* ist sehr sicher und störfest (z.B. Spread Sectrum), erfordert aber einen relativ zeitaufwendigen Modulations- und Demodulationsvorgang mit Signalverarbeitungselektronik.

b) *Schmalbandige Modulation* ist weniger störfest (z.B. Single Carrier ASK), kann aber mit synthetisierter Modulatorschaltung und schneller Hardwaredemodulation mit geringeren Verzögerungen arbeiten.

Nach dem ersten Prinzip ist mit der jetzt verfügbaren Elektronik eine ausreichend schnelle Signalverarbeitung mit vertretbarem Aufwand (Interfaceschaltung!) für CAN nicht realisierbar! Entsprechend dem zweiten Prinzip wurden bereits verschiedene Anwendungen realisiert. Den Aufbau eines Prototypentestsystems mit der Versorgungsleitung eines 230V-DC Motors zeigt Abb. 3. Zwei Busteilnehmer (BN 1 und BN 2) kommunizieren über die Versorgungsleitung. Netzteil und Last (Motor) sind über Filter an die Busleitung angeschlossen, und damit für die Datensignale HF-mäßig hochohmig.

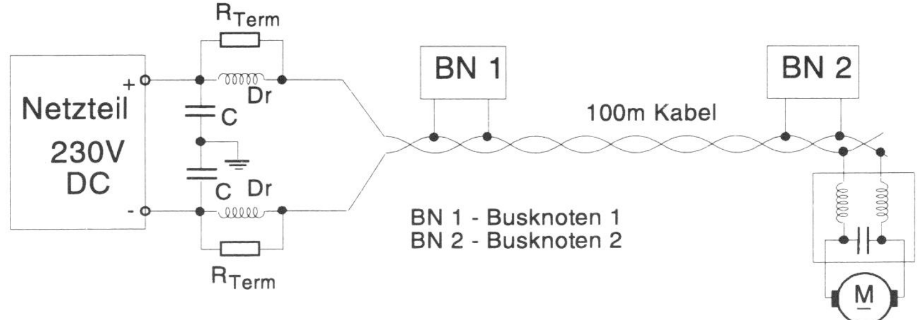

Abb. 3. Testaufbau eines Powerline-Prototypen für das Netzwerk CAN

In diesem einfachen Testsystem wurden Bitübertragungsraten von 10kbit/s realisiert. Bei Bitraten ab 30kbit/s traten mit der Prototypenelektronik Instabilitäten auf. Bei der getesteten Leitungslänge von 100m traten aufgrund des lastabhängigen Gleichspannungabfalls starke Pegelschwankungen am Leitungsende auf, die in keiner Weise die Kommunikation beeinträchtigten.

4 Ausblick

Die bekannten Lösungen und Systeme zur Realisierung von Powerline Communication in der Feldebene zeigen, daß sich diese Technik erst zu entwickeln beginnt. Unter der Nutzung von Signalverarbeitungsalgorithmen, die auf entsprechend leistungsfähigen Signalprozessoren laufen, lassen sich Breitbandverfahren (u.a. Spread Spectrum- und Frequency Hopping-Verfahren) am vorteilhaftesten anwenden (z.B. große Störfestigkeit, gute Kodierbarkeit, geringe Störstrahlung).

Da aber derzeitig Aufwand und Kosten dieser Elektronik noch unvertretbar hoch liegen, müssen noch weitere einfachere Verfahren und Entwurfswege untersucht und getestet werden. Dazu zählt auch ein Multiple Carrier-Verfahren (MCP) für das CAN-PLC-Interface. Unter der Nutzung schneller hochkomplexer programmierbarer Arrays (FPGAs), die Verarbeitungsfunktionen mit großer Parallelität zulassen (z.B. Korrelation, FFT) sind Lösungen in absehbarer Zeit realisierbar. Allen Entwurfsarbeiten zur Interface-Elektronik von Physical Layern berücksichtigen die nachfolgende Umsetzung

der Ergebnisse in einen einzigen anwendungsspezifischen Schaltkreis (ASIC) als End-
ergebnis.

Literatur

1. Schindler, Th.; Rauchhaupt, L.; Beikirch, H.: Über eine Leitung. Integrierte Übertragung von Informationen und Hilfsenergie beim Controller Area Network (CAN).
   Elektronik, Franzis Verlag, Feldkirchen, 8/1997, S. 76 - 82

2. Meusling, A.: Einsatzaspekte und Stand der Technik von Powerline Communication
   (PLC) Systemen. Congress ONLINE `99, Düsseldorf, 01. - 04. Febr. 1999,
   Congressband 1, S. C140.01 - C140.16

3. Dostert, K.: Das elektrische Energieverteilnetz als TK-Medium für die letzte Meile.
   Congress ONLINE `99, Düsseldorf, 01.-04. Febr. 1999, Congressband 1, S.
   C141.01 - C140.17

4. Kriesel, W.; Madelung, O. W.: ASI Das Aktuator-Sensor-Interface für die Automation. Hanser Verlag, München, 1994

5. Block, P.: Untersuchung der Störemission und Störfestigkeit von Feldbussystemen.
   Diplomarbeit, Hochschule Wismar, Fachbereich Elektrotechnik und Informatik,
   1997

6. Kriesel, W. u.a.: Bustechnologien für die Automation. Hüthig Verlag Heidelberg,
   1998

7. Voß, M.: Smart-Busknoten in Feldbusnetzen. Diplomarbeit, Universität Rostock,
   Fachbereich Elektrotechnik und Informationstechnik, 1997

8. Geisler, M.: Physical Layer zur busintegrierten Stromversorgung beim Controller
   Area Network (CAN). Diplomarbeit, Universität Rostock, Fachbereich Elektrotechnik und Informationstechnik, 1996

9. Reißmann, M.: Burstfrequenzmesser nach dem Korelationsprinzip. Diplomarbeit,
   Universität Rostock, Fachbereich Elektrotechnik und Informationstechnik, 1996

10. Beikirch, H.; Voß, M.: Controller Area Network (CAN). 9. Symposium Maritime
    Elektronik, Universität Rostock, 15.-17. April 1998, Tagungsband, S. 129-132

Design of Bus Media Redundancy in CAN

José Rufino[1], Paulo Veríssimo[2], and Guilherme Arroz[1]

[1] Instituto Superior Técnico - Universidade Técnica de Lisboa,
Avenida Rovisco Pais, 1049-001 Lisboa, Portugal.
NavIST Group CAN Page - http://pandora.ist.utl.pt/CAN
[2] Faculdade de Ciências da Universidade de Lisboa,
Campo Grande - Bloco C5, 1700 Lisboa, Portugal.
Navigators Home Page - http://www.navigators.di.fc.ul.pt

Abstract. Network media redundancy is a clean and effective way of achieving high levels of reliability against temporary medium faults and availability in the presence of permanent faults. This is specially true of critical control applications such as those supported by the Controller Area Network (CAN). This paper deals with the complexity of implementing bus media redundancy in CAN, discussing how this can be done through a simple scheme that uses inexpensive off-the-shelf components.

1 Introduction

Continuity of service and determinism in message transmission delays are two fundamental requirements of fault-tolerant real-time applications, that must be fulfilled through the use of some form of network-level redundancy.

Safety-critical applications would resort to full space redundancy, replicating media and network controllers to provide a broad coverage of faults and glitch-free communication [1, 3], usually at high design and implementation costs. Other approach is simple media redundancy, such as exists off-the-shelf in some standard LANs or as developed in Delta-4 [7, 13], where space redundancy is restricted to the physical - electrical signaling in the medium - level. This alternative design may lead to simpler and thus less expensive solutions, yet satisfying a wide spectrum of fault-tolerant real-time applications, with exception of those with very stringent safety and timeliness requirements [14].

Fieldbuses are in essence a technology whose area of application requires continuity of service. Fieldbuses are widely used in systems intended for real-world interfacing (i.e. integrating sensors and/or actuators) which are specially sensitive to the availability of the network infrastructure, a problem that we address in this paper, in the context of CAN, the Controller Area Network [2]. CAN is a low-cost fieldbus, with widespread acceptance in control applications, able to extremely robust operation. However, resilience to medium partitioning has to be provided as an extension to the standard specification.

Arguing with the high costs of other solutions, an existing commercial redundant CAN design uses a self-healing ring/bus architecture [4]. Though such a scheme provides resilience to open/short-circuits in the physical wiring, it does not solve the problem of CAN continuity of service efficiently: ring reconfiguration takes time (it can last as long as $100ms$) and meanwhile the network is partitioned. Bus media redundancy does not exhibit those shortcomings and do represent a natural design choice, but efficient mechanisms for its implementation were not known, until recently.

In [11], we have presented a *Columbus' egg* idea that opens room for a simple implementation of bus media redundancy in CAN. However, to be of practical use such scheme should be enhanced with mechanisms handling not only medium partitions but also stuck-at and omission failures, as discussed in [11]. The present paper addresses how those mechanisms can be efficiently implemented using off-the-shelf components, discussing how CAN properties may be exploited in the design of the different media redundancy management modules in order to keep complexity low, and explaining how these modules interface with the standard CAN components.

This work is part of a broader effort aiming at designing an embedded fault-tolerant distributed system around CAN [15, 12, 8]. An overview of the results achieved so far can be found in [10].

2   CAN Physical Layer Fault-Tolerance

Some commercial CAN *fault-tolerant transceivers* (e.g. [6]) have been designed to provide resilience to one of the following failures in the CAN communication medium: one-wire interruption; one-wire short-circuit to power or ground; two-wire short-circuit. Continuity of network operation is achieved by switching from the normal two-wire differential operation to a single-wire mode, at the expense of a reduced signal-to-noise ratio. Switching upon the detection of a failure and switching back to the differential mode when recovered, is automatic.

However, no standard mechanism supports CAN non-stop operation if both wires of the network cabling get simultaneously interrupted.

3   Bus Media Redundancy in CAN

In [11], we have thoroughly addressed CAN resilience to medium partitioning and, dismissing existing cost and complexity arguments, we have came up with an extremely simple scheme for the implementation of bus-based media redundancy in CAN. Next, we shortly discuss that *Columbus' egg* idea.

Let us assume a simplified fault model, considering only medium partitioning. Let us also assume every bit issued from a MAC[1] sub-layer is simultaneously transmitted on all redundant media interfaces. If a medium is partitioned: nodes located at the *in-partition*[2] receive a correct signal on all redundant media interfaces; nodes at the *out-partition* receive a recessive[3] signal from the (idle)

---

[1] Medium Access Control.
[2] I.e., the partition that includes the transmitter.
[3] A recessive symbol is represented by a logical one level.

failed medium. Given the *quasi-stationary* operation of CAN and its wired-AND nature, a correct *channel*[4] receive signal can be produced simply combining the redundant media receive signals in a logical AND function, as exemplified in Fig. 1 for a dual-media architecture.

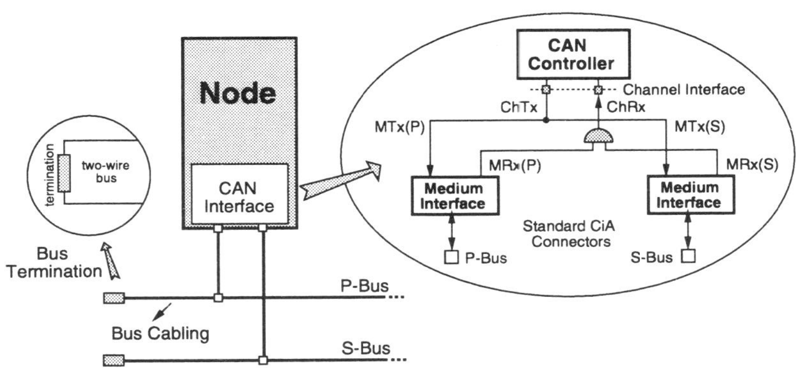

Fig. 1. A *Columbus' egg* idea for bus media redundancy in CAN

The only disadvantage of this approach is too restrictive a fault model. Only medium partitions and stuck-at-recessive faults are tolerated. However, this basic architecture can be enhanced to support a less restrictive and thus more realistic fault model [11]: stuck-at-dominant faults can be handled through a special-purpose *watchdog* timer; omission faults can be monitored and a medium exhibiting an excessive number of omission errors can be put in *quarantine* until (and if) it "behaves well" again.

4 Design of Bus Media Redundancy in CAN

This paper addresses how to support non-stop network operation, through bus-based media redundancy, in CAN. A *media selection and management unity*, to be inserted between the redundant media interfaces and the CAN controller (Fig. 2), is intended to provide resilience to crash (stuck-at/broken) and omission failures, through the mechanisms specified in [11].

Our design does not assume the use of any particular CAN controller, and multiple solutions are allowed for the physical layer: inexpensive two-wire differential cabling used together with classical (non fault-tolerant) transceivers [5] or fiber optics technology [9]. That means, apart from replication, commercial CAN components are used. All provisions for media replication management are made as extensions to the standard.

---

[4] We call *channel* to the physical path - cable medium and transceivers - used by MAC entities to communicate.

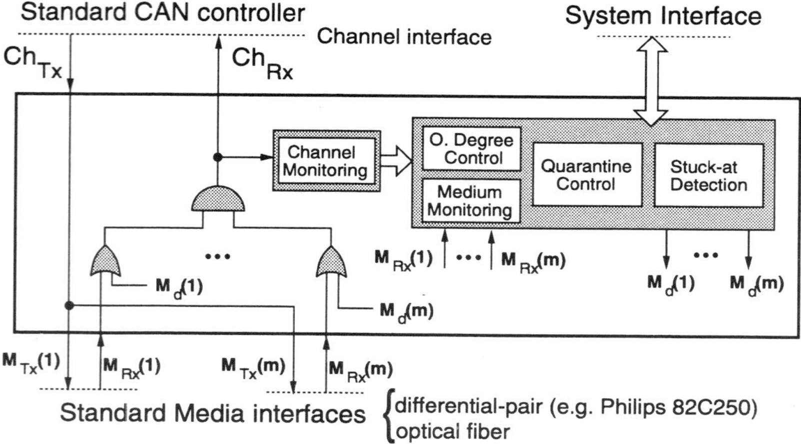

Fig. 2. CAN media selection and management unity

The architecture of the *CAN media selection and management unity* is depicted in Fig. 2. The central component of this unity is a conventional AND gate that, as defined by our *Columbus' egg* strategy, combines the values of the different media receive signals[5] to generate the channel receive signal (Ch_{Rx}), to be delivered at the CAN controller interface.

The functionality provided by this simple mechanism is the required to ensure resilience to *medium partition* and *stuck-at-recessive* medium failures, because the recessive signal of a failed medium does not disturb the result of the AND function. Nevertheless, it is relevant to detect any bus idle (recessive state) period with an abnormal duration. Suck kind of indication is of fundamental importance to the design of high-level diagnose applications aiming: to distinguish a stuck-at-recessive from a medium partition failure; to pinpoint the location of the medium partition failure in the network cable.

The *stuck-at detection* module of Fig. 2 has been designed in order to provide a low-level support for those functions, in addition to the detection of *stuck-at-dominant* medium failures. The occurrence of *stuck-at-dominant* medium failures must have a special treatment in our architecture, because an erroneous dominant value would otherwise propagate from a failed medium receiver (one of the M_{Rx} signals in Fig. 2) to the channel interface (Ch_{Rx} signal).

The contribution of a medium receive signal to the AND function is allowed to be disabled, whenever such signal assumes a dominant value that lasts beyond what it may be expected, in the worst-case, from the normal operation of the CAN protocol [2]. The disabling of a given $M_{Rx}(m)$ signal can be achieved by combining that signal with a medium disable indication[6] in a simple OR gate,

---

[5] We use $M_{Tx}(m)$ and $M_{Rx}(m)$ to denote the transmitter and receiver interfaces of medium $m \in \mathcal{M}$, the set of all media interfaces.

[6] In Fig. 2, we use $M_d(m)$ to denote the medium m disable indication.

as shown in Fig. 2. This simple method secures resilience to *stuck-at-dominant* medium failures.

The design of the *stuck-at detection* module (Fig. 2) can be optimized in several ways: the properties of the CAN protocol allow the error detection thresholds to be defined independently of network configuration parameters; stuck-at recessive and dominant failures are mutually exclusive events, and this can be exploited to reduce the complexity of the module circuitry.

The media redundancy strategies in [11] define specific mechanisms able to detect and account for frame omission errors. Short-term frame omissions usually have its origin in noise sources, that corrupt frames by electromagnetic interference. Other omission error patterns, with origin in more subtle causes such as a defective connector mount or a faulty medium interface circuit, may affect network operation during longer periods.

The *omission degree control*[7] module of Fig. 2 accounts for the frame consecutive omission errors in data/remote frame transmissions, at each medium. Network errors occurring in the interframe space are not accounted as frame omissions. The *omission degree control* module uses the indications provided by the *channel monitoring* and the *media monitoring* modules (Fig. 2), to take a decision on whether the *omission degree* of a given medium (m) should:

- be incremented by one unit, due to the detection of a frame omission with origin in that particular medium;
- maintain the same value, because the origin of the error cannot be assigned to any particular medium;
- set to zero, upon a frame has been successfully transfered in the channel, without any error being detected in medium m.

A medium exceeding a given *omission degree* bound should be considered failed. In addition, special precautions should be taken upon the occurrence of a frame omission error, whose origin cannot be localized. Usually, such kind of errors are due to single medium faults that nevertheless generate omission errors in all media. These scenarios call for fault treatment procedures where the set of "incorrect" media are temporarily disabled (*quarantined*), to allow operation to proceed with the "correct" set. A detailed discussion of CAN-oriented *quarantine* techniques, similar to those introduced in [13] for LANs, to be supported by the *quarantine control* module of Fig. 2, is out of the scope of this paper and it will be reported in a future work.

5 Conclusions

There is a demand for fault-tolerant and real-time distributed systems based on fieldbuses. Many of these systems are intended for critical control applications, where continuity of service is a strict requirement. Network media redundancy is

---

[7] Informally, the *omission degree* of a component is the number of consecutive omission errors in an interval of reference.

a clean and effective way of achieving high levels of reliability against temporary medium faults and availability in the presence of permanent faults.

In a previous work [11], we have discussed a strategy for the implementation of bus media redundancy in CAN, based on an extremely simple idea. This paper is concerned with the design and implementation of a *media selection and management unity* based on that strategy, discussing how the media redundancy mechanisms of [11] can be structured in a modular and simple architecture.

These implementation issues are extremely important with regard the integration of the *media selection and management unity* in a single, inexpensive, medium capacity Programmable Logic Device.

References

1. F. Cristian, R. Dancey, and J. Dehn. High availability in the Advanced Automation System. In *Digest of Papers, The 20th International Symposium on Fault-Tolerant Computing*, Newcastle-UK, June 1990. IEEE.
2. ISO. *International Std. 11898 - Road vehicles - Interchange of digital information - Controller Area Network (CAN) for high-speed communication*, November 1993.
3. H. Kopetz and G. Grunsteidl. TTP - a protocol for fault-tolerant real-time systems. *IEEE Computer*, 27(1):14–23, January 1994.
4. RED-CAN a fully redundant CAN-system. NOB Elektronik Product Note, Sweden. http://www.nob.se.
5. Philips Semiconductors. *PCA82C250 - CAN Controller Interface*, April 1994.
6. Philips Semiconductors. *TJA1053 - Fault-tolerant CAN transceiver*, October 1997.
7. D. Powell, editor. *Delta-4 - A Generic Architecture for Dependable Distributed Computing*. ESPRIT Research Reports. Springer Verlag, November 1991.
8. L. Rodrigues, M. Guimarães, and J. Rufino. Fault-tolerant clock syncronization in CAN. In *Proceedings of the 19th Real-Time Systems Symposium*, pages 420–429, Madrid, Spain, December 1998. IEEE.
9. M. Rucks. Optical layer for CAN. In *Proceedings of the 1st International CAN Conference*, pages 2.11–2.18, Mainz, Germany, September 1994. CiA.
10. J. Rufino, P. Veríssimo, and G. Arroz. Defining a CAN-based infrastructure for faul-tolerant real-time distributed computing. In *Proceedings of the 19th Real-Time Systems Symposium*, Work In Progress, pages 27–30, Madrid, Spain, December 1998. IEEE. http://www.cse.unl.edu/rtss98wip/proceedings.
11. J. Rufino, P. Veríssimo, and G. Arroz. A Columbus' egg idea for CAN media redundancy. In *Digest of Papers, The 29th International Symposium on Fault-Tolerant Computing Systems*, pages 286–293, Madison - USA, June 1999. IEEE.
12. J. Rufino, P. Veríssimo, G. Arroz, C. Almeida, and L. Rodrigues. Fault-tolerant broadcasts in CAN. In *Digest of Papers, The 28th Int. Symposium on Fault-Tolerant Computing Systems*, pages 150–159, Munich, Germany, June 1998. IEEE.
13. P. Veríssimo. Redundant media mechanisms for dependable communication in Token-Bus LANs. In *Proceedings of the 13th Local Computer Network Conference*, Minneapolis-USA, October 1988. IEEE.
14. P. Veríssimo. Real-time Communication. In S.J. Mullender, editor, *Distributed Systems*, ACM-Press, chapter 17. Addison-Wesley, 2nd edition, 1993.
15. P. Veríssimo, J. Rufino, and L. Ming. How hard is hard real-time communication on field-buses? In *Digest of Papers, The 27th International Symposium on Fault-Tolerant Computing Systems*, pages 112–121, Seattle - USA, June 1997. IEEE.

Programmable Process Interfaces for Smart Bus Nodes

Helmut Beikirch, Matthias Voß

Department of Electrical Engineering and Information Technology
University of Rostock, D-18051 Rostock
e-mail: beikirch@e-technik1.uni-rostock.de
phone: +49 381 498 3514, fax: +49 381 498 3608

Abstract. In Field Bus Systems intelligent functions are more and more shifted to the sensors. For the realization of these functions a powerful decentralized electronics is required directly at the sensor. At present extensive activities run to realize these intelligent functions near to the sensor. Such intelligent sensor systems in serial industrial networks are called Smart Bus Nodes. They are efficient low cost bus nodes combining an extensive software functionality with a very low power consumption. They are able to communicate on a higher OSI layer and to execute signal processing functions independently. A configuration of the process interface with the present circuit design possibilities can be done with analog multiplexers and programmable amplifiers. For the preparation of solution ways, we have in the last time the possibilities programmable integrated analog and digital Interface Circuits investigated.

1 Introduction

In distributed industrial network systems (for example Field Bus Systems) intelligent functions are more and more shifted to the sensors. For the realization of these functions (signal processing, communication capabilities) a powerful decentralized electronics is required directly at the sensor. At present extensive activities run to realize these intelligent functions near to the sensor.

Such intelligent sensor systems in serial industrial networks are called *Smart Bus Nodes* (SBN). They are efficient low cost bus nodes combining an extensive software functionality with a very low power consumption. They are able to communicate on a higher layer (OSI-Model) and to execute signal processing functions independently. The architecture of such a bus node is shown in Fig 1.

A configuration of the process interface with the present circuit design possibilities can be done with analog multiplexers (AMUX) and programmable amplifiers (PGA).

A new quality of the process interface is reached by complete flexibility of the connections and the configuration ability of the sensor interface. That means, for the connection of sensors and actuators an universal hardware is used, which adapts itself automatically to the requirements of the sensor/actuator signals (kinds of sensors, signal range, signal processing, etc). Any connected sensors will be detected by a scan. The SBN tries to adjust the required parameters for signal conditioning and data

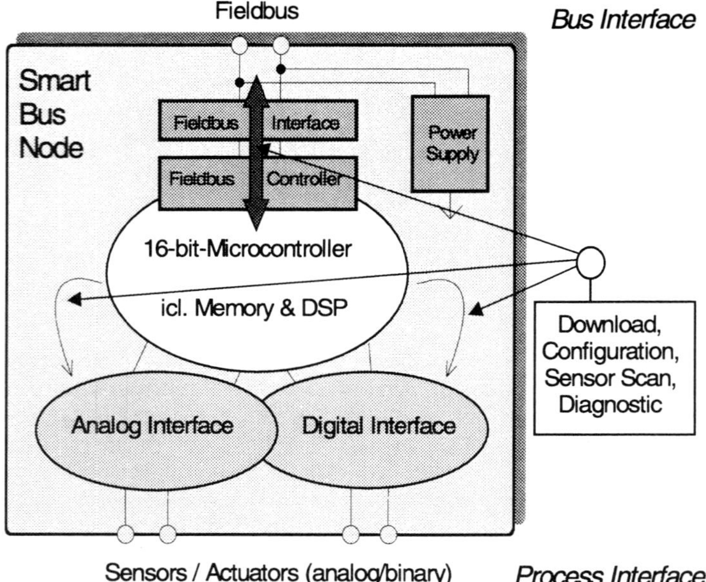

Fig. 1. Architecture of Smart Bus Nodes

acquisition optimally. The configuration of the process interface can take place through in-system-programming of the analog and digital interface components on two ways:

1. Independently decentralized by the microcontroller of the Bus Node through a global configuration plan (selection of prefabricated configuration modules) or

2. Through download of configuration files from a central service station.

The 2nd way ist comparable to the configuration principle of communication connections in advanced field bus systems.

In order ·to realize these ways in-system-programmable circuits are used (programmable in the built-in state).

2 Integrated Programmable Circuits

Digital programmable interface circuits are known for a long time. They can be realized with CPLDs (Complex Programmable Logic Devices) or FPGAs (Field Programmable Gate Arrays). These devices are already available with complexities up to 1.000.000 gates (in 1999: XILINX Virtex).

For some time there are programmable analogous circuits available. The following circuit families are known at this time:

- TRAC (Totally Reconfigurable Analog Circuit)

- EPAC (Electrically Programmable Analog Circuit)

- FPAD (Field Programmable Analog Device) and

- FPAA (Field Programmable Analog Array).

Fig. 2 shows the FPAA Gate Array principle.

The storage of the designs implemented in those analogous circuits is RAM-based like the configuration storage of the digital FPGAs. A comparison of the named analogues FPGAs is shown in Table 1. With these circuits a realization of first concepts of very flexible bus nodes is possible.

The FPAD circuits are of interest especially for signal conditioning and data acquisition from sensor signals. They contain universally usable programmable cells (amplifiers, reference sources, ADC etc.), wich are adaptable especially well to signal processing jobs. An integrated sinus-VCO enables also functions for frequency dependent signal processing.

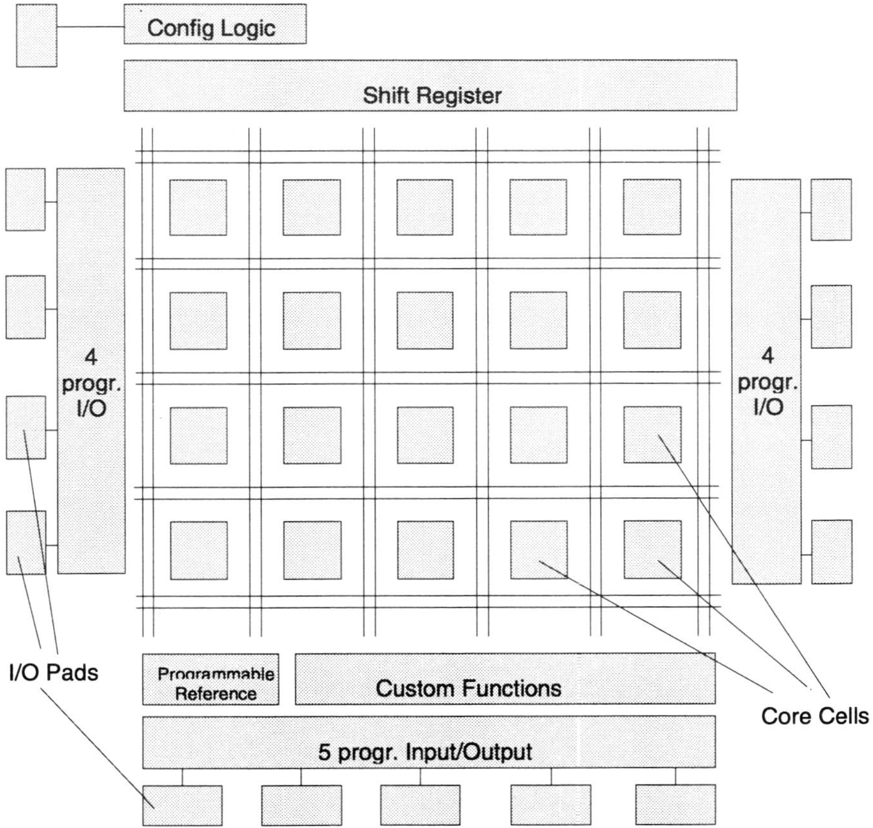

Fig. 2. Principle of Field Programmable Analog Array (FPAA)

Table 1. Comparison of programmable integrated analogues circuits (* Integrated Circuits no longer available)

	TRAC	FPAA	FPAD	EPAC
IC-Typ	TRAC020; 36SSOP/40PDIP	MPAA020-Chip; 84PLCC, 160QFP	ASB101 (ASB200) 84CQFP	50E10, 50E20 und 50E30; 28/44PLCC/QFP
Cells/Blocks	20 Analog Cells	20 Matrix Cells	3 Building Blocks + spec. Cells	18 Macro Cells (50E10)
Program-mable Circuit Functions	Addition, Logarithm, Negate, Rectifier, Auxiliary (Amplify, Differentiate, Integrate), None Inverting Pass, Off	per Cell: 1 Op-Amp, 1 Comparator, 5 programm. 8-bit-Capacitor Rows; very much Analog Switches	per Block: Basic-Cells with 6 Op-Amps. & 4 MUL; spec. Cells: Sinus-VCO, Bandgap, ChopperAmpl., Buffer, Instr.Ampl.; 16-bit-ADC (only ASB200)	program. Analog Circuits for Signal Acquisition and Processing: A-MUX, OA, DAC, TP-Filter, Comparators, T&H, Security
Switch Principle	Analog Switches; several IC cascadable	Switched-Capacitor-Principle; $\leq$ 6864 Switch.	Analog Switches; Configur. Analog Cells	Switched-Capacitor-Principle
Configuration Memory	RAM	RAM	RAM	EEPROM
Power Supply, Dynamics	Power Supply $\pm$ 2,5V Bandwith typ. 4MHz	Clock Frequ. 1 to 5MHz; OpAmp-BW $\leq$ 5MHz; V_{PS}=5V	Bandwith $\leq$ 150kHz (ASB – Analog Silicon Breadboard) V_{PS}=5V	Bandwith $\leq$ 150kHz (Turbo Mode); V_{PS}=5V
Manufacturer, Start Time	ZETEX, 1996	MOTOROLA*, 1997	Fh-IMS Dresden, 1997	IMP*, 1994

3 Example Temperature Sensor Interface

In first development steps, programmable sensor interfaces are limited to certain sensor types respectively sensor groups. That can be all kinds of sensors for temperature measurement for example. There are thermo elements, resistor thermometers and similar sensors mainly in use. As very efficient appears the sensor classification corresponding to the application profiles (f.e. motion control, environment monitoring etc.) used in field bus systems. With these profiles the applicable sensor-pool is restricted on a quite certain spectrum, so that a fully automatic sensor recognition and adaption can take place.

With the basic request, to determine a connected sensor independently from the terminal usage (randomly clamping of the sensor), to find his optimal measuring range and to monitor him during the operation (trends, diagnosis), high requirements are put to the signal processing unit to be installed decentralized. Therefore these „smart nodes" must have a fast efficient processor. For use in explosion-proof environments a power consumption as low as possible is also required.

During the scan of connected sensors recognition cycles are running, wich can be compared remotely to processes of automatic test equipment. The knowlegde base generated at the scan will be used for the optimization of the neasurement range and a subsequent sensor control.

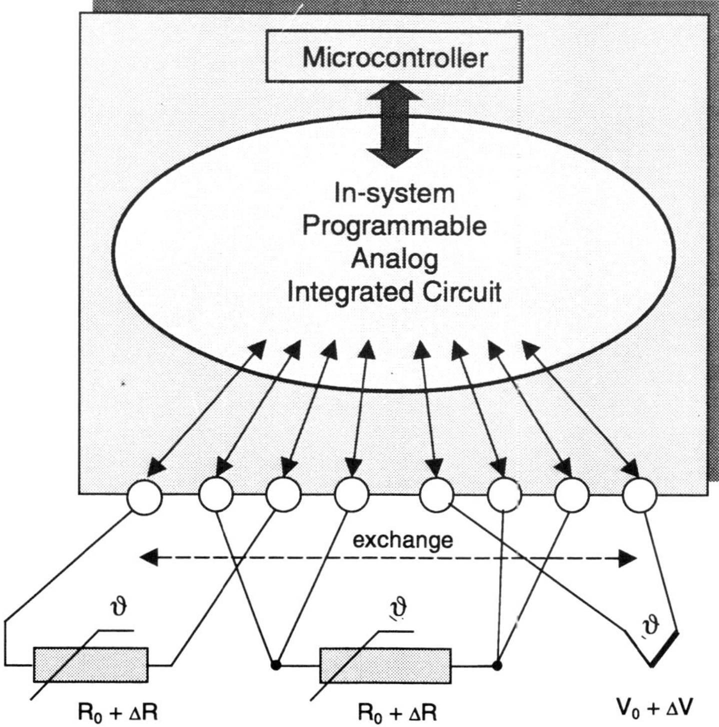

Fig. 3. Temperature sensor interface

A first breadboard circuit for the conception of a temperature sensor system as shown in Fig. 3 is under construction. The goal is the development of a bus node for the field bus system CAN (Controller Area Network). With CAN the necessary communication on higher layers (after ISO) is guaranteed, and a usage in time-critical applications is also possible.

4 Trend, Projects

The main focus of further works consists of the optimization of type, number and flexibility of the analogous functions to be implemented (amplifier, MUX, ADC/DAC, voltage/current-sources, analogues switch matrices). Target is the realization of a single-chip solution.

Another important point is the creation and continuous expansion of application profiles. As base of these profiles the existing field bus profiles can be used.

With the summary of sensor-parameters in freely available data bases and the monitoring of the connected sensors, new diagnostic and control functions can be realized in the bus nodes. The result would be an increase of security and reliability of the total system. With the use of higher-layer protocols a remote diagnosis and configuration via intranet / internet is possible too.

References

1. Voß, M.: Smart-Busknoten in Feldbusnetzen. Diplomarbeit, Universität Rostock, Fachbereich Elektrotechnik und Informationstechnik, 1997

2. Field Programmable Analog Array MPAA020. Motorola Inc., Chandler, AZ, 1997

3. Bender, S.: Rapid Prototyping – jetzt auch analog! F&M, Hanser Verlag, München, 105 (1997) 9, S.661-664

4. Analog Silicon Breadboard – ASB 101, ein frei programmierbarer Analogschaltkreis (AFPGA). Fh-Institut f. Mikroelektronische Schaltungen, Dresden, 1998

5. EPAC Device Development System D210. IMP Inc., San Jose, CA, 1995

6. Serial Programming for the IMP50E10. IMP Inc., San Jose, CA, 1995

7. Wannemacher, M.: Das FPGA-Kochbuch. Int. Thomson Publishing, Bonn, 1998

8. Beikirch, H.: Voß, M.: Smart- Architekturen im Feldbereich. Kongreß EchtZeit '97/ iNet '97, Wiesbaden, 09.-11. Sept. 1997; Conference Proceedings S. 274-280

9. Beikirch, H.: Smart Bus Nodes mit flexiblen Interfaces. Kongreß Embedded Intelligenz '98, Sindelfingen, 18.-20. Febr. 1998, Konferenzband S. 48-57, WEKA Fachzeitschriften-Verlag Poing

Eine Feldbusarchitektur mit realzeitfähiger Fehlerkorrektur

Thomas Erdner

FernUniversität Hagen
Fachbereich Elektrotechnik
58084 Hagen

Abstract. Dieser Artikel beschreibt eine Feldbusarchitektur, die speziell für realzeitfähige Anwendungen ausgelegt ist, in denen das Zeitverhalten der Übertragungsstrecken nicht nur vorherbestimmbar, sondern auch während des Prozesses stets kontrollierbar bleibt. Der prinzipielle Aufbau besteht aus einem Doppelringbus, in dem zwei Telegramme gegensinnig übertragen werden und somit Laufzeitmessungen durchführbar sind. Die Signalkodierung durch ein modifiziertes "Frequency-Shift-Keying"-Verfahren ermöglicht den Empfängerbausteinen, selbst bei starken Störungen auf dem Übertragungskanal, ein sicheres Erkennen der Eingangssignale. Durch die zusätzliche Hamming-Kodierung der Telegramme ist eine Fehlerkorrektur in jedem Empfängerbaustein ohne Übertragungswiederholung möglich.

Abstract. This article describes a fieldbus architecture particularly designed for real-time applications in which the runtime performance cannot only be determined a priori, but also controlled during the entire process duration. Basically, it consists of a dual ring bus, in which two telegrams are transmitted on both channels in opposite directions, thus making delay measurements feasible. Even in the presence of strong disturbances on the transmission channel, the signal coding by a modified "Frequency Shift Keying" method enables the receivers to securely recognize the input signals. By additional Hamming encoding of the telegrams, in each receiver an error correction is possible without requiring repetition of transfer.

1 Stand der Technik

Die Industrie fordert aufgrund von Produktivitätssteigerungen immer leistungsfähigere und damit schnellere Systeme. Aus bisherigen Anforderungen resultieren zum Teil die Nachteile der bestehenden Feldbussysteme. Da konventionelle Bussysteme für die Mehrzahl aller Anwendungen ausreichten, wurde von der Industrie das Bedürfnis an „echte" Realzeitfähigkeit nicht besonders gefördert. So vertraut man heutzutage bei Realzeitanwendungen in erster Linie auf schnelle Systeme, welche vorgegebene Zykluszeiten aufgrund ihrer Geschwindigkeit einhalten. Da diese Geschwindigkeit sehr oft durch leistungssteigernde Maßnahmen wie z.B. schnelle Prozessoren oder spezielle Speichertechniken erkauft wird, bleiben natürlich wichtige Eigenschaften wie z.B. Vorhersagbarkeit auf der Strecke.

Einer der Hauptnachteile bestehender Systeme ist die Art der Fehlerkorrektur bei Übertragungsfehlern. Hier werden bei fast allen bestehenden Systemen Übertragungsfehler zwar durch eine entsprechende Fehlererkennung (Paritätsbit, Blocksicherung oder CRC-Test) detektiert, aber die Fehlerbehandlung läuft grundsätzlich nach dem ARQ-Verfahren (error detection with automatic request repeat) [4] ab, indem Nachrichten einfach wiederholt werden. Da Übertragungsfehler in der Regel aber stochastischer Natur sind, bedeutet jeder erkannte Fehler eine Wiederholung und damit eine Verzögerung des Datenaustausches, die nicht vorhersehbar ist. Eine Übertragungszeit kann ein Hersteller aus diesem Grund nur für einen absolut fehlerfreien Datenaustausch garantieren.

Eine weitere Einschränkung im Bereich der Feldbussysteme ist die Programmlaufzeit zur Telegrammbearbeitung. In der Regel wird diese Zeit nicht oder sehr unzuverlässig [5] angegeben, zudem ist sie vom verwendeten Prozessor abhängig. Diese Zeit muß in bestehenden Feldbussystemen meßtechnisch ermittelt werden.

2 Zeitmessung im Doppelringbus

Eine Verbesserung in Hinsicht auf die Realzeitfähigkeit stellt folgender Entwurf dar. Zur Ermittlung der Datenübertragungszeiten wird ein Doppelringverfahren (s. Abb. 1) eingesetzt. Der hier verwendete Doppelring besteht aus 2 voneinander unabhängig aufgebauten Datenkanälen (Kanal 1 und Kanal 2), über die gegenläufig die gleiche Nachricht gesendet wird.

Für die Zeitmessung müssen daher diese beiden Datenkanäle vollkommen identisch sein, d.h. es muß garantiert sein, das ein bestimmtes Signal auf beiden Kanälen die gleiche Laufzeit hat. Die Einhaltung dieser Bedingung kann während des Betriebs vom Master durch Vergleichsmessungen überprüft werden. Desweiteren können durch permanente Überwachung der Laufzeit und entsprechender Korrektur temperaturabhängige Laufzeitänderungen kompensiert werden. Hierdurch kann für jedes einzelne E/A-Slave-Modul die spezifische Laufzeit vom Master-Modul bis zum Slave-E/A-Register bestimmt werden (s. Gl. 6 u. Gl. 10).

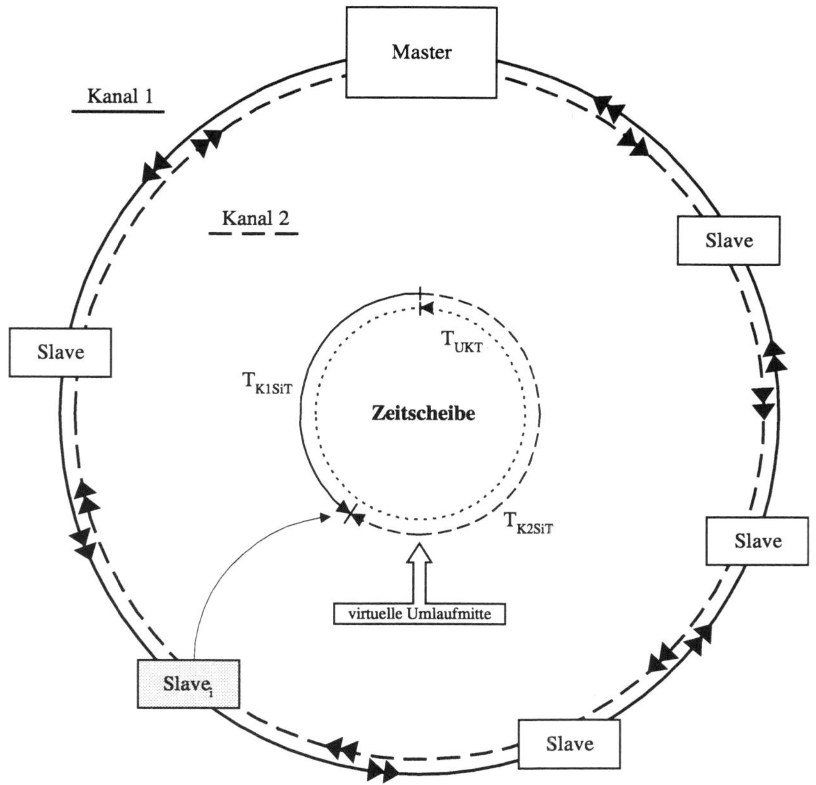

Abb. 1. Zeitmessung im Doppelringbus

2.1 Definition der Verwendeten Abkürzungen

T_{UKT} Kanalumlaufzeit eines Telegramms (identisch auf beiden Kanälen)

ΔT_{SiT} Laufzeitdifferenz des Telegramms bei Slave$_i$ zwischen Kanal 1 und Kanal 2

T_{K1SiT} Laufzeit des Telegramms vom Master zum Slave$_i$ über Kanal 1

T_{K2SiT} Laufzeit des Telegramms vom Master zum Slave$_i$ über Kanal 2

2.2 Aufstellen des Gleichungssystems aus Abb. 1:

$$T_{K1SiS} + T_{K2SiS} = T_{UKS} \qquad \text{Gl. 1}$$

$$T_{K2SiS} - T_{K1SiS} = \Delta T_{SiS} \qquad \text{Gl. 2}$$

390

aus Gl. 1 folgt:

$$T_{K1SiS} = T_{UKS} - T_{K2SiS} \qquad \text{Gl. 3}$$

aus Gl. 2 folgt:

$$T_{K2SiS} = \Delta T_{SiS} + T_{K1SiS} \qquad \text{Gl. 4}$$

Gl. 4 in Gl.3 eingesetzt:

$$T_{K1SiS} = T_{UKS} - \Delta T_{SiS} - T_{K1SiS} \qquad \text{Gl. 5}$$

$$\boxed{T_{K1SiS} = \frac{T_{UKS} - \Delta T_{SiS}}{2}} \qquad \text{Gl. 6}$$

aus Gl. 1 folgt:

$$T_{K2SiS} = T_{UKS} - T_{K1SiS} \qquad \text{Gl. 7}$$

aus Gl. 2 folgt:

$$T_{K1SiS} = T_{K2SiS} - \Delta T_{SiS} \qquad \text{Gl. 8}$$

Gl. 8 in Gl. 7 eingesetzt:

$$T_{K2SiS} = T_{UKS} - T_{K2SiS} + \Delta T_{SiS} \qquad \text{Gl. 9}$$

$$\boxed{T_{K2SiS} = \frac{T_{UKS} + \Delta T_{SiS}}{2}} \qquad \text{Gl. 10}$$

Die zeichnerisch in Abb. 1 eingesetzte virtuelle Umlaufmitte symbolisiert den Zeitpunkt, an dem die Laufzeiten der Telegramme über beide Kanäle gleich sind, bzw. ΔT_{SiT} zu Null wird. Das Telegramm, welches zuerst eintrifft, bestimmt die Lage des Slaves im System, d.h. ob der Slave vor oder hinter der virtuellen Umlaufmitte liegt.

3 Realzeitfähige Fehlerkorrektur

Die Störungen auf den Übertragungskanälen, welche normalerweise durch elektromagnetische Störfelder hervorgerufen werden, sind in der Regel Impulse, die sich dem Nutzsignal überlagern und Amplituden- bzw. Pegeländerungen hervorrufen und die übertragenen Telegramme verfälschen.

3.1 Das modifizierte Frequency-Shift-Keying-Verfahren

Aus diesem Grund wird zur Übertragung der Signale ein modifiziertes FSK-Verfahren (Frequency-Shift-Keying) verwendet [2]. Hierbei wird nicht nur die Frequenz (f_1 oder f_2) je nach Zustand gewechselt, sondern auch der Gleichspannungsanteil (s. Abb. 2b). Mit Hilfe dieser Modifikation läßt sich der jeweilige Zustand sowohl über die Frequenz als auch über den Gleichspannungsanteil auswerten. Die Auswertung der Frequenz ist daher zuverlässig bei niederfrequenten Störungen ($f_{Stör} \ll f_{Signal}$), da sich bei diesen Störungen, in Bezug auf die Dauer eines Datenbits, der Gleichspannungsanteil ändert. Andererseits ist die Auswertung des Gleichspannungsanteils bei hochfrequenten Störungen von Vorteil, da sich diese in erster Linie als Frequenzänderung bemerkbar machen können.

Zusätzlich kann man durch die fehlenden FSK-Anteile bei den Synchronsignalen (s. Abb. 2b) eindeutig zwischen Synchron- und Datensignalen unterscheiden. Eine entsprechende Auswertelogik entscheidet in den Empfängerbausteinen über den Zustand.

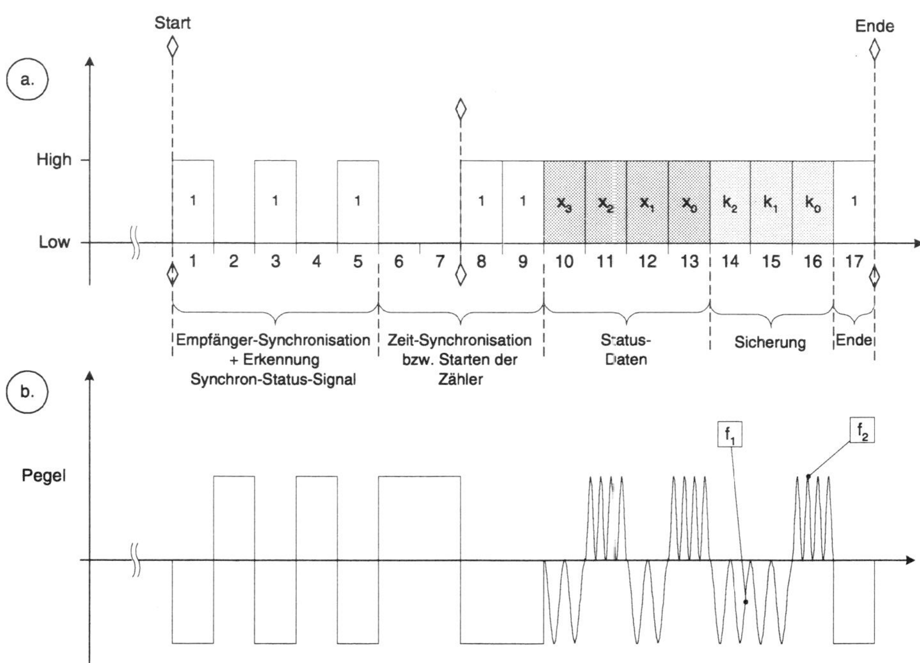

Abb. 2. Daten- und Signalcodierung

3.2 Die Hamming-Codierung

Die nächste Stufe der Datensicherung wird durch eine Hamming-Codierung [6] erreicht. Mittels der Hamming-Codierung kann man bei einem Telegramm eine vorherbestimmbare Anzahl von Datenbits sichern bzw. korrigieren. Bei diesem Entwurf wird durch 3 Sicherungsbits (s. Abb. 2a - $k_0 \dots k_2$) die Fehlerkorrektur für die 7 Daten- und Sicherungsbits ($x_0 \dots x_3$ und $k_0 \dots k_2$) auf 1 Bit und die Fehlererkennung auf 2 Bit festgesetzt. Im Gegensatz zur konventionellen Methode, also Fehlererkennung mit Wiederholung bei einem Fehler, ist die Korrektur bei der Hamming-Codierung zeitlich vorherbestimmbar. Der Nachteil liegt jedoch in der höheren Redundanz der Telegramme, wodurch natürlich die Übertragungszeit der eigentlichen Nutzdaten steigt.

3.3 Übertragung in Daten-Nibbles

Die dritte Stufe zur Fehlerkorrektur ist die Übertragung von Daten-Nibbles (4 Bit) auf beiden Datenkanälen. Durch diese Teilung der Daten läßt sich bei einer fehlerhaften Sendung das Daten-Nibble des zweiten Kanals in der Slavebaugruppe zur Korrektur heranziehen.

Zusätzlich ist es möglich, 1Bit-Fehler die sich während der Übertragung einschleusen, in jedem Empfängerbaustein auf dem Übertragungsweg zu korrigieren. Diese Möglichkeit erzielt man durch den Verbund von 4 Daten- und 3 Sicherungsbits. Ist ein solcher Datenverbund bei der Übertragung in irgendeinem Slave vollständig angelangt, d.h. die 4 Daten- und 3 Sicherungsbits befinden sich in den Empfangsregistern, so kann die Slavebaugruppe selbständig korrigieren.

Weiterhin es ist sehr einfach, sämtliche Vielfache von 4 Bit-Kombinationen zu übertragen. Hierdurch lassen sich nicht nur 8-, 16- oder 32-Bit-Kombinationen sehr einfach realisieren, sondern auch Blocksicherungen [3], bei denen die Fehlererkennung und –Korrektur noch einmal stark erhöht wird.

3.4 Telegrammform im Doppelringbus

Als Telegrammform wird für den Doppelringbus ein synchrones Summen-rahmentelegramm, wie z.B. im INTERBUS-S, eingesetzt, welches im Gegensatz zu nachrichtenbasierten Telegrammformen eine wesentlich höhere Codeeffizienz aufweist [1]. Aus diesem Grund werden die Slaves in verteilte Schieberegister aufgeteilt, welche ein gleichzeitiges Schreiben und Lesen der Slaves ermöglichen. Diese Art der Telegrammübertragung ermöglicht daher einen realzeitfähigen Voll-Duplex-Betrieb des Feldbussystems.

Die bei Linienbussystemen, wie z.B. Ethernet, auftretenden maximalen Leitungslängen spielen bei diesem Entwurf eine untergeordnete Rolle, da alle Slavebaugruppen automatisch als Repeater mit Codekorrektur fungieren. Die Buslänge spielt also nur zwischen den Slavebaugruppen eine Rolle, nicht aber für das gesamte Bussystem. Die Ausweitung des gesamten Bussystems, bzw. die Anzahl der Slaves, ist somit nur von der Schieberegistergröße der Masterbaugruppe abhängig.

Literatur

1. Baginski, Alfredo; Müller, Martin: Interbus-S, Grundlagen und Praxis, Hüthig, 1994

2. Riegelmayer, Wolfgang P.: Verkabelungskonzepte, Grundlagen und Praxis, Vogel Buchverlag, 1. Aufl., 1995

3. Schnell, Gerhard, (Hrsg.): Sensoren in der Automatisierungstechnik, Vieweg Verlag, 1993

4. Schnell, Gerhard, (Hrsg.): Bussysteme in der Automatisierungstechnik, Vieweg Verlag, 1994

5. Schwager, Jürgen: Auswahlkriterien für Feldbusse, Ludwig Drebinger GmbH, iNet '92 - Conference Proceedings, 1992

6. Swoboda, Joachim: Codierung zur Fehlerkorrektur und Fehlererkennung, Oldenburg Verlag, 1973

Safety-Oriented INTERBUS

- INTERBUS Safety -

Karsten Meyer-Gräfe

Phoenix Contact GmbH & Co., Postfach 1341, 32819 Blomberg, Germany

Abstract. The use of bus systems in automation technology has led to considerable savings, of which many companies are beginning to take advantage. This is also the reason for the increasing acceptance of fieldbus systems e.g., by automotive manufacturers, who now specify fieldbus networking as a prerequisite when planning a new system. However, one area has remained largely unaffected by these rationalization options. Safety technology is only gradually being implemented using transmission systems; however, the demand for this can no longer be ignored. It will soon be possible to implement safety-oriented applications using INTERBUS. INTERBUS is aiming to reach Category 4 (Cat. 4) according to EN 954-1.

1 Risk Analysis

To design machines and systems safely, the risks they generate must first be evaluated. Various national and international standards exist to help the user with classification. At present, the designation of risks is not consistent. DIN V 19250, for example, refers to safety classes, while EN 954-1 has safety categories. IEC 61508 uses the designation "Safety Integrity Level" (SIL). In this paper, the EN 954-1 designations will be used throughout. Fig. 1 provides a full overview of the designations used in the most important standards.

The procedure for risk assessment is very similar in all the standards mentioned. Having determined the risks for a machine, the user must take steps to reduce them. These steps can relate to its construction or its electronics. A transmission system, for example, must use particular measures, dependent on the relevant safety category, in order to detect and deal with errors which occur during data transmission. The categories which should be used for INTERBUS are shown in the following examples.

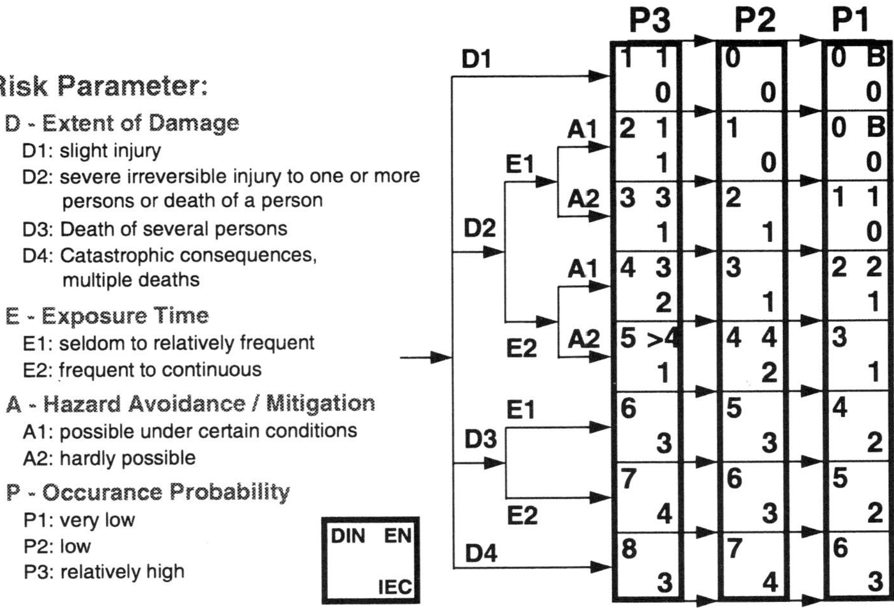

Fig. 1. Risk determination table

2 Errors and Measures on Bus Systems

All safety-oriented applications should be categorized using a risk graph. The German national standard DIN V VDE 0801 referred to above deals with computers with safety functions and computer systems with safety functions, but not with bus systems. In the area of bus systems, professional associations are working together with manufacturers to develop test principles for bus systems. In addition to the measures described in the above standards for the implementation of "computers with safety functions", which is what safe bus devices actually are, the following errors, which may occur in transmission devices, must be taken into account:

- Loss of message(s)

- Delay of message(s)

- Incorrect sequence of message(s)

- Addition of incorrect messages to the data flow

- Mix-up of safety and non-safety related messages

- Repetition of old, out-of-date messages at the wrong time

- Corruption of messages

The above transmission errors can be effectively detected and dealt with using the following measures, depending on the bus system:

396

- Consecutive number
- Time tag
- Time expectation (timeout, etc.)
- Echo (repeated transmission)
- Sender/receiver detection
- Data save (e.g., various CRC polynominals)
- Redundancy (homogenous, diverse, with comparison, etc.)

The measures for detecting each type of error are shown in the table below:

Error \ Meas.	Consec. number	Time tag	Time expect.	Echo	S/R detection	Data save	Redun. & comparison.	Data save (S ≠ NS)
Repetition	●	●					●	
Loss	●		cyclic only	●			●	
Insertion	●			●	●		●	
Incorrect seq.	●	●					●	
Corruption (of user data)				●		●	serial bus only	
Delay		●	●					
Coupling of S + NS messages				●	●			●

3 Solutions for Fieldbus Systems

A bus system can be set up in several ways in order to be able to transmit safety-oriented data. Each of the following subsections will show what options are available in principle, and which is the best for INTERBUS.

3.1 Redundancy

The simplest and most obvious way to transmit and process a message safely is redundancy. The hardware and software structure is two-channel in most cases, but can be more (≥ 3). However, the traditional, multi-channel based redundancy solution is unsuitable in many cases, such as mechanical engineering applications or small to medium-sized plants, because conventional solutions with all the necessary components are much more cost-effective. Some manufacturers use multi-channel structure systems for safety technology, incorporating all the connected components, beginning with the control systems, through the bus master and bus cables to the I/O components.

3.2 Using Safety Control Systems and Safety Protocols

The second solution is to use one or more safety control systems with I/O devices with safety functions, which communicate using a special safety protocol, which often further increases the protocol overhead because of the large amount of additional safety data, slowing down transmission and increasing reaction times as a result. In many cases, a lot of additional bytes have to be transmitted for a single safety-oriented byte. This is one of the main reasons why many safety bus systems can only transmit safety signals and not standard signals.

Suitable measures should be selected for the appropriate solution using the table above, and the manufacturer should ensure that there is an entry in each cell.

4 The New SafeGuard Concept

The SafeGuard concept is a completely different approach. It is based on the integration of an additional monitoring unit in the bus system which listens in on bus traffic and monitors the correct operation of the control system(s).

The SafeGuard solution is primarily based on the assumption that standard bus systems generally operate correctly. Safety information is transmitted over INTERBUS as a bit stream and processed by the safety-oriented devices. As the SafeGuard is in an exposed position on the INTERBUS system (directly behind the bus master), all data for the devices is routed over it. Conversely, it can read the input information of all the connected devices on the return data line which connects the last device and the master. In this way, the SafeGuard has the same process I/O image as the standard control system in which the INTERBUS master is located.

The standard control system performs both process and safety-oriented functions. The safety-oriented output components are switched off once they have met the safety condition required of them. The safety functions can be programmed by the user or by preset function block, which the user then only has to parameterize.

The SafeGuard is also capable of safety logic operations. The monitoring unit creates its safety-oriented process output image from the available safety-oriented input data. The Safeguard compares the resulting output data with the output data which the PLC or bus master is intending to sent to the safety-oriented output components and, in the event of an error, switches off the power supply, on the standard control system side, to the individual segments or the entire system. It should be noted at this point that the standard control system technology now operates so reliably, that intervention by the SafeGuard is required only seldom, and switching off the power supply is reserved for cases of absolute error.

The INTERBUS protocol is not changed as part of the SafeGuard solution, as redundancy is only applied by the monitoring unit to the safety data which passes it. It is therefore possible to adapt existing systems as well as to create new ones.

Fig. 2 shows how SafeGuard is integrated into a fieldbus system using the example of INTERBUS.

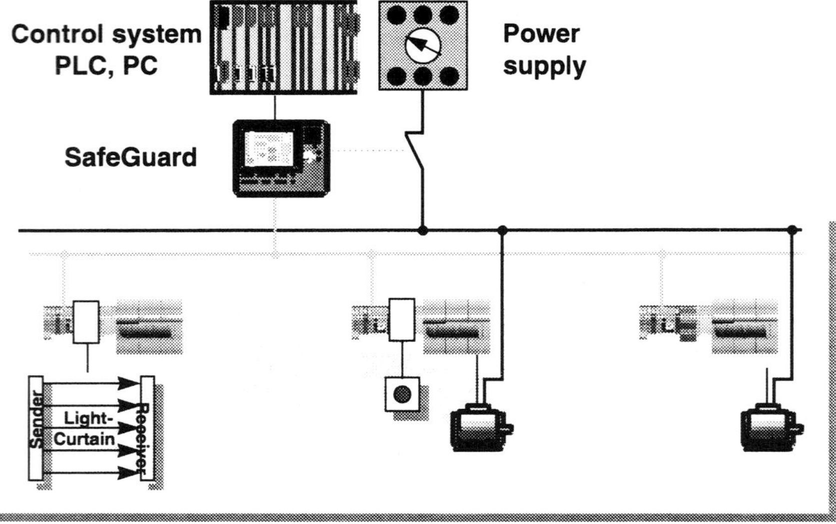

Fig. 2. The SafeGuard concept

5 Summary and Outlook

The advantage of implementing safety technology using INTERBUS Safety is that the familiar system features remain available:

- Many devices

- Long cable lengths

- Numerous transmission media are available (optical fibers, infrared, radio, etc.)

- Automatic LifeGuarding

- Synchronous data input and output

- Wide range of diagnostic options

- High degree of distribution

- Standard tools for diagnostics, monitoring and configuration

5.1 Advantages of the SafeGuard

In addition to the advantages offered by the standard technology, the SafeGuard concept can also offer the following benefits:

- Host or control system independence

- Separation of process and safety data

- Process control without technology barriers
- Fast reaction times
- Central and distributed structure possible
- One solution for all fieldbus systems
- Extremely easy test and startup processes
- Easy and short validation process
- Retrofits possible
- Global and fast error diagnostics

5.2 Further Developments in the Future

The strict separation of standard and safety technology means that users are able to select the required control system without restrictions. It is also possible to use modern PC architectures for safety-oriented tasks. Safety is ensured even if the PC or its operating system were to fail.

Ethernet (TCP/IP) and Internet technologies can be used for worldwide access. Users, no matter where they are, can scan their process data or download new programs.

The example of INTERBUS Safety points the way to pioneering developments in the future.

MultiPort RAM: a PC-Based Implementation of the Fieldbus Concept

M. Lobachov[1], T. Sauter[1], B. Plagemann[2]

[1]Institute of Computer Technology, TU Vienna
Gusshausstrasse 27-29/384, A-1040 Vienna

[2]BECK IPC GmbH,
Garbenheimerstrasse 30-38, D-35578 Wetzlar

Abstract. The rapid introduction of industrial personal computers into the manufacturing automation throughout the last years gave an increasing importance to the PC-based distributed automation systems. One such system is MultiPort RAM, a software package which implements the conventional fieldbus concepts while running on a standard personal computer architecture. The paper reviews a general structure of this system, it's functionality, operation principles, and application development aspects. The networking implementation is investigated in detail. Finally, a methodology for the MultiPort RAM integration with other field and wide area networks is discussed.

1 Introduction

The necessity to control more and more complicated manufacturing processes requires the usage of sophisticated automation systems. Traditional centralized control solutions cannot be applied any more. They are replaced by decentralized control systems which involve a number of independent nodes, running the components of a distributed application and communicating between each other. This idea, networking of the small autonomous intelligent stations, is implemented in the classical fieldbus systems.

MultiPort RAM (MpRAM) is a PC-based software implementation of the traditional fieldbus system concepts. Nowadays automation systems show the growing tendency to use personal computer based equipment instead of the traditional programmable logic controllers (PLC). The main reasons for this are lower price, higher node performance, greater flexibility and easy maintenance. MpRAM takes advantage of this approach and offers a comprehensive set of tools for creating real-time distributed control applications based on the open PC hardware platform.

2 Functionality overview

The functionality of MpRAM software meets the distributed control application development requirements. In order to split the system computational resources into several independent nodes, the application components running on these nodes should have access to the following facilities:

- interface to hardware input/output modules installed at the nodes (ADCs, DACs, FAN adapters, etc.) for controlling miscellaneous peripheral devices;

- interface to the network environment for information interchange between the components of a distributed application.

2.1 Interface to a node hardware

The interface to hardware I/Os allows to represent the node's I/O channels as ordinary software variables within the user application. During runtime the application program operates with its internal variables, whereby the corresponding I/O channels are read or written. Due to this mechanism the node applications can be developed in a completely hardware-independent way.

The actual hardware control is performed by a set of specially developed drivers, which interact with the rest of the system using a standardized open API provided by the upper software layers. This uniform interface allows to use different hardware, including newly developed boards, with the same node application program and eliminates the need for modification of the system core modules according to the hardware changes.

2.2 Communication between nodes

Physically the information interchange between the nodes is based on a *distributed shared memory* (DSM) model. This approach, widely used in parallel computing, implies the allocation of a special memory area at each node for storing the shared data. While this memory area appears as a conventional shared memory to the application programs, the system operation essentially relies on message passing between independent nodes, that access the global virtual memory pool.

On the logical (application) level, data exchange utilizes the concept of network variables, similar to many other fieldbus systems (LonWorks, CAN, EIB, etc.). For each node a subset of variables is defined (network variables), which becomes globally accessible to all the other nodes in the network both for reading and writing. All network variables in the system are placed in the shared memory, forming a distributed application *process image*. The system software running at each host automatically updates local copies of this image. Updates are performed in the background and do not depend on the user application (the relaxing consistency DSM principle [5]).

Variables that represent hardware I/Os can be also declared as network variables and allocated inside the process image. This provides the possibility to access and control the remote hardware exactly the same way as local I/O channels.

3 System architecture

The conceptual software framework of an MpRAM node is shown in Fig. 1. The system software is implemented as a set of memory-resident programs (TSR) running under the ROM-DOS operating system.

The main functionality is carried out by two kernel modules. The *data storage and hardware control module* is responsible for storing and handling the local copy of a process image. It also implements mapping of hardware I/O channels into the shared memory and provides an API to the hardware drivers. The *networking core module* implements the MpRAM real-time data communication protocol which is discussed later in this article.

Network support is built on top of the standard IPX or TCP/IP protocol stack and a NIC driver. Hardware I/O modules are controlled by a set of *hardware drivers* loaded as separate TSRs.

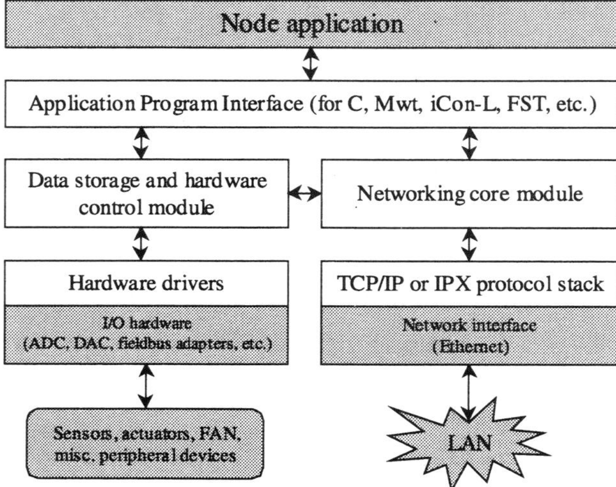

Fig. 1. MpRAM software architecture

The *application program interface* layer provides an interface between the node application program and a system kernel. Which API module is loaded, depends on the programming language used for developing the node application. Numerous different API modules are available [3].

In general, the MpRAM design pursues a structured approach, where each of the independent subsystems performs a certain dedicated function. Communication between different modules is based on the established well-defined APIs. This provides system extensibility and allows for the easy integration of new features in the future.

4 Networking

Since the nodes use the traditional personal computer platform which, basically, is an Industrial PC (IPC), network support is solely based on standard LAN/WAN protocols, such as IPX or TCP/IP, and makes use of a standard PC LAN – Ethernet. All Ethernet extensions (Fiber Optic Cable, Radio Ethernet, etc.) can be used as well.

4.1 Networking architecture

In terms of ISO/OSI reference model, the MpRAM network core module implements the 5th and 6th layers; layers 2-4 are represented by the IPX or TCP/IP protocol stack, and layer 1 functions are performed by Ethernet [4]. MpRAM is designed in such a way, that virtually any physical communication media can be used at layer 1 (for example, the support of RS232C interface as a physical data transmission environment can be implemented in the future).

This approach has a number of obvious advantages, but also introduces a nasty problem: most of the physical layer protocols used in PC LANs do not offer a strict time-based bus sharing algorithm. One of the typical examples is Ethernet with a CSMA/CD bus access method (also known as "LBT" – "Listen Before Talk") [7]. Message delivery time in this case is not predictable, which is incompatible with the basic real-time requirements – a prerequisite for the majority of manufacturing control applications. Thus, in order to eliminate bus access conflicts and achieve a deterministic behavior of the system, a separate software bus-sharing algorithm based on the virtual token passing principle was implemented.

During the system configuration phase each node in the MpRAM network is assigned a unique numeric ID, the node address. This ensures that within a token passing cycle all participants on the bus transmit their data in a strictly specified order. At a given moment of time one of the nodes, considered to be the master node, broadcasts it's current process image contents over the network. All other nodes, considered to be slaves, listen to the bus and alter their local process images according to the broadcasted data. Subsequently, the node with a nearest higher address becomes master and gets the permission to occupy the bus. Interpacket gaps, time windows between the data transmissions, are introduced in order to allow newly connected nodes to announce themselves on the network. The obvious advantage of this decentralized access method is that the system is not affected by failures of individual stations.

In order to meet real-time requirements, telegram length limitations have to be implied as well. Since complex automation systems can operate with an extremely large process image, the whole shared memory content can not often be transferred in one data packet (the data field within one packet is limited to a maximum of 200 bytes). These considerations where the reason for applying a *time slice model* (Fig. 2).

The whole process image is split into three *time windows*. Each of the windows is divided into a certain amount of *time slices* (window 0 always consists of exactly one time slice, windows 1 and 2 can contain multiple slices). The data transmitted in a telegram comprises one sequentially picked time slice from each window. Since window 0 contains only one time slice, the variables allocated to this window are transferred in each packet and have the maximum priority. Configuring the number and length of the corresponding time slices, the user can alter priorities of the variables allocated to the other time windows [1].

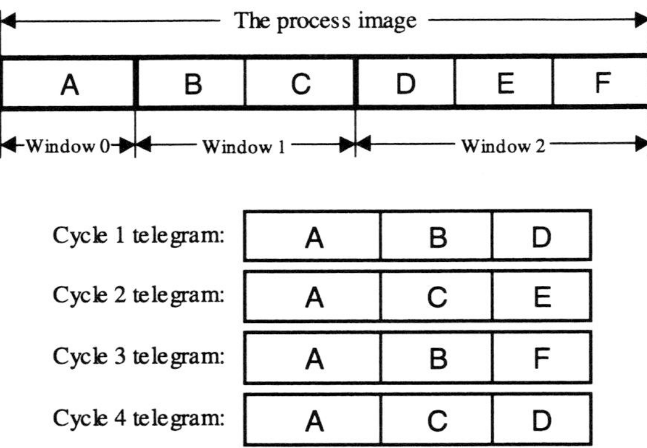

Fig. 2. MpRAM time slice model

The timing characteristics of the described algorithm are fixed and depend only on the number of nodes. Thereby the data transmission within a strictly limited period of time is guaranteed.

Apart from data transmission MpRAM protocol also includes the real-time aware error handling procedures, capable of recovering from live network reconfigurations (e.g. node connections and disconnections). Additionally the SPC security mechanism provides user applications with the information on the current network status, thus allowing the user-developed components to react on the particular node failures.

4.2 Interconnection approach

The protocol ensures predictable and limited response times, but only if the network segment is isolated. It means, that there must be no other bus users except those that support the token-passing protocol. Therefore, a direct connection of the MpRAM network to other networks (WANs, LANs) by means of the ordinary IP or IPX routers is impossible. In the simplest case, the additional PC-based nodes (e.g. computers, running monitoring software) can be easily integrated into the MpRAM network by installing the MpRAM driver set.

The most general approach for establishing a connection between MpRAM system and other networks is to use application gateways. Gateways can be applied for data exchange between MpRAM and TCP/IP network (WAN) or between MpRAM and field area network (FAN).

The principal structure of MpRAM ⇔ TCP/IP application gateway is shown in Fig. 3. On one side the gateway operates as an ordinary MpRAM node which supports the token-passing protocol and doesn't break the real-time behavior of the system. The gateway is also able to keep track of all the information in the MpRAM network (it has access to the process image). On the other side (TCP/IP network), the gateway operates like all other wide area network hosts. It accepts TCP connections from the client applications and cyclically exchanges data with them, using a request-response protocol. It is possible to serve a number of client applications with a single gateway. Each of these applications, depending on the access authorization rules, can gain both read and write access to the MpRAM data [2]. However, the information exchange through the wide area network does not fulfill the real-time requirements and is supposed to be used mostly for monitoring purposes. Security considerations also have to be taken in to account. Using the additional data encryption software is recommended.

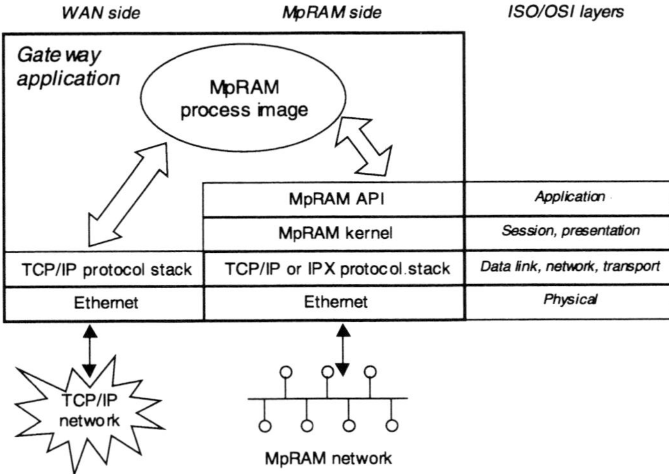

Fig. 3. MpRAM ⇔ TCP/IP gateway architecture

The architecture of an MpRAM ⇔ fieldbus application gateway is similar. On the field area network (FAN) side, a corresponding FAN adapter is installed, which is controlled by means of appropriate hardware drivers. The application, which implements the gateway itself, is usually task-specific and must be developed for each particular system. As shown in Fig. 4, a wide range of supported fieldbus adapters and convenient methods of WAN connection allow using MpRAM as a top-level solution, integrating different networks into a single control system.

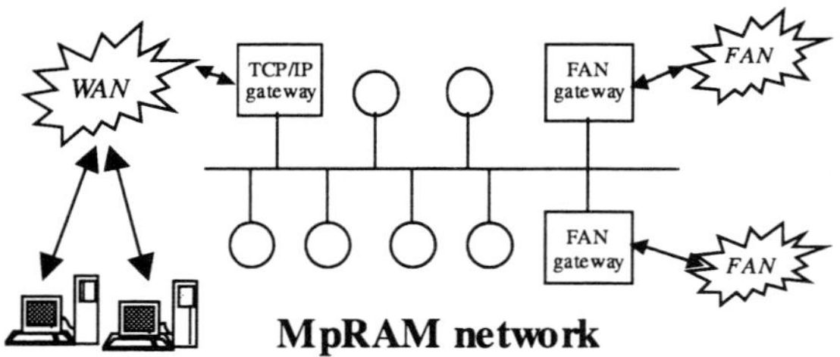

Fig. 4. Example of a heterogeneous environment

5 Programming approach

One of the most important MpRAM aspects is a wide variety of supported programming languages and development environments. Components of a distributed application running on the nodes can be developed using standard high-level programming languages (such as C, C++, Pascal), classical PLC programming languages (AWL, KOP) or IEC 1131-3-compliant PLC programming languages (SFC, ST, IL, LD, FBD) within the Multiprog/wt development environment. The latter ensures compatibility with the widespread IEC 1131-compliant automation technologies [8]. In order to run the PLC-based applications on the IPC a special PLC emulation environment is used. For instance, applications developed in Multiprog/wt are executed under the Programmable Controller Operating System (ProConOs) environment, which communicates with the MpRAM kernel and provides a transparent process image access for the PLC software. In heterogeneous networks different nodes can run applications written in different languages.

The top-level development tool, MpRAM Configurator, is used for describing the general system architecture and creating particular configurations for each node in it. The whole distributed application project is represented as a tree of boards (hosts) with the hardware modules attached to them. The hardware module configuration contains the description of the I/O channels provided by the unit. The board configuration includes a list of node network variables with data type specification, the desired programming language and other relevant information. A number of parameters, such as the time slice layout, are configured globally for the whole system. As a result of the project compilation the node configuration files are created. They are installed to the nodes along with the user-written applications [1].

The ability to integrate a user API into different development environments on the one side allows the developer not to be "tied" to a single programming language, and on the other side simplifies porting existing applications to the MpRAM platform.

6 Summary

MultiPort RAM (MpRAM) is a software implementation of the fieldbus system architecture. It realizes the concept of distributed application and provides easy-to-use and robust methods for hardware control and data exchange between nodes in an automated control system, built on top of an open-platform and cost-effective hardware. The underlying networking environment is very widespread, well supported and based on the open-standard protocols.

Further developments of the concept focus both on software and hardware aspects. The system software will become cross-platform (for instance, porting MpRAM to a 32-bit real-time operating system, vxWorks, is currently in progress). The IPC hardware will be also improved by integrating all necessary functionality on one single IC together with a hardcoded TCP/IP stack and MpRAM kernel modules.

References

1. MpRAM Konfigurator Handbuch. BECK IPC GmbH, 1998.

2. MpRAM Gateway, the system manual. BECK IPC GmbH, 1998.

3. Aufbau und Anwendung des MpRAM Programmerinterface. BECK IPC GmbH, 1998.

4. International Standards Organization (ISO): Basic Reference Model for Open Systems Interconnection, ISO 7498, 1983.

5. Protic, J., Tomaevic, M., Milutinovic, V.: Distributed Shared Memory. Concepts and Systems. Los Alamitos, CA: IEEE Computer Society, 1998.

6. Solomon, M., Kramer, J.: Distributed systems and computer networks. Prentice-Hall Inc., 1987.

7. Institute of Electrical and Electronics Engineers, Inc. (IEEE): Carrier sense multiple access with collision detection (CSMA/CD) access method and physical layer specifications, IEEE 802.3, 1998.

8. International Electrotechnical Commission (IEC): Programmable Controllers Programming Languages, IEC 1131-3, 1993.

Fieldbus Physical Layer in the Application

Udo Fetzer

R. Hirschmann GmbH &Co., Produktmanagement, D-72606 Nürtingen.
e-mail: ufetzer@nt.hirschmann.de

Abstract. A major characteristic of automation technology is a strong growth in the market for network components which are responsible for the fault free transmission of data. The range of network components for fieldbus applications pass from the simple copper connector over fiber optic fieldbus repeaters up to hubs for ETHERNET. This article will show the advantages of new technologies for communication and connection. The two major technologies which are discussed are fiber optic fieldbus networks and high speed connector systems. High speed means in this case very short installation times. Also developements for the future like ETHERNET for industrial networking are shown.

1 Fiber optic technology for failsafe data transportation

More and more machines and plants are connected via the network. Different types of networks are used in industrial applications. The simple RS 232 point to point communication, redundant PROFIBUS networks or ETHERNET; all these data transmission systems are using fiber optic communication if there is a strong demand in interference free communication or long-distance communication. Also the growing demand for bandwith enlarges the applications for fiber optic communication.

Nowadays a rapid progress in new developments for the connection technology is taking place. The fast installation technology for copper systems influences also the demand for fiber optic fast installation. Today fiber optic connectors are considered to be expensive and time wasting. High speed connections are now also available for fiber optic systems by using new assembling technologies.

The fastest fiber optic connectors are those for plastic fibers. This step index fibers have a core of 980 µm diameter, a cladding of 1000 µm diameter and an outer diameter including the insulation of 2.2 mm. This dimensions enable connectors less precise than the connectors for the standard glass fibers with a core between 9 and 62.5 µm diameter. Figure one shows an example for a connector system for plastic fibers. This connector system makes possible a very fast assembling because the outer insulation of the standard plastic fiber cable does not need to be stripped. Also no special tools are necessary for mounting. The fiber is fixed without glue by a defined pressure on the outer insulation. If a high optical quality is needed for distances of more than 20 m the fiber surface should be polished with standard polish paper.

Because of saving time during installation the plastic fiber connector systems will be used more and more for automation projects. For the future it is expected that fiber optic systems will replace a significant part of the classical copper cabling systems because of reduced engineering costs by using fiber optics and reduced cable and installation costs.

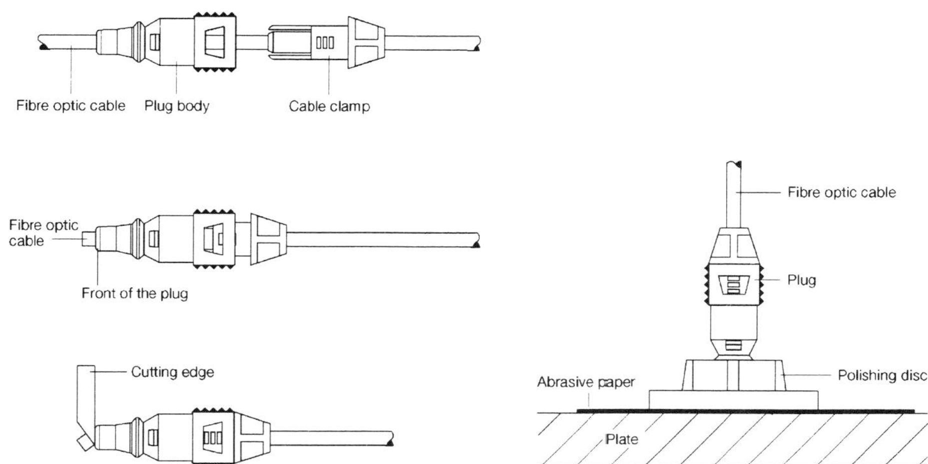

Fig. 1. Example of a fast assembled connector for plastic fibers

Fieldbusses which are using fiber optic technology – for example PROFIBUS – are used not only for factory automation but also in applications with a very high demand for safe processes. An example for extreme safety requirements are nuclear power plants. But also other applications like traffic control systems or the automation of airports require a high reliability of data communication. In this case fiber optic redundancy concept fulfills the demand for safety regarding break down of a fiber optic cable. Figure 2 shows the fiber optic ring redundancy for PROFIBUS. If any of the fibers between two repeater modules breaks down automatiaclly the ring changes to a normal bus or line topology. So the connected devices can continue their communication without interruption.

2 Fast copper installation technology in IP 67

New contact and installation technologies are now also available for the classical copper connections. With those technologies time and therefore money is saved during installation. Connectors which are mounted without soldering or screwing to the unshielded sensor cable are an example. This connectors are using the IDC (insulation displacement connection). With IDC it is even not necessary to strip the insulation of the wires. Now there is also a fast installation technology for the shielded fieldbus cables available. The task is to have the IP 67 protection class for a shielded connector system for two copper wires. The first applications where this technology is used are the IEC 61158-2 fieldbuses (PROFIBUS PA, Foundation Fieldbus). Figure 3 shows this technology. The central component is the T-housing with three connection ports for direct bus connection or M12 connectors. The contact element guarantees the connection between the three ports by IDC. If required also the bus terminators can be integrated in this contact element. By this it is possible to connect very fast and comfortably the devices to the fieldbus.

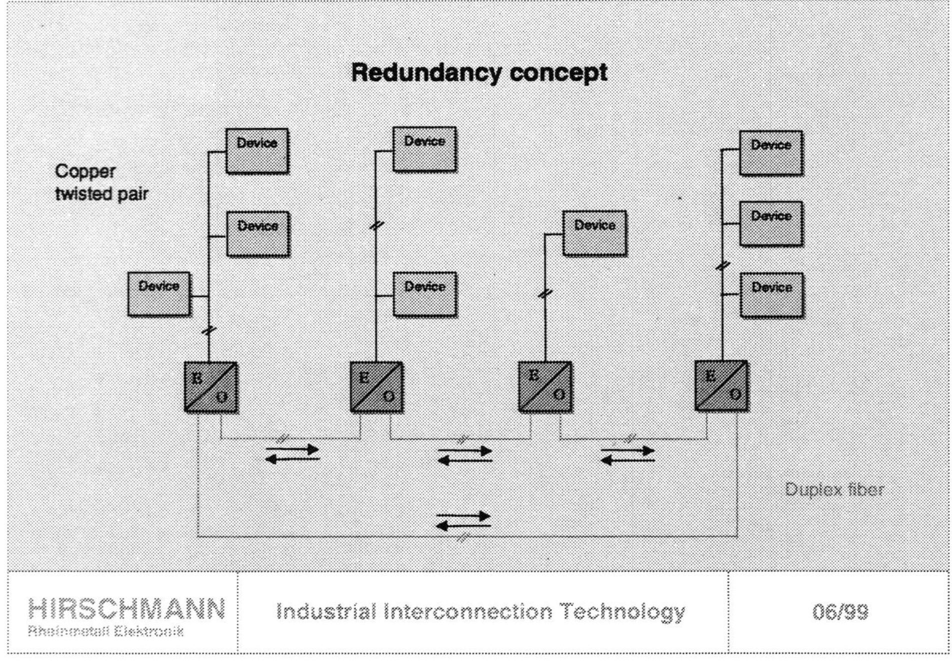

Fig. 2. Redundancy in a fieldbus network with a fiber optic ring

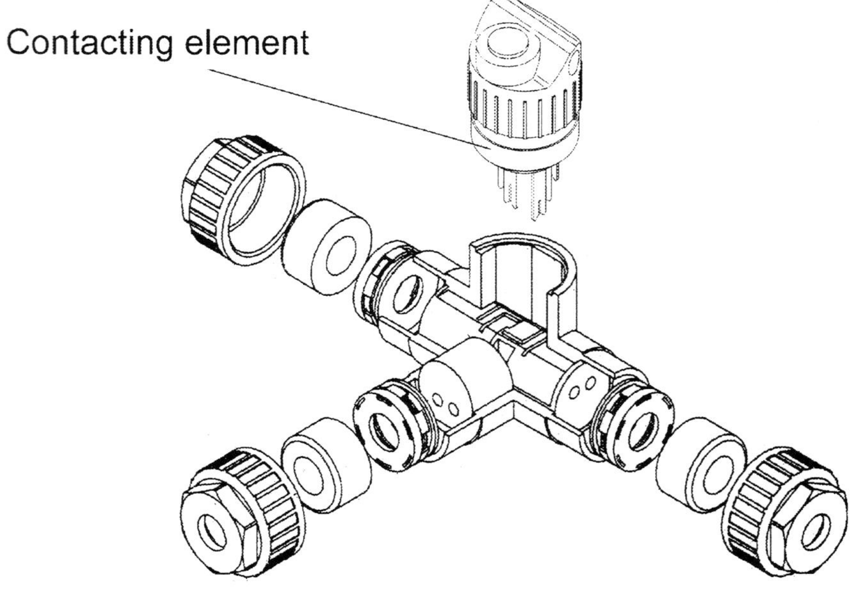

Fig. 3. Connector system for fast installation in the field

3 Future developments

With worldwide many millions installed nodes fieldbusses are the industrial networks of these days. Based on deterministic protocols and guaranteed real time capability PROFIBUS for example has a data speed of 12 Mbit/s. More and more the bus cabling replaces the classical star cabling where each simple device was connected point to point to the control unit. The bus topology reduces installation and service costs and the demand for the project design.

Fieldbus devices nowadays have usually integrated the RS 485 interface as bus communication port. The number of devices with already integrated fiberoptic ports is steadily increasing. Integrated fiber optic ports allow the direct integration of a device into a fiber optic network With the plastic fiber technology the user has a connection and communication technology which is easier and faster to handle than the classical copper technology. Even considering the total costs plastic fiber technology is competitive to the copper technology.

With external fiber optic repeaters devices which have a RS 485 interface can be easily integrated into a fiber optic network. It is not necessary to make an adaption in the device itself. So standard products can use all the advantages of the fiber optic technology.

As a supplement to the existing fieldbuses ETHERNET will increase its market share in the near future. ETHERNET exists since the early 80´s and is today with its open TCP/IP protocol the worldwide standard for office networks. For automation networks ETHERNET is used for the control level in industrial networks. By this for example the link could be made between the office world and the production control in a factory.

Switching technology and 100 Mbit/s ETHERNET allows a "quasi real time functionality". In applications where "hard real time" is not demanded – for example the control level – ETHERNET could be used. Normaly a network consists of a 100 Mbit/s fiber optic backbone with switched 10 Mbit/s or in the future 100 Mbit/s ETHERNET segments. For the hard real time processes the fieldbus networks e.g. PROFIBUS are integrated into a mixed network consisting of an ETHERNET and PROFIBUS or any other fieldbus.

The hardware for such a vertical and horizontal integrated industrial network is available today. But some parts of the software have to be developed. It is necessary to model the relevant fieldbusses independent of special manufacturers into ETHERNET. Also Windows CE will be adapted for industrial applications.

For service and diagnostic tasks the INTERNET is an excellent technology. There are future developments to make network links between industrial networks and the INTERNET. The target is to have the worldwide access into an automation application. Of course the requirements for the safety against prohibited access via the INTERNET are extremely high.

Chapter 11: NOAH Esprit 26951 Project

ESPRIT Project NOAH - Introduction

Udo Döbrich*, Patrice Noury**

*Siemens AG, A6D GT 3, Rheinbrückenstr. 50, D-76181 Karlsruhe
Tel.: 0721 595 2550, Fax: 0721 595 6728, udo.doebrich@khe.siemens.de
**Vice-president technical marketing
ALSTOM technology, 5 Av Newton, 92142 CLAMART France
Tel: +33 1 46 29 13 93, Fax: +33 1 46 31 75 45, patrice.noury@techn.alstom.com

Abstract. The ESPRIT Project Network Oriented Application Harmonisation (NOAH EP 26951) specifies and validates important parts of on open platform for distributed automation systems.

1 Introduction

In the domain of industrial automation, the classical paradigm based on a central device to perform control has moved to a new one using distributed devices with communication facilities, even if the engineering of devices doesn't provide tools for distribution. In entering the world of distributed world, industrial automation has had to reinvent its tools and methods. IT (information technology) is providing technologies of great hope to help industrial automation to set its issues.

The field communication systems are now well known to be able to transfer more than the straightforward measured value. Microprocessors embedded in field devices implement added value functions, which were the privilege of the control devices. Thank to Fieldbus capabilities the control functions themselves can be de-localised to be closer to the field. System aspects such as synchronisation of tasks, time distribution, time triggered application should be supported in a fully distributed environment. This adds an other level of modularity and scalability to the systems in order to match the exact requirements of the process.

The information Technology (IT) latest development provides mechanisms to describe data, provides convenient features to describe technical information such as physical shape, dimensions, electrical connections, behaviour, documentation, performances. Some of these methods may be good candidates for describing automation world elements. The major issue is that all these methods have narrow focus to serve a specific purpose. It is then obvious that they overlap, compete and does not provide a complete solution when put all together. What is lacking is a reference model to position each of them. The model should be able to perform the better rational approach than the OSI model has done for communication 15 years ago. The model was good but incomplete as it left open the relationship between application and communication and did not specify a model for the application functions. This has led to multiple approaches that have to be harmonised now.

An additional dimension should be considered, is that an industrial application has to follow a very specific lifecycle. An automation application has to engineered, tested, validated, installed on site, commissioned and maintained and modified according to the process evolution. An adequate tool but specific now often supports each phase. Then data has to be moved from one phase to the other in a consistent manner. This state is not achieved yet, and data should often been manually re-entered.

Although a straightforward data transfer may be achievable, there is no guaranty that data may be computable at the next step of the project lifecycle. This result in a loss of data and know-how in between project steps and also in reuse of know-how through out projects. This is a major factor of project costs, which is difficult to solve.

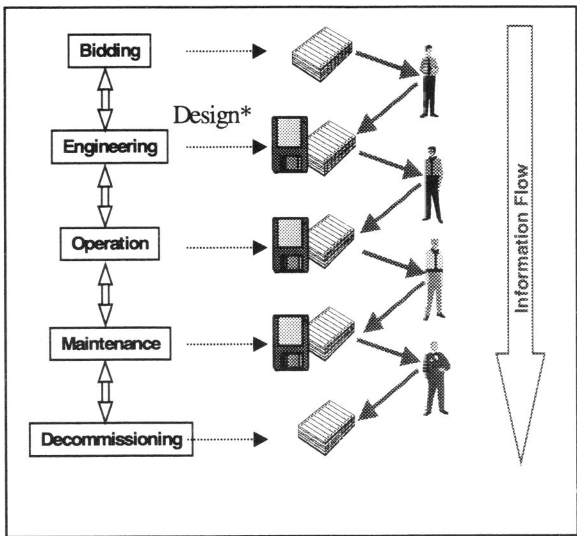

Fig. 1. Project Flow

The ESPRIT NOAH Project EP 26951 "Network Oriented Application Harmonisation" intends to solve some of the issue in providing an open platform for distributed industrial application.

2 The ESPRIT NOAH Project

2.1 Consortium

Twelve members have put forward the project:

- (Co-ordinator)   ALSTOM Industrie
- (Associated)   LORIA – INPL (Institut National Polytechnique de Lorraine)
- (Partner)   ENEL
- (Associated)   Foschungszentrum Informatik Karlsruhe (FZI)
- (Associated)   Institut für Automation und Kommunikation (Ifak)
- (Associated)   Politecnico di Torino

- (Partner)   PROCES-DATA A/S
- (Associated)   Schneider Automation SA
- (Partner)   Schneider Electric
- (Partner)   SIEMENS A&D
- (Partner)   Softing
- (Associated)   Università degli Studi di Catania

2.2 Project Summary

- The project start is the first of January 1998.
- The Duration is 24 months and the effort is 329 Person Month
- The Budget is 4.400 kEuro
- 65 Deliverables

2.3 Project Objectives

The project objectives are structured into three technical main aspects:

- Application Independent Communication Interface
- Device Harmonisation
- System Integration

These three aspects are then put together in a common validation phase on a pilot project.

The first aspect, the Application Communication programming interface, should include not only the operational services but also the services necessary to set up and maintain the network called in the communication jargon the "system and network management". NOAH continues the work outlined by previous project and completes it. The scope covers the communication solutions, which are included in the EN50170 at the date the project was started.

NOAH develops an overall device model allowing the representation of the various device aspects serving different purposes (operational, configuration trouble-shooting...). Device information is then split in two categories. The first, which is common to devices in a given type, will be standardised; the second which is either device specific or non real time specific will be described according to a description method which should be machine readable. Description of the different device types such as drives, motor controllers, sensors will be gathered into harmonised profiles so that devices are functionally compatibles.

Finally, the integration of devices in an open, distributed automation system will be managed through an information system based on a single logical data repository. This repository will provide the access needed by the tool in charge of a given project step. NOAH intends to specify the various "facets" supporting project phases but not the internal model which is left to the system integrator know how. The objective is that the flow of data is transparent throughout the project development

The specification produced in the project will be subjected to prototype implementation in order to validate their relevance within a pilot project. The pilot will be fitted which each of the three solutions included in the EN 50170. Each Fieldbus will be linked to an automation system, which will be in charge of controlling a section of the pilot

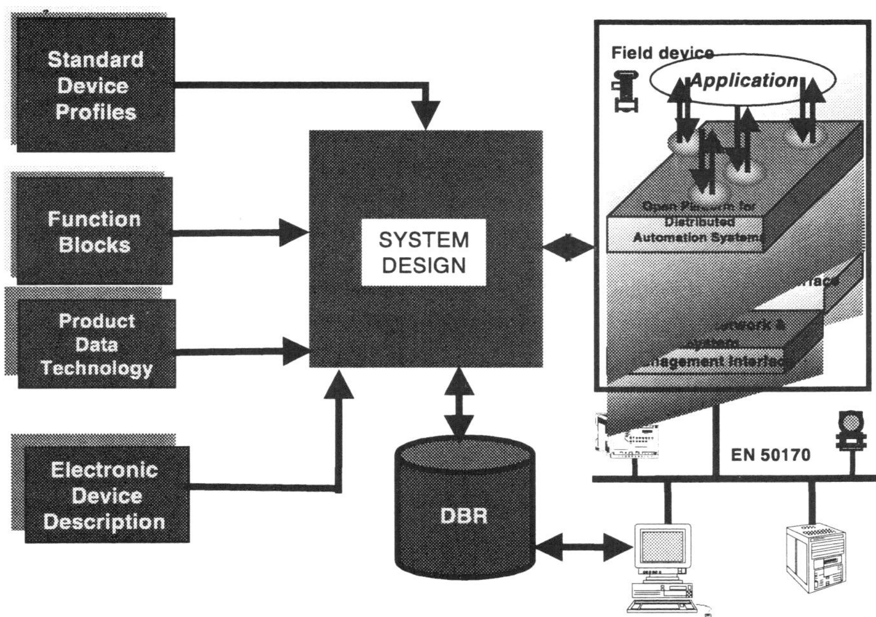

Fig. 2. Open Platform

The validated specifications will be then transferred to the appropriate standardisation bodies in CENELEC and IEC.

2.4 Project Structure

The project is structured in seven working packages (WP) following three threads:

- Technical Activities   (WP 1– 5)

- Project Management   (WP 6)

- Exploitation and Dissemination   (WP 7)

The technical activities were structured in 5 working packages following the three different technical main objectives and aspects:

Each working package is sub-structured in tasks. The tasks of the technical working packages follow the principles:

- 1st Task - User Requirements

- 2nd Task - Specification(s)

- 3rd Task - Implementation(s)

- 4th Task - Validation/ Test / This task will collect all project results and integrate them in a stepwise approach

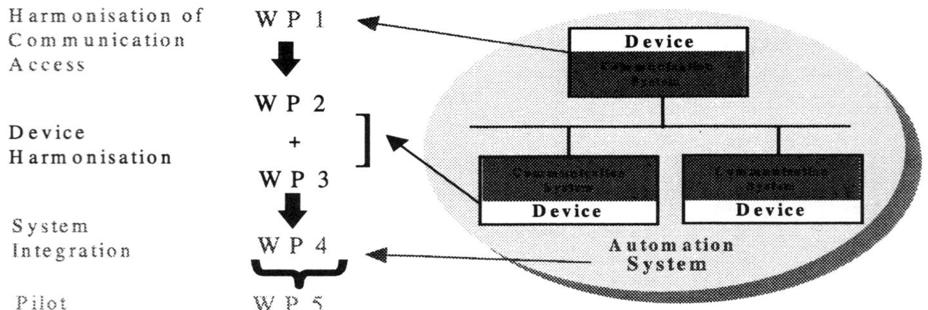

Fig. 3. Project Structure

3 Project Expected Results

3.1 Application Independent Communication Interface

The Application Independent Communication Interface to the Fieldbus systems is the first step to an open platform for industrial automation. Today the application programs and models are directly bound to a specific communication system. The use of a same application program on different communication systems is not possible without an adaptation more or less complex.

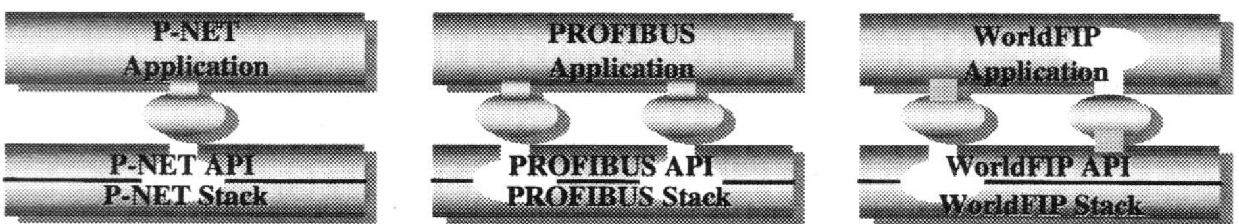

Fig. 4. Problems with the different Communication Systems

The idea of the ESPRIT Project RACKS (EP 20468) was to specify an application independent interface for the Fieldbus systems of EN 50170 (P-NET, PROFIBUS and WorldFIP). The RACKS project has specified the model of the control and data flow and the communication objects. The network management has not been considered.

NOAH extends now this communication interface to the network and system management. Network and system management objects are defined, and their access method and behaviour specified. The interface will be implemented as a prototype for the three volumes of the EN 50170 standard and validated in a pilot plant.

3.2 Device Harmonisation

The starting point of the device harmonisation is the existing profiles or companions standards accompanying the EN50170 Fieldbuses. As they are fully-fledged specifications they cannot be merged in the straightforward manner. They have been developed on different device models. Then NOAH has been driven to develop a common device model for the purpose to build a common reference for understanding.

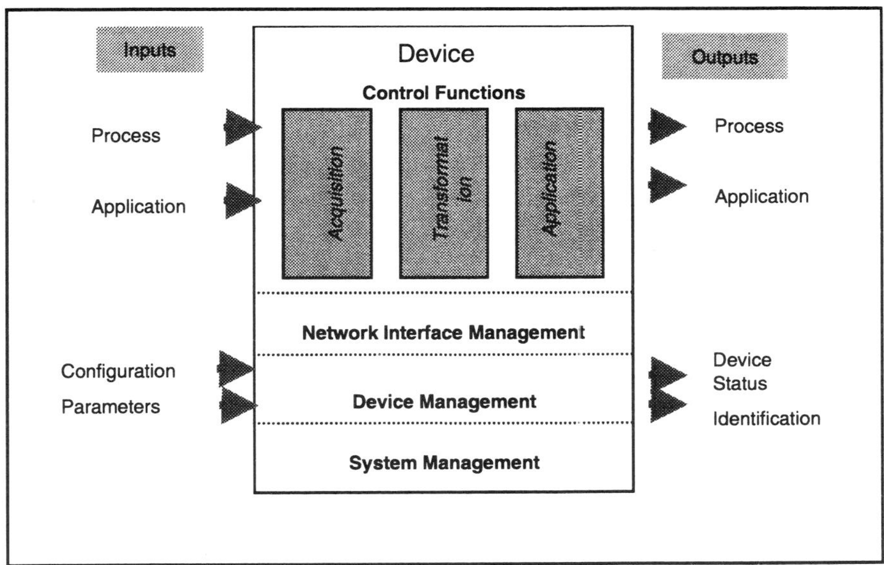

Fig. 5. Device Model

The device model describes the components of a device, defines the relationships between them. It also defines device global behaviour towards the outside world. Agreed rules for writing harmonised profiles were used to perform the detailed investigations of each device type for each particular profile and companion standard. The results are the NOAH profiles, which cover a common device structure (shown in the figure) and detailed specialisation for each device component.

Specialisation means the block structure was extended in a hierarchical way, where every new level and element in the hierarchy add new properties and behaviour. NOAH profiles are available for the following device types:

- Motor Control Centre (MCC)
- Variable Sped Drives (VSD, different types)
- Temperature Transmitter
- Pressure Transmitter
- Differential Pressure Transmitter
- Modulating Actuator
- Simple I/O
- On/Off Actuator

The NOAH profile let the manufacturers the freedom for additional features resulting from product innovations. This is done by defining the NOAH profiles on a conceptual level. Conceptual means, that the implementation details have to be mapped to a specific platform. Based on these NOAH profiles devices, controller proxies, tool libraries and user training can be done.

Real products on the market have manufacturer-specific and/or technology-specific features, which intentionally are not part of profiles. For a non-interrupted information flow during the life cycle of the system and especially for the devices an electronic readable description of the device functions and variables is necessary. Therefore one main effort of NOAH was to agree to a common device description technology. This was done by an iterative analysis process of the existing languages and the cross check with the NOAH device model. No existing language on the market was found which completely cover the NOAH device model and system requirements; i.e. the most of the languages and methods are written as "stand alone" solutions. NOAH selects the required parts from existing languages and modified the solution according to the identified requirements. The implementation of the tool chain for the production of "Electronic Device Descriptions" for the device manufacturer, the system integrator and user in the plant is still in progress.

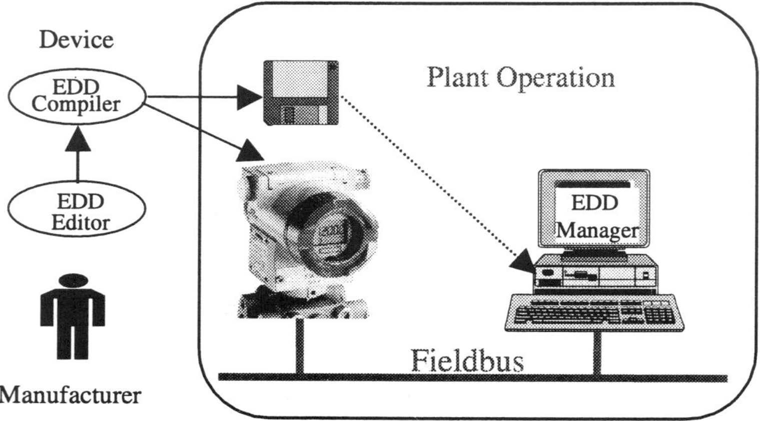

Fig. 6. Device Description Technology

Based on the device description technology the device manufacturer is able to specify its devices in a very precise way, which gives him the ability to simulate the device in advance for the purpose of engineering the system and verifying its capabilities. Additionally, commissioning tools are supported because of the machine-readable information of functions and variables contained in the device. The all over data base repository (DBR) is able to provide information about the specific details of each instantiated device.

3.3 System Integration

A distributed automation systems binds together the information and functions of the devices to a higher level. For example putting together a sensor, a controller, a control valve to form a loop means more than the sum of the components. The loop has a distinct name, specific performance, behaviour, properties, which doesn't reside in any of the component. This is the motivation to build a data repository (DBR) to store this information which is essential to understand how the control function will achieve the process stability. The database is controlled by a management application, which transfers the data from and to the devices MIB. It also provides access to the data for the various tools through adequate facets of the stored objects.

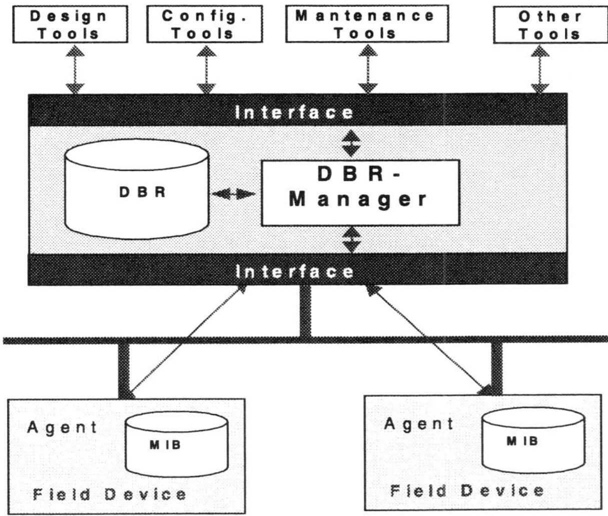

Fig. 7. System Integration

3.4 Validation of Results trough a Pilot

For the access of information from the DBR information a "Generalised Interface" will be implemented, which allows anybody (who is authorised) to access the whole plant information via a WAN (Wide Area Network). Password and access protection will be obviously provided.

In NOAH a WEB-server will provide such an access to the database. Additionally, a video camera will give a real plant view from a remote station. The concept is to connect plants world wide together, because more and more production, maintenance and diagnostic are done globally. The NOAH results are cross-checked with the activities from the EIAMUG ESPRIT project. The implementation of parts of the NOAH system approach will be implemented in the IAM (Intelligent Actuating and Measurement) pilot plant in Milan/Italy. The pilot has three different automation systems and networks (P-NET, WorldFIP, PROFIBUS).

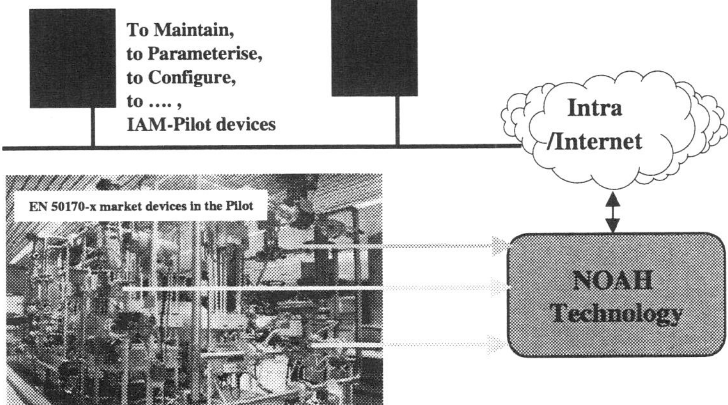

Fig. 8. IAM Pilot

4 Relationship with Standards

The ESPRIT project NOAH focuses on some critical issues the open automation platform. Other projects have already developed some work: Brite/Euram ACORN, Esprit RACKS EP 20468, EIUMAG and IAM. These work are complementary.

To encompass these efforts, a CENELEC Invited Project NOAH was created to convert the results of these projects into normative documents. The target standardisation bodies may be different TCs within IEC, ISO, and CENELEC. The CENELEC IP NOAH will act as a co-ordination project between the different standardisation TCs.

5 Benefits

The benefits will be fruitful for both the user and the system integrators while keeping control of manufacturers expenses.

Users will find a reduction in the total system cost of ownership:

- Reduction of system delivery
- Improve quality and functional safety
- No over-design
- Control of risk of not meeting the required performances

System integrators will find a better control of their development which an improvement of the reuse of previous effort:

- Reduction of engineering effort
- Use validated module based on experience
- Single data entry, multiple use
- Forecast performance before integration

Devices manufacturers will find a common method to document their device while reducing the cost of variants:

- Electronic data sheet
- Description of capabilities
- Single application for multiple communication technologies

References

1. NOAH point of information: Web-page: http://noah.loria.fr

Fieldbus Profile Harmonization –

Approach of NOAH ESPRIT 26951 Project

NOAH Working Package team

H-P. Otto, Siemens WP Leader Germany; Ch. Borodino, Alstom France; T. Bregulla, FZI Germany; Ch. Diedrich, ifak Germany; U. Döbrich, Siemens Germany; J. Hörger, Siemens Germany; J. Szymanski, Alstom France; Ch. Verney , Schneider France;

Address

Ifak Magdeburg, Germany, Steinfeldstr. 3, D39179 Barleben, chd@ifak.fhg.de

Abstract. The NOAH (Network Oriented Application Harmonization) project harmonizes device profiles of the EN50170 fieldbusses WorldFIP, P-Net and PROFIBUS among others. Within the project a certain subset of field devices are under consideration. The investigations cover transmitter for temperature and pressure, on/off and modulated actuators, simple I/O, motor control centers and variable speed drives. The range of devices shows, that basic device classes with very different scope and functionality are chosen to find fieldbus and profile independent application structures, functions and parameters. The project follows a well defined device model for the harmonization. The common parts of all profiles are presented and the consequences are discussed.

1 Introduction

The fieldbus standard EN 50170 covers a large number of potential applications for industrial control, supervision and use in the field as well. To co-ordinate application functions between simple I/O, transmitter, actuators, variable speed drives and controllers as well as visualization and operator terminals, the definition of the application functions and their variable/parameter syntax and semantics have to be done. This is the main topic of a profile. A profile is independent of the underlying communication system. This profile is known as device profile. It provides functions and parameters for co-operation with other profile devices or control, diagnosis, visualization, supervision, maintenance, and technical management purposes. The mapping of application functions and variables/parameters to a specific fieldbus protocol has to be defined in addition of specifications. This mapping specification (also known as communication profile) contains the use of services for application function co-ordination, the used communication subset and range of protocol parameters.

NOAH device profiles are: temperature, pressure and differential pressure transmitter, modulated and on/off actuator, simple I/O, variable speed motor drives, and motor control centers. The general resulting device structure is shown in Figure 1 [1].

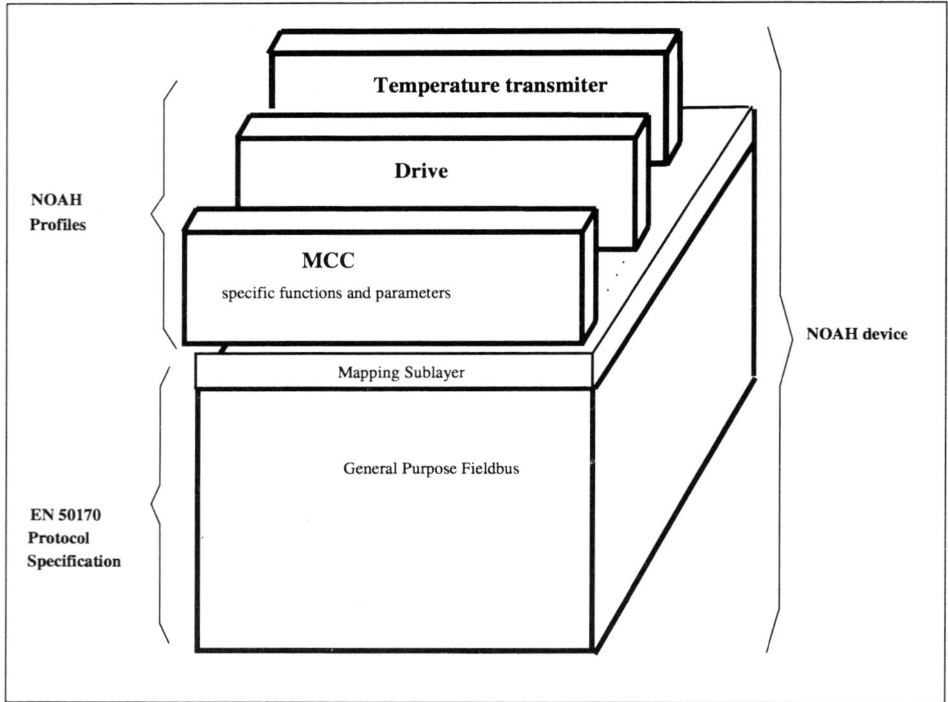

Fig. 1. Noah device structure of the specification documents

The profile offers the possibility to standardize the reactions (behavior) of devices produced by different manufacturers in one system. It makes the interactions between the device functions and the accompanying control, maintenance and diagnosis functions much easier.

All profile specification work is based on the profiles and companion standards of WorldFIP [2], P-Net [3] and PROFIBUS [4].

2 Device Classification

The scope of NOAH is focused on devices for information detection and use of information for the direct influence of the process, i.e. field devices. Devices are classified using a certain number of attributes which identify a device by a defined combination of attribute figures. These attributes are relations to the process (measurement, actuation devices), signal type (analogue, binary, discrete), measurement and actuation type (temperature, pressure, ... modulated actuation, variable speed drive, ...) measurement principles (e.g. resistor temperature measurement, thermocouple temperature measurement) and others more. NOAH range of selected devices covers a broad application area.

3 Common NOAH profile definitions [5]

3.1 NOAH block model

The NOAH profile harmonization work defines a common device model at first. This device model is a point of reference for all profiles and companion standards of the different fieldbusses under consideration. The main idea of the device model is the separation of device functionality with different purpose to different blocks. A block is an encapsulation of functions and variables, which hide the detailed data flow between the functions within the block from an external point of view. Visible from outside are input and output variables. The main reasons are to ensure technology innovation of measurement, actuation and control principles and to hide manufacturer specific know how. This approach is well known as function block technology.

NOAH defines a Device Management Block, a Technology Block, an Application Block, a Network Interface Block and a System Management Block (see Fig. 2).

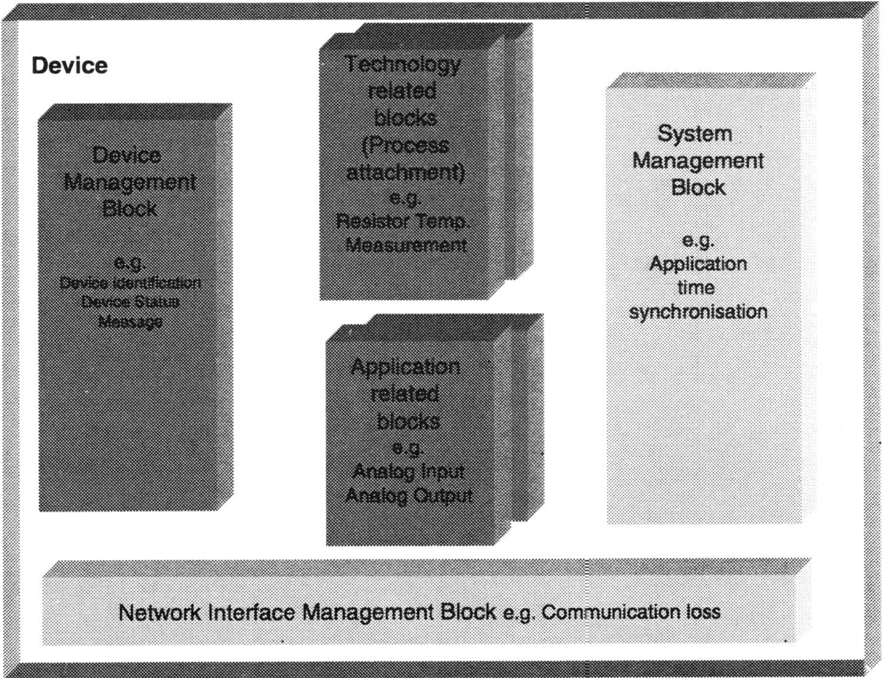

Fig. 2. Block structure of a NOAH device

Each block can be composed of several blocks or hierarchies of blocks. Not all blocks are relevant for all device types, i.e. block may be empty. For instance: simple I/O may not have knowledge of the status of the communication interface (Network Interface) or need coordination within the system (System management). In this case no specification of the Network interface and System management has to be done.

3.2 Device Identification

Devices have to be unambiguously identified during their life cycle. This identification have to cover the needs of the device vendors, the system design and commissioning and the operation and maintenance. Therefore for device identification the following parameters shall be supported:

- DEVICE_VENDOR
- DEVICE_MODEL
- DEVICE_REVISION
- DEVICE_SER_NO (For identification of multiple devices of the same type)

Because of the desired non-interrupted information flow between all involved tools these identification parameter should available for the following steps of the life cycle:

Step 1   Identification of the device for ordering
Step 2   Mapping identification of the device in the manuals
Step 3   Identification of devices for representation in the data base of the end users
Step 4   Identification of the device regarding the valid Electronic Device Description
Step 5   Self-identification of the device via communication access
Step 6   Mapping identification of the device in the proxy

The degree of support of these steps by the device vendors, tool vendors and end user facilities determines the degree of non-interrupted information flow between the tools.

3.3 Standard Status for measurement

It is a piece of information attached to every measuring variable to assist the user of measurement data (typically control functions) in assessing its utility. It may assume the following values:

- Secure obtained from redundant faultless sensors.
- Valid : no fault present (single sensor)
- Impaired: less accurate measurement with respect to nominal.
- not-valid : measurement
- (invalidated) : no validation functions implemented

This abstract definition has to be mapped to concrete data types and codes. One example is shown in Table 1.

Table 1. Mapping of abstract status to IEC61131-3 status data

NOAH Standard Status	PROFIBUS-PA Status
Secure obtained from redundant faultless sensors	manufacturer specific
Valid	Good
Impaired	Uncertain
Not-Valid	Bad
No validation	manufacturer specific

3.4 Device Status

The device synthesized status is devoted to assist the user of a device (in particular a Control operator/algorithm) in assessing its remaining capabilities and to adapt accordingly its strategies. All NOAH device follow the same device state chart (Fig. 3), which is based on the Harel state model [6]. Not all states are mandatory for the all devices. In minimum a NOAH device has to detect network faults, application faults and the application have to be switch from automatic to manual.

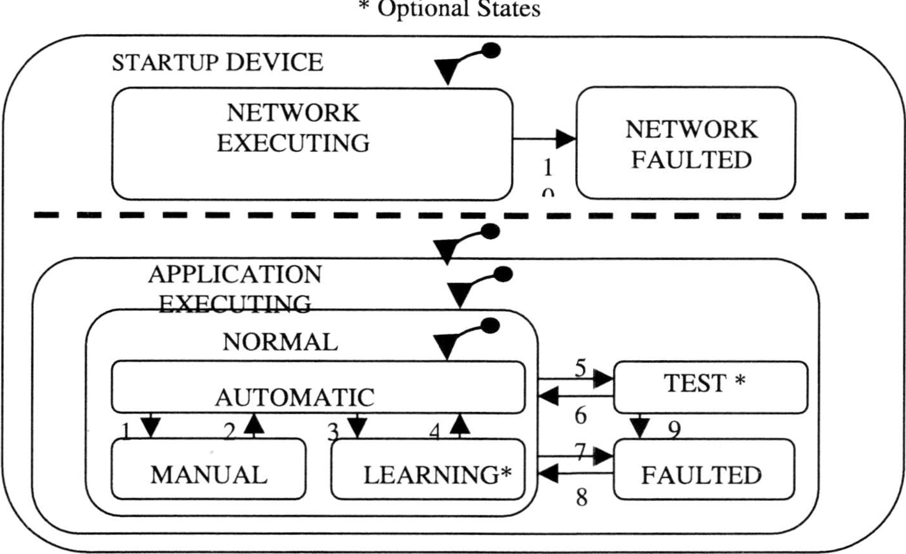

Fig. 3. Device status state chart

3.5 General profile definition

All profiles define their certain block structure. Each function of each block is described verbally. At least all variables/parameters are summarized in a table containing some of their attributes (see Table 2).

Table 2. Variable description table (example), Legend* - For future use

Variable/Parameter Name	Description	Data type	Read/ Write	Default value *	Retain *	Class M/O/C
MAIN_MEASURE	Pressure/differen- tial pressure value	Numeric	R	Data type specific	Volatile	M
MEASURE_STATUS	Bad, Good Uncertain	List of Boolean	R	Vendor defined	Volatile	M
...						

The profile definition remains on an abstract, let say conceptual level, i.e. it isn't an implementation specification. A detailed mapping to data types, enumeration codes, have to be done for specific application fields. The profiles provides a view for the handling and the functional structure of the next generation of field devices.

4 Summary and outlook

NOAH is ready to start no less then standardizing the measurement and actuation functionality from the field device point of view. The NOAH project has been started with a common device model and some basic device functions for instance block structure of all functions, tightly connected measurement variable and it's status, device identification parameters, device status and variable descriptions. The profiles don't define how the profile implementations look like. It is possible to implement NOAH devices, proxies within PLCs which provide NOAH like function block interfaces or parameterization and configuration tool with NOAH like HMIs. That means, that device manufacturer are able to use the specification for future field device design and implementations and tool manufacturer can design their tool architecture and libraries according to the NOAH device model. One pre-requisite is of course the acceptance of the NOAH device model and profile specification within the community of manufacturer.

This is a step forward to a common functional behavior, handling, architecture and interface to intelligent field device functionality. To reach full benefits of the profiles, implementation specifications for certain application areas have to follow. The way to fully interoperable and interchangeable devices of different manufacturer is visible. The NOAH project integrated the know how and concepts of main players in the automation arena together.

References

1. EN 50170: General Purpose Field Communication System, CENELEC 1996.

2. WorldFIP Companion Standards, WorldFIP, Paris 1999

3. P-NET Standardised general purpose channel types, October 1998, Process Data

4. PROFIBUS-PA Profile for Process Control Devices, PNO Karlsruhe 1999

5. NOAH Project, EP 26951, Deliverables D230 to D238, May 1999

6. Science of Computer Programming – Vol. 8 "Statecharts: A Visual Formation for Complex Systems" by D. Harel, 1987, p. 231-274

Electronic Device Description

René Simon[1], Claudio Demartini[2]

[1]Institut für Automation und Kommunikation e.V. (ifak)
Steinfeldstraße 3 (IGZ), D-39179 Barleben
Tel.: 49-39203-81060, Fax: +49-39203-81100
E-mail: *rsi@ifak.fhg.de*

[2]Politecnico di Torino
Dipto. di Automatica e Informatica
Corso Duca degli Abruzzi 24, I-10129 Torino
Tel.: +39-011-5647010, Fax: +39-011-5647099
E-mail: *demartini@polito.it*

Abstract. Engineering costs are becoming more and more dominating in the complete live cycle of automation systems, especially under a true cost-of-ownership principle. Electronic Device Descriptions are a good means to reduce the engineering costs. The following report presents the history of Electronic Device Descriptions and describes the method and technology, developed by the ESPRIT Project "Network Oriented Application Harmonisation" (NOAH).

1 Introduction

Electronic Device Descriptions (EDD) are becoming more and more important for the engineering and commissioning process of Distributed Control Systems (DCS). They will replace traditional approaches based on paper documents, which are more error-prone and therefore time and resource consuming.

The EDD Language is used to describe a broad range of devices, from simple, i.e. sensors / actuators, to complex, i.e. hard- and software modular devices. Devices from different vendors are described using the same language. EDD is not restricted to field devices, it is suitable for any (hardware) components with communication interfaces. The descriptions are used by software tools, which adapt themselves (communication interface, HMI interface) to the specific device they interact with, to fulfil e.g. configuration, commissioning, and supervision tasks. That means, the development, distribution, and installation of new drivers is not necessary to integrate a new device into a commissioning tool because the tool is able to gain the information from the EDD file.

The ESPRIT Project "Network Oriented Application Harmonisation" (NOAH) addresses several fields, such as a harmonized application interface of communication systems (WP1), a harmonized function block model (WP2), an Electronic Device Description (WP3) and a system information repository (WP4). All technical specifications, prototype tools and components will be validated in a special pilot implementation. NOAH results will be used in standardization activities (e.g. CENELEC).

Within the Work Package 3 of the NOAH project, an Electronic Device Description and accompanying support tools will be developed. Task 3.2 creates two items, a Language Study Report and a Language Specification. This work is based on results of Work Package 2 (the device model) and a requirements specification (Task 3.1) respectively. Task 3.3 will develop a toolkit, which will be validated using a pilot implementation in Task 3.4. This paper presents the results of the tasks 3.1 and 3.2, finally giving a short hint about 3.3.

2 Overview about NOAH EDD

2.1 Requirements

The requirements to the Electronic Device Description (EDD) can be summarized as follows:

- An EDD (file according to the EDD Language specification) must be delivered by the device vendor together with the device.

- EDD is used in the engineering process of the distributed control system, supporting planning, commissioning, operation, diagnostics and maintenance.

- An EDD file represents a static description, i.e. the declarative part of the device, therefore only the external interface/ behaviour of the device has to be described (no interest in internal code).

- EDD is used within the context of a determined architectural model and supported by a corresponding chain of software tools (editor, compiler, interpreter) with standardized interfaces.

- There are two different EDD presentations:
    - source (human readable and computer interpretable)
    - binary (optional , created by compiler, interpreted by interpreter)

- The EDD can be stored both, on disc and within the device itself (transport via fieldbus).

- The EDD is independent from the underlying fieldbus system.

- The EDD is used to describe information identifying items and defining relationships between them (hierarchical, relational)

- The EDD offers language elements for presentation within a HMI and for communication access.

2.2 Background

Digital information processing in field devices and digital fieldbus communication leads to a change in the handling of automation systems in manufacturing and process control. The field devices provide now much more information than the 4-20mA signal. In addition, they carry out some functions which are/were originally programmed within the PLC or DCS. These field devices are also known as smart devices. The consequence is a distributed system. The tools for commissioning, maintenance as well as for the design and programming of such automation systems need an exact description of the data and functions in the field devices. The way from the design of the description to its use is called device description technology.

The description contains fieldbus configuration parameter values as well as ranges, engineering units, default values, limits, labels and help information. The device description focusses mainly on the configuration and parameterization processes. The configuration adjusts the component set of a device which is composed for a specific purpose. The parameterization adjusts the device to the process specific measurement or actuation purposes.

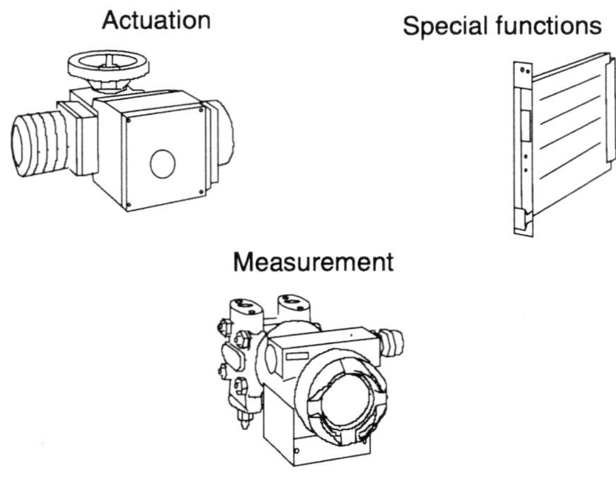

Fig. 1. Field devices demanding device description

The configuration and parameterization of the field devices may be seen from another point of view. The communication configuration chooses those features from the possible capabilities of the devices, which fulfil the requirements from the system design. The application parameterization relates to the functions and data performing the main purpose of the device. Most of the field devices become more and more complex. Therefore a device specific guidance of the operator for configuration and parameterization is necessary. The operator will be guided by special parameter labels, help texts, menu hierarchies and consistency checks. All these aspects are covered by a device description. However, not all devices need descriptions of all aspects at the same amount. The content of the descriptions is a result of the device complexity, the fieldbus features and the number of configurable functions and data.

The interactive handling of field devices is nothing new on the market. The device manufacturers provide local operator panels and/or PC based tools if necessary for years. The new feature is the interaction of one tool with devices from different manufacturers. The end user wants to handle the field devices as simple as 4-20mA ones. In addition, smart devices have to fit in all phases of the engineering life cycle.

2.3 Study of Existing Approaches

To compare the existing approaches, a "meta model" was developed, which outlines the different language aspects. The meta model covers both, technical and non-technical aspects. 9 language studies were done, which were all trying to cover the specified language aspects. The following languages (technologies) are considered:

> Hart, DeviceNet, PROFIBUS – DDL, PROFIBUS – GSD, IEC 61131, ISO 10303 (STEP), IEC 61915, WorldFip, Fieldbus Foundation

Comparing the different languages it seems appropriate to divide them into 3 groups.

The first group contains just the PROFIBUS – GSD. This very simple, special purpose language (one separate key word for each feature) has not the flexibility to be enhanced according to the defined requirements, so it cannot be used within the project. But it is necessary to provide a migration path from PROFIBUS – GSD, in order to use the big amount of device information, already existing in electronic form already.

IEC 61131 and ISO 10303 (STEP) can be put into the second group. Both are very interesting either from a application (IEC 61131) or implementation point of view (ISO 10303). The devices to be described by EDD are increasingly programmed using IEC 61131. STEP is used in many areas for product description purposes. For device descriptions, there is a need of a specific semantic which is not found either in IEC61131 or ISO 10303. But it must be recognized that influencing the IEC 61131 standardization process is beyond the possibilities of the project. This is true also for ISO 10303, providing a new Application Protocol is not possible.

The third (and most important) group covers Hart, PROFIBUS – DDL, IEC 61915, DeviceNet, WorldFip and Fieldbus Foundation. Basically, all these approaches follow more or less the ideas first developed by Hart. They all offer basic language elements to describe device information, parameters and how to access them, i.e. provide device description specific semantic. They fulfil more or less the mentioned requirements. So the starting points for the Language Specification in NOAH are the languages of this group 3.

3 Language Overview

The following figure shows the main scope and structure of the EDD.

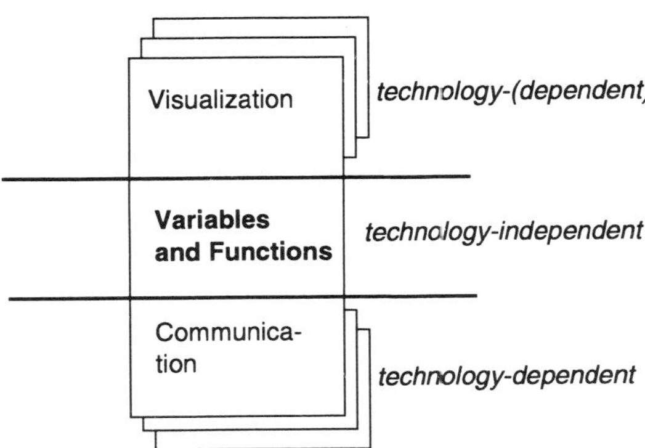

Fig. 2. Scope and structure

The kernel is composed of the variables and functions (methods). These structures describe the structure of the device itself (as set of variables) and its behaviour (the methods processing the variables). The kernel is technology-independent, as far as there is no further requirement on the device where the variables reside or the host where the methods are executed.

The visualization provides elements for the user of the EDD to interact with the device via menus and tables. This requires certain (very simple) technology to be implemented, but this technology is easily provided by all common "windows like" operating systems.

The communication part is more complicated. It describes how variables from the device can be accessed by a host via a communication system. Because of the great diversity of the available (fieldbus) communication solutions it is difficult here to find an overall solution.

Maybe, it is necessary to lay out, that different "kinds" of communication (addressing) are necessary to get data out of a field device, not all of them being within the scope of EDD:

-   Device addressing: The device must be addressed within a communication network or a set of networks, so a name space of devices is necessary. This is not part of the EDD, which deals only with single devices but not with their arrangement within projects (systems).

-   Services: Services are needed e.g. to read or write variables. EDD knows very abstract services (called transactions) within its command language element.

- Addressing within a device: Each variable must be addressed within a device, i.e. below the name space shared by all devices within a project. This is provided by the EDD Language within its command language element (absolute and relative addressing schema).

- Data types: Different data type systems are used within the device (depending on the programming language used to implement the device), from the communication system and even in the host device. The EDD Language implements its own data type concept, but does not describe how to map these data types to other technologies.

Within NAOH EDD all the language constructs are put into one of two groups: Basic and HMI elements. These groups contain:

Basic elements

- Variables, blocks, methods, commands, collections & arrays, some other

HMI elements

- menu bars, structured tables

Each of these constructs has a set of attributes associated with it. The attributes are used to define each construct. For example, a menu has two attributes: item and label. A specific menu is defined by defining each of these attributes. Attributes can also have sub-attributes, which refine the definition of the attribute and hence the definition of the construct itself.

An Electronic Device Description source is developed using the EDD language syntax. Examples of the information specified in an EDD source are:

Parameter definitions

These definitions identify all available device variables used from applications of control and maintenance systems and other devices. For instance, a process variable and operation mode variable can be described. Parameter definitions include simple variables, or structured ones.

Relationships

Relationships among parameters are common in devices, EDD descriptions include refresh relationships, and unit relationships.

Human interface support

Menus, help text, and display formats are available as hints to the interface developer. These facilities provide the device developer with a degree of control over the presentation of the device to the end user. Enough information is provided to implement a menu driven interface for a small-display device, or a simple full screen display.

Blocks

These items describe the parameters, parameter lists, and associated menus, edit displays, relations etc. of a block. The various types of blocks are described in Function Blocks Specification, e.g. profiles or companion standard of fieldbus systems.

4 NOAH EDD Toolkit

Basically, there are 2 tasks the EDD Toolkit is supporting. As part of the development process of the field device, an EDD (as ASCII-File) is produced using the EDD IDE. This IDE consists of a HMI (P-Producer) and the EDD Manager. Although this can be done using a simple ASCII editor, it makes sense to support the developer with more sophisticated tools to manage complex EDDs. The EDD is consumed by engineering and commissioning tools (HMI (C-Consumer) for example). Furthermore, the EDD Manager provides instance data information, doing so, it needs access to the field device (via the WP1 Interface, productive). Access information for the network is provided by the Fieldbus Console.

Please keep in mind, that the following figure shows the viewpoint of WP3, not a general NOAH architecture.

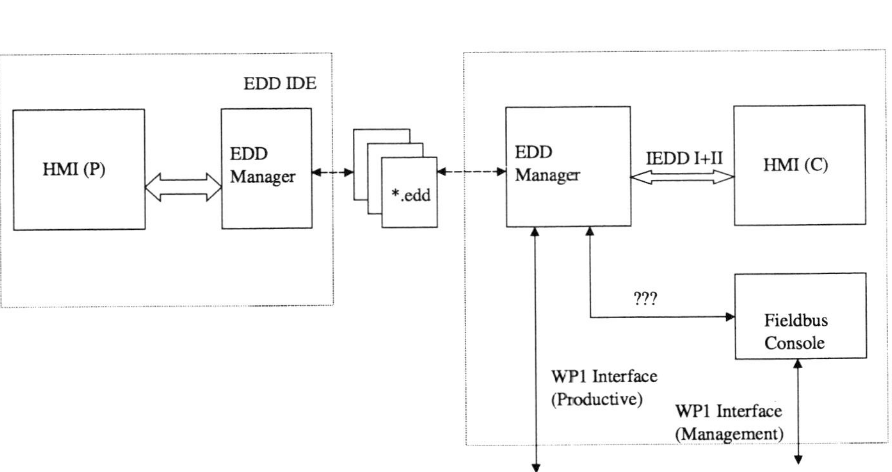

Fig. 3. Toolkit architecture

The modules of the EDD Toolkit can be characterized as follows:

HMI (P)   Interactive Human Machine Interface for creation, editing and managing EDD (usually done by the device vendor). Uses the functionalities provided by the EDD Manager Interface.

EDD Manager	Contains functionalities for reading / writing EDD files as well as instantiating, populating, and browsing them.
EDD Manager Interface	Provides the functionalities of the EDD Manager.
EDD Files (*.edd)	According to language specification.
HMI (C)	Interactive Human Machine Interface for e.g. engineering, commissioning, and supervision of field devices.
WP1 Interface	Fieldbus interface developed in WP1. Management Interface used by the Fieldbus Console. Productive Interface used by the EDD Manager.
Fieldbus Console	Resolves the access path to the device (network information).
??? Interface	Provides the access path to the device (network information).

5 Summary and Prospects

This paper presents an overview about Electronic Device Description for field devices.

The use of such a language needs both, software tools for an effective development and management of EDD files and software components which enable a fast plug-in into software tools used in the engineering and commissioning process. This chain of software tools will be specified and implemented as EDD Toolkit within Task 3.3.

To establish a basis for the wide use of the EDD Language, formal standardization is necessary. Therefore NOAH should provide this specification to different standardization bodies such as CENELEC or fieldbus user organizations.

References

1. Device Description Language, InterOperableSystems Project Foundation, 1993

2. NOAH documents, not yet in the public domain

A DBR-Based Approach for System Management

C. Demartini[1], R. Iosif[1], C. Raibulet[1], J.P. Thomesse[2]

[1] Dipartimento di Automatica e Informatica, Politecnico di Torino,
corso Duca degli Abruzzi 24, 10129 Torino (Italy)
Phone: +39-011-564 70 10
Fax: +39-011-564 70 99
E-mail: *demartini@polito.it, iosif@athena.polito.it, raibulet@athena.polito.it*

[2] LORIA, Campus Scientifique - BP 239
54506 VANDOEUVRE-lès-NANCY CEDEX
Phone: +33-03 83 59 30 00
Fax: +33-03 83 27 83 19
E-mail: *thomesse@loria.fr*

Abstract. This work describes the Database Repository (DBR) as specified within the ESPRIT Project "Network Oriented Application Harmonization" (NOAH). The DBR provides a uniform solution to support the control, configuration, management and maintenance of industrial plants. The basic aim of this approach is to define a general information model that can describe any distributed automatic system. The Database Repository specification unifies the actual representation of the devices within complex systems allowing heterogeneous devices and their links to be described according to a common approach, regardless of their specific implementation.

1 Introduction

The Database Repository specification provides a structured description of information associated with specific parts of a system life cycle. The basic aim is to provide a common view of the system in such a way that relevant applications can be developed by focusing on all system composing elements. The capability of ensuring a common view for any application provides a unified perspective for several activities including installation, configuration, maintenance and technical management.

A further advantage offered by the definition of a common and general information model is represented by having a unified approach to the problem of describing complex distributed systems. This improves the reuse of previous results and specifications so that the engineering costs can be greatly reduced.

Section 2 introduces the application concepts used for the database repository specification. Section 3 describes the mapping of the application concepts onto a subset of the AP 212 [1] objects. The time constraints related to the dynamic aspects of data are presented in Section 4. Section 5 contains an example for the DBR model validation.

2 A Model for the DBR

A uniform information model is the best way to ensure a common view of the plant components. That view should be the result of the analysis of the basic activities emphasized by users: Control, Configuration, Maintenance and Technical Management. This approach derives from the natural observation that, a data driven description of the plant should be preferable to a functional approach. In other words, a complete description of the information model provides system integrators and vendors with the capability of developing their own engineering tools, each one accessing its specific view of the common database describing the whole plant.

The model includes both static and dynamic aspects of data related to an industrial plant. Static description is concerned with the logical and physical structure of the plant, while dynamic aspects regard the evolution of the system in time. In this work a static database model is provided using STEP EXPRESS [1, 2], a standard language for the definition of product data. Dynamic aspects that mainly regard the time constraints that must be respected along the system life cycle, are modeled in SDL [2].

The STEP standard [1,2] has been chosen as the most convenient standard proposal currently available for the implementation aims of the DBR static model. Within the STEP standard there are several Application Protocols (AP) available, e.g., ISO/IEC 10303 Part 212 DIS [1] which has been taken as a basis for this work because it includes all necessary concepts present within the plant description such as devices, interfaces, networks, functions, variables, parameters, etc.

The DBR model uses a subset of AP 212 to describe the main application concepts and their relationships.

2.1 Application Concepts

The relevant application concepts represented in the DBR were classified according to the user requirements in order to achieve the degree of generalization which permits the common description of a plant:

Function block: it is a software functional unit comprising an individual, named copy of a data structure and associated operations specified by a corresponding function block type [2] (the function block type specifies the characteristics of that type). It has associated algorithms described by an external reference (e.g., a state chart representation). The input of a function block is specified by zero or more input parameters. The output of a function block is specified by zero or more local variables used by its algorithm. It is implemented by one or more physical devices [2].

Local variables and parameters: they are software entities associated with a function block that keep the values used by the function block's algorithm. Local variables are defined within the scope of a function block. Parameters are input/output events that do not carry associated data (i.e., boolean values) or input/output variables [2]. From a physical point of view they are considered input/output ports.

Logical structure: it represents the way in which function blocks are connected within the system. The connections are specified in terms of connected parameters. There are two types of connections: one-to-one (a direct connection between an output port and an input port) and one-to-many (a direct connection between an output port and two or more input ports). Connections are implemented by a physical networks.

Localization: it describes the physical position and arrangement of the physical entities, in particular, devices and network segments. In addition, the physical position of a device group that forms a logical entity can be represented.

Physical structure: it describes the way in which devices are connected within the system. One-to-one and one-to-many connections are specified in terms of ports.

Allocation: it describes aspects regarding the intended use of the equipment within the system, expressing relationships between function blocks and devices, between logical and physical structures, etc.

2.2 ISO 10303 AP 212 - Subset Definition

The following are the units of functionality (UoF) selected for the DBR model design:

Allocation – defines information that specifies the relationship between the objects designated in both the function_structure UoF and the product_structure UoF, describing the allocation of function blocks to devices.

Classification – describes concepts needed to categorize the packages of information contained in the product data.

Course – specifies the data used to describe a three-dimensional path to specify the spatial connection of equipment items and routes between them.

Designation – describes concepts used to identify equipment items and function blocks.

External_reference – provides a way to specify information not included in AP 212.

Function_structure – specifies the concepts used to describe the functional structure of the system. The functions may be either primitive or composed from other functions.

Functional_connectivity – specifies all those concepts that are required to specify the logical connectivity of the system, i.e., connectivity among function blocks.

Network_allocation – describes information that specifies the allocation of functional connectivity to physical connectivity established by the equipment used in the system.

Physical_connectivity – specifies information required in order to describe the connectivity of the equipment used in the system.

Product_structure – specifies the concepts needed for the description of the equipment used in the system implementation.

Properties – specifies a generic concept needed to assign technical or management data to the information packages that make up product data.

Site – specifies information regarding the position of the equipment within the system. Position can be specified in terms of GPS coordinates or local Cartesian coordinates.

3 The Mapping of the DBR Concepts onto AP 212 Elements

This subsection presents the mappings of the function block and the devices onto AP 212.

3.1 Function blocks

A function block is represented as an association between the function_definition, representing the function type and a function_unit, representing the instance of the function type. The association is represented by a functional_composition_relationship object. Note that more than one instance is allowed to share the same function type.

Every instance has to have a unique designator in order to identify it within the system. The designator is represented by the item_designation object associated with the function_unit object by means of its id attribute. Note that item_designation is an abstract entity, therefore its non-abstract subtype object_reference_designation must be used instead.

Every function_definition has to specify version information by means of the function_version entity related with the function_definition by its version attribute. For a function_version, the only mandatory attribute is the base_function which relates the function_version with a functionality entity, which describes the information associated with the function block algorithm i.e., the name (acquisition, linearization, etc.) and optionally, an informal description.

Fig. 1 shows the EXPRESS-G diagram [2] of the function block model.

3.2 Devices

A device is represented by the single_device entity which is a subtype of the abstract entity device.

The device object must be associated with a unique identifier, represented by the item_designation entity which is associated with the device by means of its id attribute. Note that item_designation is an abstract entity, therefore its non-abstract subtype object_reference_designation must be used instead.

A device definition has to be specified using the design_discipline_item_definition entity. The definition carries information related to the device type. More than one device can share the same definition that is, more than one device of the same type can be present in the system. The design_discipline_item_definition entity specifies the version information by means of the item_version object associated with it. An

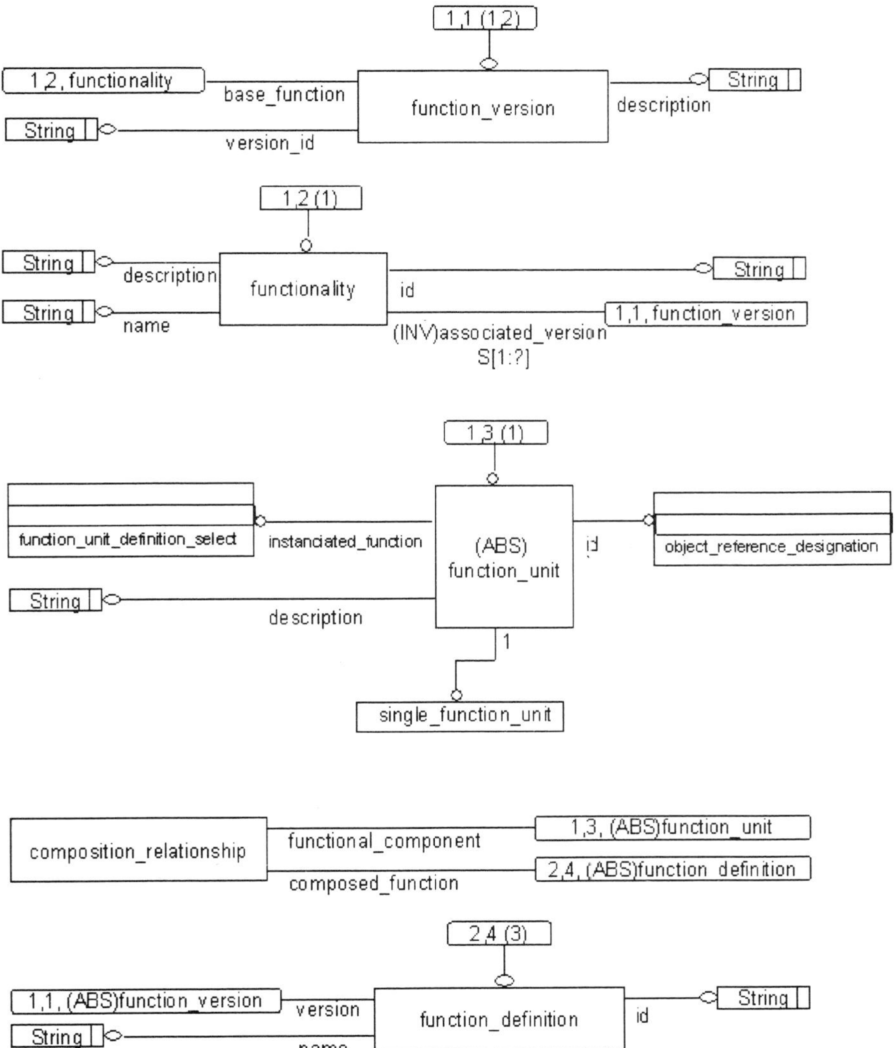

Fig. 1. EXPRESS-G diagram of the function block model

item_version has to have an associated item object whose unique identifier is represented by the item_identification object related to the item by means of its id attribute.

Fig. 2 shows the EXPRESS-G diagram of the device model.

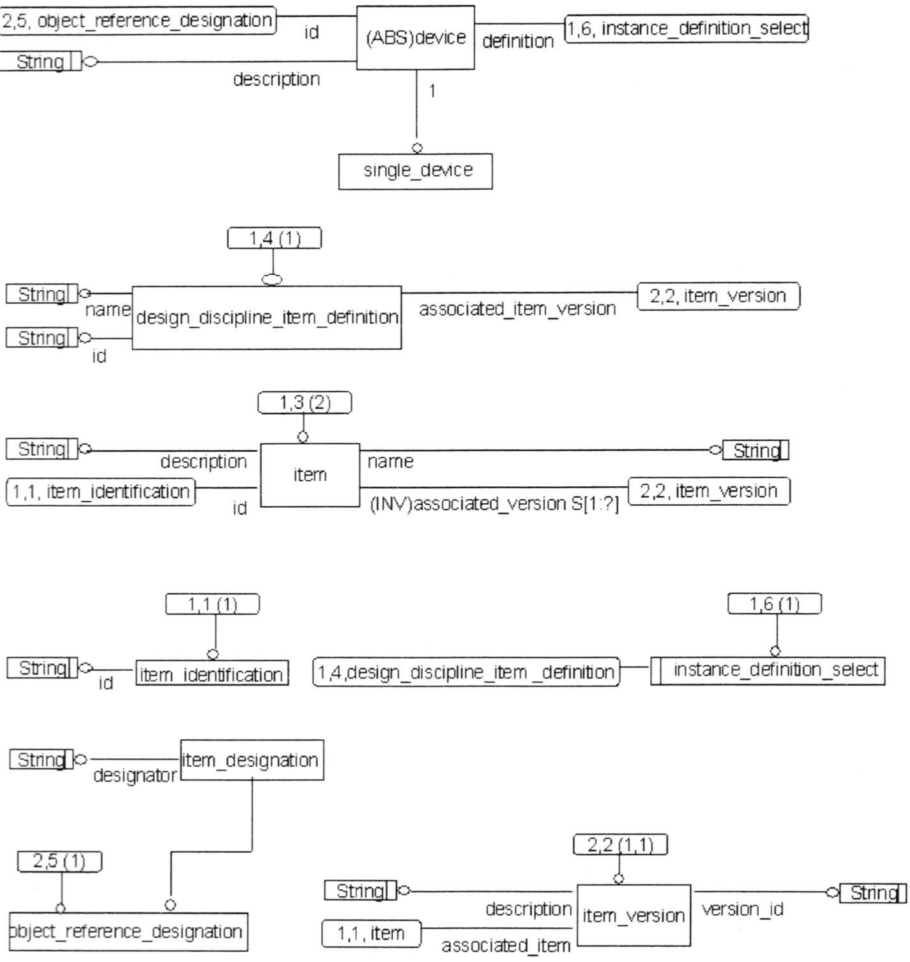

Fig. 2. EXPRESS-G diagram for the device model

4 The Dynamic Description of the DBR

Real-time constraints are related to the behavior of the system in time. They refer not only to the logical order of some actions or operations, but also to the duration and to the moments when actions take place. Within the DBR model, the time constraints are expressed only on an event or on a set of events. This approach is valid only if the behavior of the system can be modeled as a trace of occurrences of events. And this is suited for the DBR model.

All the end user constraints are translated into constraints on events. For example, a time constraint called "simultaneity" of data may be expressed on a constraint on the "simultaneity" of the events called "productions of data". A deadline on a task is a constraint on the event "end of the task". A maximum response time is a constraint on the duration between the "begin" and "end" events.

The real-time constraints are considered a type of object and they are described as any other object in the model. The time constraints can be also described using the STEP/EXPRESS language that has been chosen as the specification language for application elements of the DBR model.

5 The Validation of the DBR Model

The DBR model validation was demonstrated with a real system example: a Temperature Loop in the IAM-Pilot laboratory at ENEL. Water temperature is kept constant by modulating the valve position while the heat exchanger has a constant heat transfer rate, the controlled value is measured by temperature transmitter after the heat exchanger. The system is composed of devices (Temperature Transmitters, Control System and Actuator) connected via a fieldbus. Fig. 3 shows a generic scheme of the temperature control system loop.

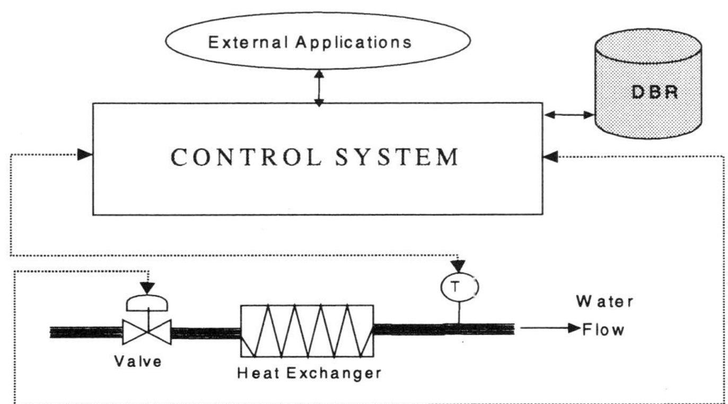

Fig. 3. A simple control schema

A functional view of the control system is presented in Fig. 4. The system consists of a temperature transmitter (TT) connected to a PID which commands the actuator. The actuator finally commands the valve.

In this example it was considered only a detailed description of the temperature transmitter. The description was done in accordance with the Temperature Transmitter Profile. Further, the PID and the Actuator were considered to have just one function block with no parameters. The mapping of the TT onto the DBR can be taken as a generic example on the device modeling. It consists from three function blocks logically connected in cascade: acquisition, linearization and measure management.

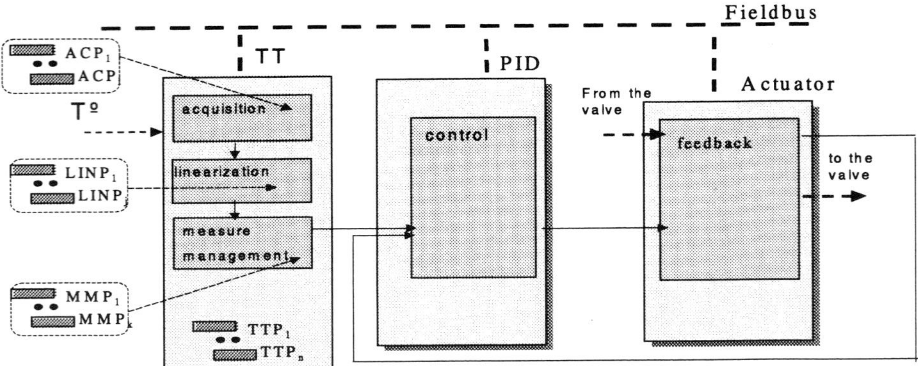

Fig. 4. A detailed view of the control block

The generation of the DBR for the above presented system, was done according to the generic rules described in the previous subsections. The system was first decomposed into function blocks, then for every function block a number of parameters was created, representing the function block's internal parameters. The next steps were the modeling of the three devices and the creation of the physical link representing the fieldbus. Finally, allocation relationship were created in order to relate function blocks to devices, ports to terminals and the logical networks to the physical one.

6 Conclusions

This work presents the database repository (DBR) specification developed within the ESPRIT Project "Network Oriented Application Harmonization". The DBR is designed to provide a uniform solution to describe all the relevant information of an industrial plant. Application concepts have been identified and mapped onto a set of entities and relationships derived from the ISO 10303 – AP 212 international standard. The validation of the DBR model was successfully done on a real system example.

References

1. ISO/IEC DIS 10303-212: 1998, Product data representation and exchange, Application Protocol: Electrotechnical Design and Installation.

2. NOAH documents, not available for public.

NOAH Project: an Example of Application

Antonio Mella, Francesco Russo

ENEL-Ricerca, Via Volta 1. 20093- Cologno Monzese (MI), Italy

Abstract. A NOAH demonstrator is planned in the IAM Pilot based at ENEL Ricerca's premises in Milan. The IAM Pilot is the result of ESPRIT projects funded by the European Commission, and is a cooperative action of ENEL with the User support Groups for automation based on CENELEC EN50170. It is a permanent platform used to investigate research results, standards and market offerings for distributed automation. The paper describes how the IAM-Pilot laboratory, already equipped with EN50170-compliant instrumentation, has been upgraded to implement an example of the NOAH architecture. This experimental site is intended to enable us to bring together some key concepts of the NOAH project (which are explained in the other papers presented at this Conference) and will permit us to show, concretely, how main NOAH concepts may be brought to bear to obtain the expected results. The paper recalls the IAM-Pilot structure and investigation approach and describes the general demonstration scenarios developed for NOAH. Particular attention is paid to the practical aspects of the possible use of the NOAH concepts and to verification of the real benefits end users and system integrators can derive from the Noah approach. Another key expected result from a real practical investigation is the possibility of checking the key aspects of the NOAH specifications in order to submit verified specifications to CENELEC with a view to drawing up a suitable standard.

1 Introduction

The NOAH project aims at developing an "open automation platform" in order to harmonise operations that must be performed in an automation plant. In the systems currently in use, many kinds of incompatibilities exist, and these severely limit the use of different communication systems and/or devices, supplied by different manufacturers. This problem complicates the task of System Integrators, who must cope with many different problems and must possess detailed knowledge in several different areas.

The Noah project aims to contribute towards solving these problems through the development of a harmonisation approach that operates at three different levels:

- Communication harmonisation: the goal is to provide a harmonised communication interface that will enable applications to work irrespective of the specific network used. This means that an open, communication-independent interface will be specified in order to adapt the various services provided by the different networks.

- Devices harmonisation: devices from different manufacturers use different semantics, so that, even if two devices perform the same task and possess the same features they are not be interchangeable. The harmonisation of the semantics is one of Noah's main tasks. Moreover, a device often possesses specific manufacturing functions that must be known to System Integrators. So the use of an "Electronic Device Description" will be of great help during both the engineering stage - it will enable the designer to acquire information about devices - and the runtime stage - to acquire information about the parameters and device-options of the plant. Function blocks will be used to describe the behaviour of the varying functionalities in the plant.

- System harmonisation: each system provider has his own concepts of system operation, and this makes co-operation between different parts of the plant difficult. Noah drives system harmonisation through a Management Information Base and a Data Base Repository to provide access to open functions.

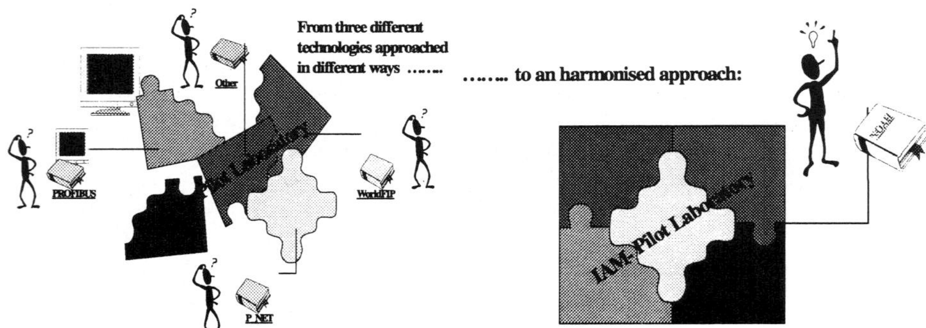

Fig. 1. NOAH's aim: a harmonised approach

Noah, instead of generating specific software or implementations, will provide concepts and approaches. These must be carefully tested before coming into use. A theoretical study is not possible, nor could it be detailed enough to provide ensure results. For this reason the development of a real test environment is of primary importance to the project. The test environment chosen for Noah is the IAM Pilot based at ENEL Ricerca. The presence of three different fieldbusses (Profibus, WorldFIP and P.NET) and of many devices from different manufacturers provides a heterogeneous environment, where NOAH has harmonisation problems to cope with.

2 The IAM-Pilot

The IAM-Pilot is the result of the series of the IAM ESPRIT projects funded by the European Commission. It represents joint action between ENEL and the User support Groups for automation based on CENELEC EN50170. It is a permanent platform for investigating research results, standards, and distributed automation available on the market. For each fieldbus specification the IAM-Pilot is instrumented (**Fig.2**) with market devices and systems in such a way as to permit investigation of market support for the CMM (**C**ontrol, **M**aintenance, technical **M**anagement) and IAM (**I**ntelligent **A**ctuation and **M**easurement) user requirements.

Fig. 2. The IAM-Pilot

The IAM-Pilot WEB site (http://pc551.pea.enel.it/IAM-Pilot/IAM-home.htm, see also other WEB references in the acknowledgements) gives details of the actual composition of the systems installed.

The IAM-Pilot physical process consists of a closed circuit of treated water (Fig.3) implementing four loops: a pressure regulation loop, a level regulation loop, a flow regulation loop, and a temperature regulation loop.

2.1 The IAM-Pilot field devices for NOAH.

The field devices taken into account for the NOAH demonstration are:

> EN50170-2 (Profibus) devices: level and temperature loops instrumented with a number of devices (two differential pressure transducers and one pneumatic actuator) using the PA extension of the protocol specification.

448

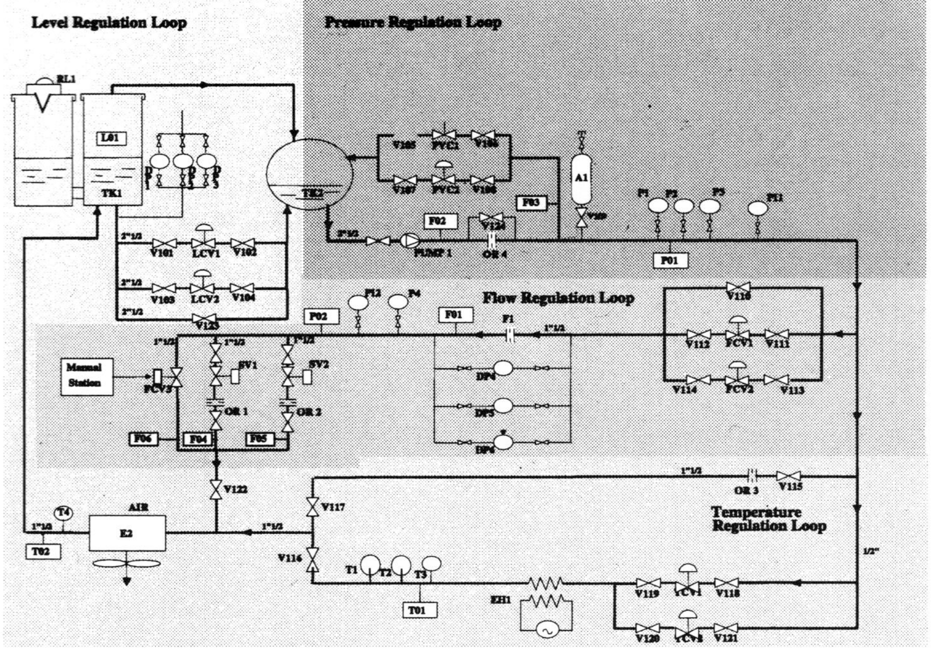

Fig. 3. The IAM-Pilot scheme

<u>EN50170-3 (WorldFIP) devices</u>: flow and temperature loops, instrumented with a number of devices (in particular a prototype device implementing a NOAH-compliant management information base (MIB) able to make it visible over an Internet connection).

<u>EN50170-1 (P-NET) devices</u>: flow and temperature loops, instrumented with a magnetic flow transducer and a multiplexer for Hart instruments.

2.2 The IAM-Pilot NOAH Demonstrator

Fig. 4 shows the NOAH system architecture. For experimental purposes this architecture is developed and installed as a demonstrator able to support the performance of testing.

This instrumentation corresponds to the lower part (shown in green) of Fig.4, where the three separate fieldbus systems are represented. In other words it represents the field part of the complete automation. In our investigation we consider these as different sections of a plant provided, within separate bids, with different market technologies.

The figure shows all the relevant aspects of NOAH's architecture, which are also taken into account in the Pilot demonstrator, intended to test and to show major aspects of the NOAH approach.

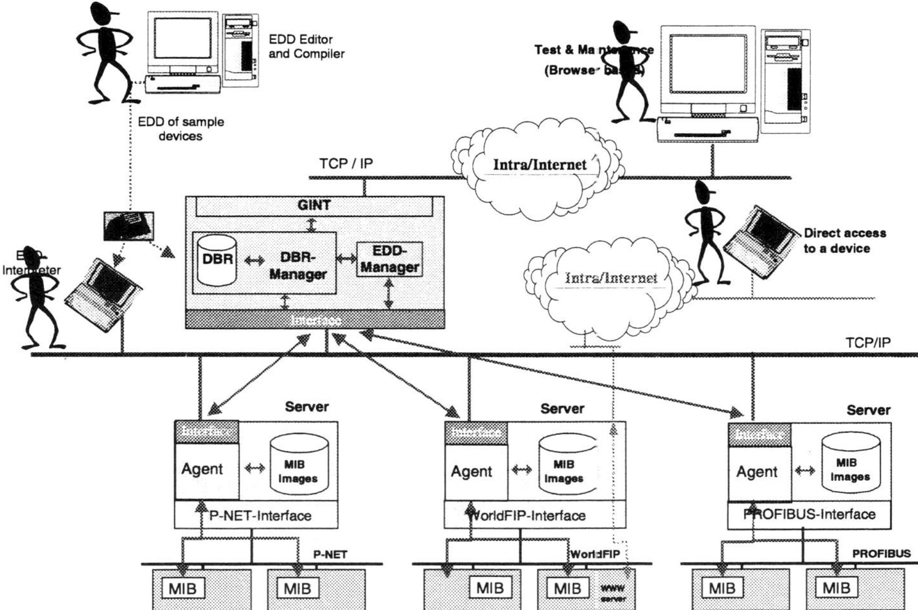

Fig. 4. The IAM-Pilot NOAH Demonstrator and its testing interfaces

The green boxes represent the field instrumentation and the control part of the automation (PLCs). The management information present in each field device evidenced as the Management Information Base (MIB).

The second level from the bottom consists of the NOAH management adaptation layer which, for the sake of clarity, is implemented in the IM-Pilot by three separate PC's (referred to in Fig.4 as servers). This adaptation layer presents consistently to the upper interface all management data of the underlying sub-system. This level is crucial to the success of the NOAH approach and is thus subjected to testing, both at specification level and at the subsequent implementation level. This level also hosts the MIB's image data base, which may hold images of data in the devices mapped on NOAH objects, or just pointers should the device already implement these objects.

The machine, which is connected to the 'servers', contains the Data Base Repository (DBR) [2]. This is seen as a plant-wide, high-level data base that includes all life-cycle information concerning the whole of the plant automation from the design and commissioning to the run-time and maintenance information.

The DBR also contains data or pointers to data contained in the MIB's. It is in fact a task of the DBR manager and of the agents in the 'servers' to provide a consistent picture of all data stored in the distributed data base formed by DBR, MIB images, and MIBs in the field devices.

Tools resulting from the ACORN Brite-Euram European project [6] are used to populate the DBR with a description of the IAM-Pilot automation.

Another point of practical interest in the NOAH project is the definition of an electronic device description language (EDD) [1]. In the IAM-Pilot we are able to investigate both the process of creation and the practical use of example EDDs by using the prototype tools provided by NOAH partners and prototype field devices. The EDD description of devices makes it possible consistently to access devices implemented by different manufacturers and with different technologies. Device parameterisation will be performed in the NOAH demonstrator using this interface.

The access to all plant-wide information is made available via the Generalised Interface (GINT) which for purposes of the NOAH demonstrator is implemented using HTTP protocol technology. This enables us to extend the IAM-Pilot demonstrator's purpose to that of a living demo of NOAH results publicly available over the Internet.

In order to test User Requirements, a test application (featuring basic data retrieval functions useful for Maintenance applications) is implemented. The access from this interface to any plant information, thanks to NOAH technology, is completely transparent with respect to the underlying specific technology.

An example of a direct connection to a field device, implementing a NOAH-compliant MIB, through the management adaptation interface, is performed, thus giving a practical demonstration of the differences between access to the subsystem (Management Information Base, MIB) and access to the DBR.

The network configuration of the complete system is outside the scope of the project and, for purposes of NOAH testing, the systems will be considered already properly configured and running (that is, configured with their own configurators).

3 NOAH Test Scenarios and Methodological Approach

The methodological approach used for tests in the IAM-Pilot is carried out as described in[3]. References for the analysis and test definitions are the user-requirement documents [4] which have been defined at the beginning of the NOAH project by assembling all experience acquired by the partners in previous ESPRIT projects (such as PRIAM, EIAMUG, IAM-PILOT) and from their professional experience.

Subjects of analysis and tests will be:

- Paper documentation (specifications) and

- Prototypes provided

This approach makes possible full analysis, since implemented prototypes will probably only partially implement the complete specification, and since not all details are accessible via a real interface.

The study of documentation will also be a prerequisite for defining an appropriate test for the prototypes implemented.

<u>Subjects of On-Paper-Analysis are:</u>

- *Uniform Management Model specification documentation*

- *Device Model specification documentation*

- *Harmonised Profile specification documentation*

Basic items for On-Paper-Analysis are:

- *Uniform configuration Management*

- *Uniform fault management*

- *Identification*

- *Parameterisation*

- *Measure Actuation (Validation)*

- *Tests on demand*

- *Device Status*

- *Access Rights Management*

<u>Subjects of Practical investigation are:</u>

- *All devices and prototypes installed in the IAM-Pilot NOAH demonstrator according to the stated parts of the specifications implemented (which will be a sub-part of the complete specification). Practical tests will only address the requirements relevant to the implemented specification.*

Example scenarios for practical investigation are:

Electronic Device Descriptions:

- *Editing*

- *Compilation*

- *Human readable device description (functions, parameters, timing aspects, etc.)*

- *Machine readable device description (example of configuration, calibration.)*

Uniform, local and remote, Access to:

- *Application structure*

- *Network structure*

- *Mapping of application to managed objects*

- *Common access to application objects*

Several kinds of test are planned in the pilot, the aim being to verify a number of specific concepts.

As an example, from a NOAH console a value can be read from devices placed on different networks. Moreover, function blocks in different devices located in different Fieldbusses could be started from the console, thus demonstrating how it is possible to operate on the plant, irrespective of the network where the devices are connected.

Another kind of test concerns the use of the electronic device description, which can be read from the console, thus enabling some parameters of the device and their setting to be displayed. A remote console, through Internet could perform some of these operations.

4 Conclusions

The IAM pilot is used as a recognised European tool for testing NOAH concepts in a real plant in order to demonstrate their general validity and their applicability in a wider process control context.

This will enable us to submit NOAH specifications to CENELEC as verified specifications with a view to drawing up a suitable standard.

The IAM-Pilot laboratory NOAH implementation will be made available for free access over Internet as a living demo of the project results achieved.

5 Acknowledgements

This paper is mainly supported by work done in the IAM-Pilot ESPRIT PROJECT 23525 (1997-1998) (*http://pc551.pea.enel.it/IAM-Pilot/IAM-home.htm*, and also *http://www.prosoma.lu/cgi-bin/show.py?page=result&id=2477*) and currently in the NOAH ESPRIT PROJECT 26951 (1998-1999) (*http://noah.loria.fr*).

References

1. R. Simon "Electronic Device Description" FET '99 Magdeburg 1999.

2. Demartini ET al. " A DBR based approach for system management in distributed process control" FET '99 Magdeburg 1999

3. Mella, T. Bangemann, P. Chatelet "Fieldbus: User requirements (IAM) Conformance Analysis " ANIPLA '98 Milano 1998.

4. Capetta L., ET al (96). "Fieldbus communications: expectations and experience of an end user." ANIPLA Seminar, Torino, Jan. 1996.

5. Russo "Standards for fieldbus and Automation" ANIPLA '98 Milano 1998.

6. Russo ET al." Distributed automation based on fieldbus: ACORN integrated toolkit." ANIPLA '98 Milano 1998.

SpringerTechnik

Dietmar Dietrich,

Herbert Schweinzer (Hrsg.)

Feldbustechnik in Forschung, Entwicklung und Anwendung

Beiträge zur Feldbustagung FeT '97, Wien, Österreich,
13.–14. Oktober 1997

1997. XVI, 443 Seiten. 233 Abbildungen.
Broschiert DM 128,–, öS 896,–
ISBN 3-211-83062-6

Feldbussysteme sind im Zusammenwirken mit LANs (Large Area Network) und WANs (Wide Area Network) als Nervensystem anzusehen, das in Systeme aller Bereiche unseres Lebens eingebettet wird. Die Automatisierung erlebt dadurch zur Zeit eine revolutionäre Weiterentwicklung. Die Tagung FeT '97 in Wien hat sich zum Ziel gesetzt, aktuelle Fragen zu diskutieren.

Das Buch beinhaltet Aufsätze, die anläßlich dieser Tagung eingereicht und vom Programmkomitee bestbewertet wurden. Damit ist ein guter Überblick geboten, welche Schwerpunkte in der Feldbustechnik gegenwärtig bearbeitet werden. Der Aufbau des Buches erfolgt in zwei Teilen: der erste Teil beinhaltet Beiträge aus Forschung und Entwicklung, der zweite Teil Beiträge mit Produkt- und Anwendungsorientierung. Innerhalb dieser Teile gibt es eine Gliederung nach den Anwendungsschwerpunkten Industrietechnik, Gebäudeautomation und bereichsübergreifende Beiträge.

SpringerWienNewYork

Sachsenplatz 4–6, P.O.Box 89, A-1201 Wien, Fax +43-1-330 24 26, e-mail: books@springer.at, **Internet: http://www.springer.at**
New York, NY 10010, 175 Fifth Avenue • D-14197 Berlin, Heidelberger Platz 3 • Tokyo 113, 3–13, Hongo 3-chome, Bunkyo-ku